D0849912

Carbanions
in
Organic
Synthesis

CARBANIONS
IN
ORGANIC
SYNTHESIS

JOHN C. STOWELL

University of New Orleans
New Orleans, Louisiana

A WILEY-INTERSCIENCE PUBLICATION

JOHN WILEY & SONS
New York • Chichester • Brisbane • Toronto • Singapore

Library of Congress Cataloging in Publication Data

Stowell, John Charles, 1938-
    Carbanions in organic synthesis.

    "A Wiley-Interscience publication."
    Includes bibliographical references and indexes.
    1. Carbanions. 2. Chemistry, Organic—Synthesis.
I. Title.

QD305.C3S76   547'.1'372   79-373
ISBN 0-471-02953-X

Printed in the United States of America

10 9 8 7 6 5 4 3 2

Preface

The variety of synthetically useful carbanion types has expanded phenomenally in the past decade. The charge-delocalizing effects of many different groups allow preparation of carbanions which in turn are used to make complex molecules containing a wide choice of functionality. In this volume a broad selection of carbanions, new and classical, is presented, with illustrative examples of the more frequently used reactions of each. A thorough coverage of all the specialized reactions and limitations of these reagents would be unthinkable in a single volume, so a practical general overview is given. The historical origins of the reactions are not traced, but the cited references may be consulted for more examples as well as earlier material.

The first two chapters cover the general methods of preparation and many subsequent uses for carbanions. Chapters 3 to 6 concentrate on the carbon-carbon bond-forming reactions and are arranged in approximate order of increasing degree of delocalization of the negative charge.

This book should be useful to researchers engaged in the synthesis of natural products, analogs, drugs, and other complex molecules. Students of organic synthesis at the graduate level may find this a useful starting point as well.

JOHN C. STOWELL

March 1979
New Orleans, Louisiana

Contents

Abbreviations

Ac	Acetyl
aq.	aqueous
Bu	butyl
DABCO	1,4-diazabicyclo[2.2.2]octane
DME	1,2-dimethoxyethane
DMF	dimethylformamide
DMSO	dimethyl sulfoxide
Et	ethyl
HMPA	hexamethylphosphoric triamide
LDA	lithium diisopropylamide
Me	methyl
Ph	phenyl
Pr	propyl
rt	room temperature
THF	tetrahydrofuran
TMEDA	N,N,N',N'-tetramethylethylenediamine
Ts	p-toluenesulfonyl
°	degree centigrade

Carbanions
in
Organic
Synthesis

CHAPTER 1

Preparation of Carbanions

$\mathbf{T}$HE GENERAL TERM "carbanion" refers to substances in which part of a unit negative charge resides on a certain carbon atom. The charge may be small if there is largely covalent bonding to the accompanying metal cation. It may be very small when π conjugation allows the charge to reside largely on more electronegative atoms as we find in ketone enolate anions. Furthermore it may be dispersed by π conjugation to other carbon atoms, by inductive effects of adjacent heteroatoms, or by d orbital overlap as in ylides. The very concentrated, localized carbanions require more energetic processes for their formation, and the more dispersed, stabilized ones are made under mild conditions with less potent reagents. The unifying factor in this wide range of materials is their common ability to give useful products by formation of new bonds to the carbon atom.

A negative charge on carbon may be developed by either of two general reactions. In a redox process, a metal may donate electrons to a reducible substrate which may then lose a halide or hydride ion leaving a carbanion with a cation partner. Alternatively, in an acid-base reaction, a proton may be abstracted from a carbon by a negatively charged base leaving a negative charge on that carbon. The practical variations on these methods are the subject of this chapter.

1.1 REACTION OF ALKYL AND ARYL HALIDES WITH METALS

1.1.1 Lithium. A wide variety of primary, secondary, and tertiary halides (Cl, Br, I) will react exothermally with lithium wire or ribbon in a range of organic solvents such as benzene or hexane. These give solutions of alkyllithium reagents in good yields (Reviews: Schöllkopf, 1970; Wakefield, 1974). Ether may be used as solvent if it is kept cold, but it is slowly attacked by butyllithium and rapidly attacked by secondary and tertiary lithium compounds. The exceptional methyllithium is insoluble in hydrocarbon solvents, but it gives stable ether solutions. *tert*-Butyllithium may be prepared in pentane using lithium dispersion containing 2% sodium with *tert*-butyl chloride at reflux (89% yield) (Owens et al., 1960).

Vinyllithium compounds may be prepared stereospecifically using lithium dispersion (Eq. 1) (Linstrumelle et al., 1976; Millon et al., 1975). The more

$$\text{\textbackslash\textbackslash}Cl + Li (1\% Na) \xrightarrow[10^\circ]{ether} \text{\textbackslash\textbackslash}Li \qquad (1)$$

3

reactive allylic halides tend to react with the lithium reagents in situ and give coupling products; therefore the less reactive allyl phenyl ether is used instead. This is cleaved in tetrahydrofuran at $-15°$ by lithium metal to give allyllithium in 65% yield (Eisch et al., 1963).

Aryllithium compounds are prepared from the corresponding bromo compounds in ether or tetrahydrofuran. Phenyllithium slowly attacks ether; thus it is often kept in ether-benzene mixed solvent. Starting with primary, secondary, and tertiary alkyl chlorides, the corresponding lithium reagents may be made instantaneously at $-78°$ by using a THF solution of the deep-blue radical anion from lithium and p,p'-di-*tert*-butylbiphenyl instead of the free metal (Freeman et al., 1976).

All of these reagents are sensitive to moisture and oxygen and so must be prepared and used under an inert atmosphere, usually nitrogen or argon.

1.1.2 Magnesium. The less reactive magnesium requires a complexing ether solvent such as diethyl ether or tetrahydrofuran for reaction with organic halides. Primary, secondary, and tertiary halides (Cl, Br, I) as well as aryl bromides and iodides form the well-known Grignard reagents (Reviews: Nützel, 1973a, Kharasch et al., 1954). The reaction does not always start quickly after the reagents are combined or heated; therefore a small crystal of iodine or a drop of 1,2-dibromoethane may be required to initiate reaction. A trace of water may prevent initiation of the reaction. Once started, it is quite exothermal, so most of the organic halide should be reserved in an addition funnel and added at a rate which gives controlled refluxing. For example see Eq. 2 (Cason et al., 1955).

$$\text{\Large \char`\^}\!\!\!\!\!\!\text{Br} \ + \ \text{Mg} \ \xrightarrow[\text{reflux}]{\text{ether}} \ \text{\Large \char`\^}\!\!\!\!\!\!\text{MgBr} \qquad (2)$$

Allyl chloride and bromide give Grignard reagents that are only slightly soluble in ether but can be used very well as a slurry in subsequent reactions (Benkeser, 1971). Vinyl halides give mostly acetylenes when treated with magnesium in ether, but in THF they give high yields of useful vinylmagnesium halides (Normant, 1960a).

Organomagnesium halides are in equilibrium with diorganomagnesium plus magnesium halide in solution. When pure diorganomagnesium reagents are desired, the addition of an equimolar amount of dry dioxane will precipitate the magnesium halide from ether solution (Huston et al., 1941; Nützel, 1973b).

1.1.3 Zinc. Zinc is most frequently used in the Reformatzki reaction of α-haloesters where the organometallic reagent is consumed in situ as described in Section 5.3.3. Alkyl iodides as well as benzyl and allyl bromides will react readily with powdered zinc metal in THF to give organometallic reagents of low reactivity (Nützel, 1973c). An activated zinc powder prepared by potassium metal reduction of zinc bromide will react with alkyl and aryl bromides also (Rieke et al., 1973).

1.2   ALKALI METAL REDUCTION OF C–H COMPOUNDS

Active metals can donate electrons to sufficiently reducible compounds to generate hydride anions and the organometallic compound. For example lithium reacts readily with fluorene (Eq. 3) (Gilman et al., 1958). Other

$$\text{(fluorene)} + 2\text{Li} \xrightarrow{\text{THF}} \text{(9-Li-fluorene)} + \text{LiH} \qquad (3)$$

71% yield determined as the carbo-
nation product

hydrocarbons that have been metalated directly include indene, cyclopentadiene, and acetylene. Sodium metal readily converts ethyl acetoacetate and diethyl malonate to the enolates in benzene or ether (Review: Schmidt, 1963). The more electropositive cesium will even react with toluene to give benzyl cesium plus hydrogen gas (Postis, 1946).

Many of these reactions may be facilitated by first forming a homogeneous solution of the radical anion sodium naphthenide, which will rapidly reduce active C-H compounds to the carbanions (Eq. 4) (Normant, et al., 1960b).

$$\text{(naphthalene)} + \text{Na} \xrightarrow[30°]{\text{THF}} \text{(naphthalenide)}^{\cdot-} \text{Na}^{\oplus} \xrightarrow{\text{Ph}_2\text{CH}_2} \text{Ph}_2\text{CH}^{\ominus}\text{Na}^{\oplus}$$

$$\xrightarrow[\text{2. } H_2O, \, H_2SO_4]{\text{1. solid } CO_2} \text{Ph}_2\text{CHCOOH} \qquad (4)$$

70%

1.3 HALOGEN-METAL AND METAL-METAL EXCHANGE

In cases where direct reaction between an organohalide and a metal does not occur owing to low reactivity or the presence of interfering functionality such as phenolic groups, the halogen is often exchanged for a metal using another organometallic reagent such as methyllithium, wherein the by-product is bromomethane (Reviews: Wakefield, 1974; Schölkopf, 1970b; Jones, 1951). The reversible equilibrium is most favorable for exchange when the metal is transferred to a carbon of greater electronegativity, for example from an alkyl carbon to an aryl or vinyl carbon or from a tertiary carbon to a primary carbon. This is evident in the selective exchange of n-butyllithium with only the aryl bromine in β-(o-bromophenyl)ethyl bromide (Parham et al., 1976a; Hergrueter et al., 1977). tert-Butyllithium will exchange with primary iodides at −78° (Corey et al., 1977b). The exchange is fast at very low temperatures where other reactions such as attack on nitriles is slow. The advantage of this process is exemplified in Eq. 4 where a nitrile survives a ketone addition (Parham et al., 1976b). A variety of other functional groups survive, including tert-butyl esters.

(4a)

86%

In cases where the by-product RBr from the alkyllithium reagent may interfere by subsequently alkylating the carbanion, tert-butyllithium is preferred. When two equivalents tert-butyllithium are added to an aryl or vinyl bromide, the aryl- or vinyllithium reagent is formed and the resulting tert-butyl bromide is dehydrohalogenated rapidly by the second equivalent of tert-butyllithium (Eq. 5) (Seebach et al., 1974a; Corey et al., 1972b).

$$\text{(structure with Br)} + 2t\text{-BuLi} \xrightarrow[-78^\circ]{\text{THF}} \xrightarrow{\text{CH}_3\text{I}} \text{(structure with CH}_3\text{)} \tag{5}$$

76.5%

Certain vinyllithium reagents that are unstable toward elimination or rearrangement at ordinary temperatures have been prepared and used at low temperature by the exchange method (Masure et al., 1976; Parham et al., 1977). Terminal vinyllithium reagents can be prepared stereospecifically from 1-iodo-1-alkenes (Cahiez et al., 1976a).

Once the carbanionic reagent is formed, the associated cation can also be exchanged to give reagents of reduced or modified reactivity. Classically, Grignard reagents have been converted to zinc and cadmium reagents by reaction with the anhydrous metal chlorides (Review: Shirley, 1954).

Vinyllithium reagents may be prepared from the corresponding tin compounds. Phenyl- or butyllithium will exchange with tetravinyltin to give the simple vinyllithium (Seyferth et al., 1961). Polyfunctional reagents have been prepared from the hydrostannation products of acetylenes by exchange with n-butyllithium (Eq. 6) (Corey et al., 1977b; Wollenberg et al., 1977).

$$n\text{-Bu}_3\text{SnH} + \text{HC}{\equiv}\text{C} \text{(structure)} \xrightarrow[90^\circ]{\text{AIBN}} n\text{-Bu}_3\text{Sn} \text{(structure)} \tag{6}$$

$$\xrightarrow[\text{THF, } -78^\circ]{n\text{-BuLi}} \text{Li} \text{(structure)}$$

72%

Diorganomagnesium compounds can be prepared from diorganomercury compounds by exchange with magnesium turnings (Nützel, 1973b).

Lithium reagents may be exchanged with cuprous halides to give organocopper compounds. The synthetically most useful reagents are those formed when two equivalents of lithium reagent are combined with one of cuprous halide (usually the iodide). Alkyl-, aryl-, and vinyllithium reagents may be exchanged, and the products are known as lithium diorganocuprates or diorganocopper lithium reagents (Eq. 7) (Review: Posner, 1975). The dialkyl cuprates

$$2\text{CH}_3\text{Li} + \text{CuI} \xrightarrow[-20^\circ]{\text{ether}} (\text{CH}_3)_2\text{CuLi} + \text{LiI} \tag{7}$$

derived from secondary and tertiary alkyllithium reagents are thermally un-stable. They have found some use at low temperatures when stabilized as phosphine complexes, but the stable mixed reagents such as phenylthio(alkyl)-cuprates are preferred (Posner et al., 1973). These are prepared from thiophenol as shown in Eq. 8 for the *tert*-butyl case (Posner et al., 1974). The alkyl group is

$$\langle \bigcirc \rangle\text{--SH} + \text{CuO} \xrightarrow[\text{reflux}]{\text{ethanol}} \langle \bigcirc \rangle\text{--SCu} \quad 91\%$$

$$\langle \bigcirc \rangle\text{--SCu} + t\text{-BuLi} \xrightarrow[20°]{\text{THF}} \langle \bigcirc \rangle\text{--S--}\underset{\underset{\text{Li}}{|}}{\text{Cu}}\text{--}t\text{-Bu}$$

(8)

selectively transferred when these cuprates are used in subsequent reactions such as the conversion of acid chlorides to ketones. Because lithium divinyl cuprate is difficult to prepare, another mixed cuprate derived from copper *tert*-butylacetylide is of value (Eq. 9) (House et al., 1973b). Lithium divinylcuprate itself may be prepared using dimethyl sulfide, which gives a soluble complex

$$+\text{C}\equiv\text{CH} \xrightarrow[\substack{\text{2. CuI, ether, }12° \\ \text{3. CH}_2=\text{CHLi, ether, }-49°}]{\text{1. CH}_3\text{Li, ether, }12°} +\text{C}\equiv\text{CCuCH}=\text{CH}_2 \atop \quad\quad\quad |\atop \quad\quad\quad\text{Li}$$

(9)

with the cuprous halide. The cuprous halide must be carefully purified or one may use the conveniently purified colorless crystalline complex Me_2SCuBr with excess dimethyl sulfide (House et al., 1975; Clark et al., 1976).

In each of these reagents the formula $RR'CuLi$ represents only stoichiometry, not structure. They are likely to be aggregates of this composition.

1.4   REACTION OF C–H ACIDS WITH BASES

1.4.1   Equilibrium Acidities. Carbanions that are stabilized by resonance or in-ductive effects are most frequently generated by removal of a proton from their conjugate acid by a sufficiently stronger base. The acidity of these C-H acids covers a range of over 30 pK units. The extensive literature on the methods and results of the pK measurements are reviewed by Jones (1973). Table 1 lists the equilibrium pK values of selected C-H acids in dimethyl sulfoxide as measured by Bordwell et al. (1975-1977). A general idea of the relative carbanion-stabilizing effects of various groups and combinations of groups may be obtained from

TABLE 1 EQUILIBRIUM ACIDITIES IN DIMETHYL SULFOXIDE SOLUTION

Acid	pK	Acid	pK
$CH_3SO_2O\underline{H}$	1.62	$PhC\underline{H}_2SO_2C\underline{H}_2Ph$	23.9
$C\underline{H}_2(CN)_2$	11.1	$PhCOC\underline{H}_2CH_3$	24.4
$CH_3CO_2\underline{H}$	12.6	$CH_3CON\underline{H}_2$	25.5
$CH_3COC\underline{H}_2COCH_3$	13.4	$CH_3COC\underline{H}_3$	26.5
$CH_3C\underline{H}_2NO_2$	16.7	$CH_3CH_2COC\underline{H}_2CH_3$	27.1
CH_3NO_2	17.2	$PhC{\equiv}CH$	28.8
$PhCOC\underline{H}_2Ph$	17.7	$PhC\underline{H}_2SOCH_3$	29.0
		$Ph_3C\underline{H}$	30.6
(cyclopentadiene structure with H H)	18.1	$PhSC\underline{H}_2SPh$	30.8
		$PhSO_2C\underline{H}_2CH_3$	31.0
$PhC\underline{H}_2CN$	21.9	$CH_3SO_2C\underline{H}_3$	31.1
		CH_3CN	31.3
(fluorene structure H H)	22.6	$CH_3SOC\underline{H}_3$	35.1
		$PhC\underline{H}_3$	42^a

^{a}Extrapolated value, since toluene is less acidic than DMSO.

the table. Since most of the reactions, of the less acidic tabulated examples, are carried out in aprotic solvents, the order found in dimethyl sulfoxide should be reasonably valid for generalization. The hydroxylic solvents give extra hydrogen-bonding stabilization to those anions which have charge delocalized on hetero-atoms.

Many reactions require only a low equilibrium concentration of carbanion where formation and reaction of the carbanion are carried out in one operation. In these cases the weaker bases are used; for example, the cyclization of 1,4-diketones requires only a 2% solution of sodium hydroxide (Eq. 10) (Hunsdiecker, 1942). The amide anion-type bases are capable of essentially complete

$$\text{(diketone)} + NaOH \xrightarrow[\text{reflux}]{H_2O\text{-alcohol}} \text{(cyclopentenone product)} \qquad (10)$$

92%

proton removal from compounds up to pK about 30, and even higher when HMPA is present. The stronger bases are also required when 1,3-dianions are prepared, since it is energetically unfavorable to concentrate two unit negative charges close by in a molecule. Once formed (Table 2), these dianions undergo reaction first at the site of the least stabilized of the two charges (Review: Kaiser, et al., 1977).

TABLE 2 SOME SYNTHETICALLY USEFUL DIANIONS

Dianion	Reference
$\overset{O}{\underset{\ominus}{\|}}\overset{O}{\underset{\ominus}{\|}}$ (pentane-2,4-dione dianion)	Hampton et al., 1973
$\overset{O}{\underset{\ominus}{\|}}\overset{O}{\underset{\ominus}{\|}}OCH_3$	Huckin et al., 1974
$\overset{O}{\underset{\ominus}{\|}}\overset{O}{\overset{\uparrow}{S}}{-}Ph$	Grieco et al., 1974a
$CH_2{-}C{\equiv}C$, $\ominus\ \ominus$	Bhanu et al., 1975
$\overset{O}{\underset{}{\|}}$ $^{\ominus}CH_2CO^{\ominus}$	Krapcho et al., 1977
$\overset{O}{\underset{}{\|}}$ $CH_3SCHCO^{\ominus}$, $\ominus$	Trost et al., 1975a
$\overset{O}{\underset{}{\|}}$ $CH_2{-}C{\equiv}C{-}C{-}O^{\ominus}$, $\ominus$	Pitzele et al., 1975

1.4.2 Kinetic Acidity. The kinetic acidity of very weak C-H acids such as toluene is frequently determined by measuring the rate of slow deuterium exchange under reversible conditions. Among the more acidic carbonyl compounds, kinetic acidity is preparatively important at low temperatures where reactions are still fast but essentially irreversible. Under nonequilibrium conditions the more rapidly removed proton may leave an anion which is not thermodynamically the more stable. This is exploited in the alkylation of ketones where the sterically more accessible proton is selectively removed by hindered bases at low temperatures and then an alkylating agent is added to give the product of substitution on the least substituted side (Section 5.1.1). Under equilibrium conditions the opposite product usually predominates.

The lower the strength of the base, the greater the kinetic selectivity. When the nitrile ester in Eq. 11 was treated with LDA, a mixture of anions α to the nitrile and α to the ester were formed; but with lithium hexamethyldisilazane only the former was observed, as indicated by alkylation products (Kieczykowski et al., 1975). In the following sections we consider some of the more commonly

$$
\underset{\substack{|\\ \underset{\underset{O}{\|}}{CH_2-COCH_3}}}{CH_2=CH-CH-CN} \xrightarrow[\text{THF, }-78^\circ]{\text{LiNR}_2} \underset{\substack{|\\ \underset{\underset{O}{\|}}{CH_2COCH_3}}}{CH_2=CH-\overset{\ominus}{C}-CN} \quad +
$$

$$
\underset{\substack{|\\ \underset{\ominus}{CHCOCH_3}\ \underset{O}{\|}}}{CH_2=CH-CH-CN}
$$

(11)

used bases and their applications.

1.4.3 Hydroxide Bases. Sodium, potassium, and barium hydroxides in alcohol solution or in water have been used frequently on the more acidic substrates, for example ethyl acetoacetate will dissolve as the anion in aqueous KOH and can then be alkylated with dimethyl sulfate (Henecka, 1952a). Similarly nitro-alkanes will dissolve in aqueous or aqueous methanolic KOH (Erickson et al., 1977).

Carbanions can be generated and used in alkylation or other reactions under surprisingly mild conditions using phase transfer catalysis (Review: Dehmlow, 1977). Aqueous NaOH solutions can be used in contact with an organic phase where a catalyst with salt and organic character promotes transfer of components from the water phase to the organic phase (Dockx, 1973). Phenylacetonitrile can be alkylated rapidly in the presence of the far more acidic water when benzyl-triethylammonium chloride is used as a catalyst (Eq. 12) (Makosza et al., 1976a).

$$
\text{[Ph]}-CH_2CN + C_2H_5Br + \text{aq. NaOH} \xrightarrow[28^\circ \text{ to } 40^\circ]{\text{catalyst}} \text{[Ph]}-\underset{\underset{C_2H_5}{|}}{CHCN} \quad (12)
$$

78 to 84%

One proposed explanation of this catalytic activity is that the quaternary ammonium hydroxide ion pair is extracted into the organic phase where it acts as a base to generate the carbanion which then attacks the alkyl halide. Another possibility is that the carbanion is formed by reaction with aqueous OH⁻ ion at the interface between the aqueous and organic phases and then the carbanion enters the organic phase paired with the quaternary ammonium ion. This ion pair may then attack the alkyl halide (Makosza et al., 1977). Overall, these conditions are far simpler than the homogeneous reactions using strictly anhydrous conditions and stronger bases. Other quaternary salts with good catalytic activity include "tricaprylmethylammonium chloride" and hexadecyltributylphosphonium bromide (Starks, 1971).

Similar phase transfer effects are found with macrocyclic "crown" polyethers (Gokel et al., 1976a; Johns et al., 1976) and polypode ethers (Fornasier et al., 1976), where the cationic metal is made lipophilic by the surrounding ligand. The resulting large cation and the associated hydroxide ion can then transfer from aqueous solution or even from solid NaOH to an organic solution; for example see Eq. 13 (Fornasier et al., 1976).

$$\langle\!\!\langle\bigcirc\rangle\!\!\rangle\text{-}CH_2\overset{\overset{\displaystyle O}{\|}}{C}CH_3 + \textit{n-}BuBr + \text{aq. NaOH} \xrightarrow[20°]{\text{catalyst}} \langle\!\!\langle\bigcirc\rangle\!\!\rangle\text{-}\underset{\underset{\displaystyle \textit{n-}Bu}{|}}{C}H\overset{\overset{\displaystyle O}{\|}}{C}CH_3$$

                                                                            78%

catalyst: R_2N —[triazine ring]— NR_2, NR_2 $R = (CH_2CH_2O)_4 \textit{n-}C_8H_{17}$

(13)

Anhydrous hydroxide solutions or suspensions can be prepared by adding water to potassium *tert*-butoxide. This is more nucleophilic than hydrated hydroxide as shown by the cleavage of nonenolizable ketones under mild conditions in ether (Gassman et al., 1967).

1.4.4 Alkoxide Bases. Alkoxide bases have found wide application particularly where a low concentration of carbanion is sufficient for reaction, as for example when sodium ethoxide is used in the Dieckmann condensation. Carbanions stabilized by conjugation with two heteroatoms as from diethyl malonate

(Henecka, 1952a) or nitroalkanes may be prepared in high concentration using sodium ethoxide. Potassium *tert*-butoxide has higher basicity and lower nucleophilicity and is the choice for many reactions (Review: Pearson et al., 1974). An indication of the base strength limitations of alkoxides is found in the alkylation of α,β-unsaturated ketones. An equivalent amount of potassium *tert*-butoxide in *tert*-butyl alcohol converts a fraction of the α,β-unsaturated ketone to the anion which then attacks the un-ionized ketone in a Michael condensation which reduces the yield of simple alkylation. When a 36-fold excess of *tert*-butoxide in *tert*-butyl alcohol is used, a greater portion of the ketone is ionized and the yields of alkylation are improved (Ireland et al., 1963). In the aprotic solvent DME, only a slight excess of potassium *tert*-butoxide is required to give essentially complete ionization of simple ketones (Gersman et al., 1971). For carbanions of this basicity or greater, it is now more common to use stronger bases such as the amide anions (Section 1.4.5).

The alkoxides are made by dissolving (caution: exothermal) the metal in excess of the dry alcohol or a solution of the alcohol in an inert solvent such as benzene (Schmidt, 1963).

A new, very hindered alkoxide, lithium 1,1-bis(trimethylsilyl)-3-methylbutoxide, is able to selectively (kinetically) generate enolates from methyl ketones, acetates, lactones, and even anhydrides in the presence of aldehydes to give good yields of addition of those anions to the aldehydes. The ions are trapped rapidly enough by the aldehydes to prevent equilibration among the various α-H sites, and this avoids self-condensation of the aldehyde or the other component (Eq. 14) (Kuwajima et al., 1976a; Minami et al., 1977).

$$(14)$$

86%

1.4.5 Lithium Dialkylamides. Classically many carbanions have been prepared by treatment of nitriles, terminal acetylenes, ketones, and so on with sodamide. The introduction of sterically hindered relatives of sodamide has allowed extension of this process to the preparation of carbanions from many substrates which might be attacked nucleophilically by sodamide itself. Moreover the condensation of the resulting carbanion with another molecule of starting material has been suppressed by carrying out this step at Dry-Ice temperatures. For example simple esters can be converted stoichiometrically to the corresponding α-anion and subsequently alkylated at temperatures where the Claisen condensation is avoided (Section 5.3.1). This simple sequence involves fewer steps than the older route via diethyl malonate alkylation and further allows preparation of α,α,α-trisubstituted acetates (Cregge et al., 1973). Many other anions were formed similarly as will be seen in Chapters 4 and 5.

Lithium diisopropylamide, the most commonly used base of this kind, may be prepared by adding dry diisopropylamine to an equimolar amount of methyllithium in ether at $0°$ (Sasson et al., 1975). Vigorous methane evolution is complete after a few minutes. It has been prepared in 1,2-dimethoxyethane solvent at $-20°$; above $0°$ the base attacks this solvent (Gall et al., 1972). It can be prepared in mixed hexane-toluene at $0°$ (Schlosser et al., 1969). It can also be prepared in THF by adding n-butyllithium to the amine at $-74°$ and warming to $25°$ (Albarella, 1977). Perhaps the most convenient form is a stable supersaturated solution in pentane-hexane, which may be standardized and stored for weeks at $25°$ (House et al., 1978). Bipyridyl or triphenylmethane may serve as an indicator in the subsequent reaction of the reagent. In the strongly basic solution it is deep red, and if a ketone is added, the stoichiometric endpoint for formation of the enolate may be determined by disappearance of the color.

The amide anions from dicyclohexylamine, diethylamine, and N-cyclohexylisopropylamine have also been used. Lithium N-isopropylcyclohexylamide is soluble enough to form $1M$ solutions in hexane at $-78°$ (Rathke, 1973). In cases where the reactants are especially susceptible to nucleophilic attack by these bases, the extremely hindered lithium 2,2,6,6-tetramethylpiperidide gives higher yields, for example see Eq. 15 (Olofson et al., 1973).

(15)

89% isolated yield

Lithium *N-tert*-butylcyclohexylamide gave the best yield in the removal of the activated benzylic proton from the ester shown in Eq. 16. Appreciable amide formation from attack at the carbonyl was obtained with LDA, lithium *N*-cyclohexylisopropylamide, and even lithium 2,2,6,6-tetramethylpiperidide (Loewenthal et al., 1976):

$$\text{81\%}$$

When lithium diethylamide is prepared from the amine and lithium metal in benzene plus hexamethylphosphoric triamide, it is even stronger as a base "hyperbasic" and can be used on hindered imines and hydrazones where the usual amide anions give no reaction (Cuvigny et al., 1976). A similar solution in THF generates the anion from phenyl vinyl sulfide (Cookson et al., 1976).

1.4.6 Bistrimethylsilylamide Bases. Lithium and sodium bistrimethylsilylamides can be prepared from hexamethyldisilazane and phenyllithium or sodamide in 80 to 100% yield. These bases are stable crystalline, distillable substances with good solubility in hydrocarbon and ether solvents (Wannagat et al., 1961). They can be measured for reaction by weight and are particularly useful for preparation of salt-free solutions of Wittig reagents (Bestmann et al., 1976).

1.4.7 Lithium and Magnesium Reagents. Some of the strongest bases are themselves carbanions. Lithium reagents, for example butyllithium, are used to remove protons from other C-H compounds where there are no groups which would be attacked nucleophilically. Such substrates include phosphonium salts, fluorene, alkynes, sulfoxides, sulfones, and others (Review: Wakefield, 1974). These reactions are often referred to as "metalation."

Increasing alkyl substitution at the carbanionic site in the base increases the basicity. For example, *sec*-butyllithium but not *n*-butyllithium will quantitatively remove a proton from allyl ethers in THF to give the allylic carbanion (Evans et al., 1974b). *tert*-Butyllithium in pentane will even remove a proton from methyl vinyl ether to generate α-methoxyvinyllithium (Baldwin et al., 1974).

The highly hindered mesityllithium is low enough in nucleophilicity to serve as a base in the formation of kinetic enolates from ketones (-78°), which may subsequently be acylated with acid chlorides (Beck et al., 1977).

Grignard reagents have been used in the same manner to generate acetylide anions, imine enolate anions, and others. Isopropylmagnesium chloride is especially useful for preparation of α-anions from α-arylacetic acids and other compounds of similar acidity (Blagoev et al., 1970).

An aryl proton may be removed if it is ortho to a substituent that is able to complex the metal cation (Reviews: Wakefield, 1974; Slocum et al., 1974). Such ortho-directing groups include ethers, amines, carboxamides, sulfonamides, oxazolines (Gschwend et al., 1975), thioamides (Fitt et al., 1976), and fluorine. For examples of addition reactions of such aryl carbanions see Eq. 17 (Ranade et al., 1976) and Eq. 18 (House et al., 1977b).

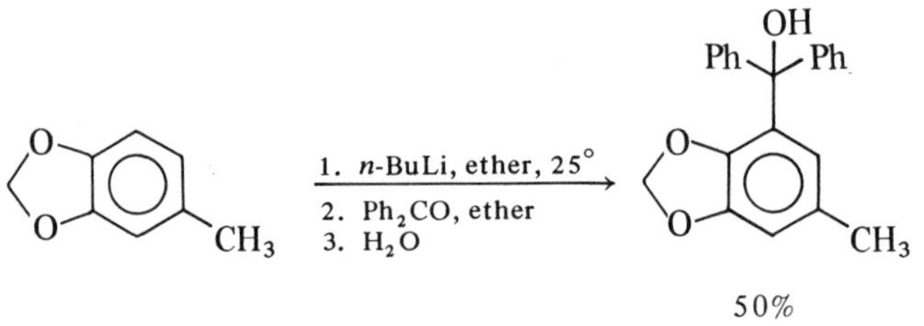

$$\text{(17)}$$

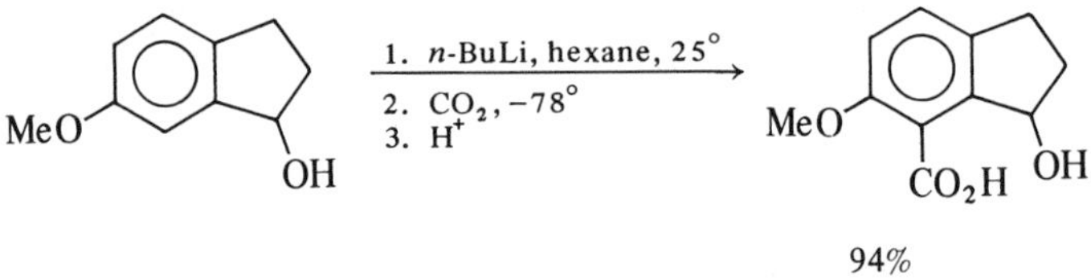

50%

$$\text{(18)}$$

94%

Arylsulfonyl hydrazones and *sec*-butyllithium give initially an α,*N*-dianion, but elimination then occurs to give a vinyl carbanion which may be treated with various electrophiles (Eq. 19) (Chamberlin et al., 1978).

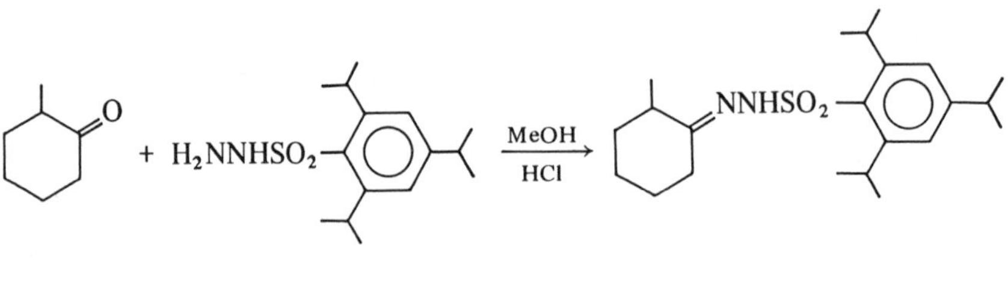

$$\xrightarrow[\text{hexane}, -78° \text{ to } 0°]{\textit{sec}\text{-BuLi}} \quad \text{[structure]} \; Li^+ \; + \; N_2 \quad \xrightarrow[\text{rt}]{DMF} \quad \text{[structure with CHO]} \tag{19}$$

63%

The basicity of lithium reagents is greatly increased by complexing agents such as TMEDA which chelate the cation, leaving a greater negative charge on the carbanionic site. The n-butyllithium-TMEDA complex is capable of rapidly removing protons from benzylic and allylic positions and even an aryl proton from benzene to generate the corresponding new carbanion plus butane (Eq. 20) (Akiyama et al., 1973; Cardillo et al., 1974; Wilson, et al., 1975).

$$\text{[structure]} \; + \; n\text{-BuLi-TMEDA} \quad \xrightarrow[-78° \text{ to } 25°]{\text{ether}} \quad \text{[structure]} \tag{20}$$

For reviews see Rausch et al., (1974) and Smith (1974). HMPA can be even more effective, as may be seen in Table 3 for the preparation of 1-(ethylthio)vinyl-lithium (Oshima et al., 1973a).

Tertiary benzamides normally acylate lithium reagents, but if sec-butyllithium-TMEDA is used in THF at $-78°$, ortho metalation occurs in yields of 88 to 95% (Beak et al., 1977).

TABLE 3 PREPARATION OF 1-(ETHYLTHIO)-VINYLLITHIUM WITH VARIOUS BASES

Base	Percent Yield[a]
n-BuLi, THF	2-3
n-BuLi, TMEDA, THF	10-20
n-BuLi, HMPA, THF	68
sec-BuLi, THF	72
sec-BuLi, HMPA, THF	90

[a]The yield is for the overall synthesis of 2-decanone by alkylation of the carbanion and hydrolysis; some of the effect of HMPA may therefore be in the alkylation step.

1.4.8 Miscellaneous Bases. The methylsulfinyl carbanion, prepared from dimethyl sulfoxide with sodium hydride, is a stronger base than Ph_3C^- and is particularly useful for generating phosphorous ylids in DMSO solvent (Corey et al., 1962; Greenwald et al., 1963; Corey et al., 1977a).

Potassium hydride reacts rapidly with esters and nitriles to give high yields of condensation products. High yields of permethylated ketones are obtained by using potassium hydride to generate ketone enolate ions in the presence of methyl iodide (Millard et al., 1978). It is also highly reactive toward alcohols and amines giving the alkoxides and the potentially valuable potassium dialkylamides (Brown, 1973).

Potassium triphenylmethide gives high yields of enolate anions from ketones and is easily prepared by the reaction of triphenylmethane with potassium hydride activated by a small amount of DMSO (Huffman et al., 1977a).

Finally, the unusual base potassium graphite (C_8K) may be used to generate the α-anions of esters and nitriles in THF at $-60°$, which may then be alkylated (Savoia, 1977). A suspension of sodium adsorbed on amorphous carbon gives ketone enolates which may be selectively monoalkylated (Hart et al., 1977).

1.5 REDUCTION OF UNSATURATED COMPOUNDS

α,β-Unsaturated ketones may be reduced by a solution of lithium in liquid ammonia in the presence of a proton donor such as tert-butyl alcohol or water to afford a specific enolate anion without equilibration to other enolate anions. These anions may then be protonated, alkylated (Eq. 21) (Caine et al., 1977), or sulfenylated (Gassman, et al., 1977a). The product in Eq. 21 was accompanied

$$(21)$$

45 to 49%

by some 3-methylcyclohexanone but no 2,5-isomer and no more than 2% diallylated product.

Lithium tri-sec-butylborohydride gives reductive alkylation with α,β-unsaturated esters as well as ketones, as shown in Eq. 22 (Fortunato et al., 1976).

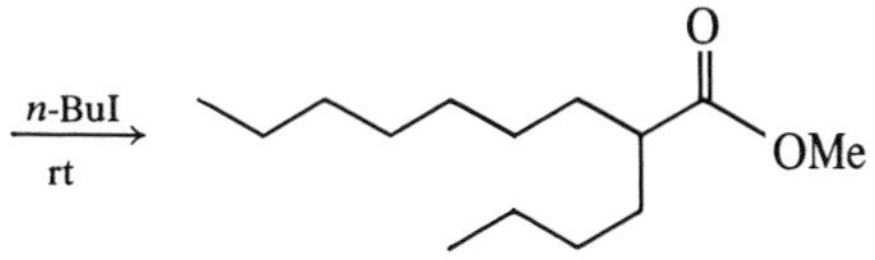

$$\xrightarrow[\text{rt}]{n\text{-BuI}}$$

63%

(22)

The carbanionic intermediate in the Birch reduction of benzoic acids can also be alkylated (Eq. 23) (Marshall et al., 1977a; see also Taber, 1976; House, 1977b; and Bachi, 1969).

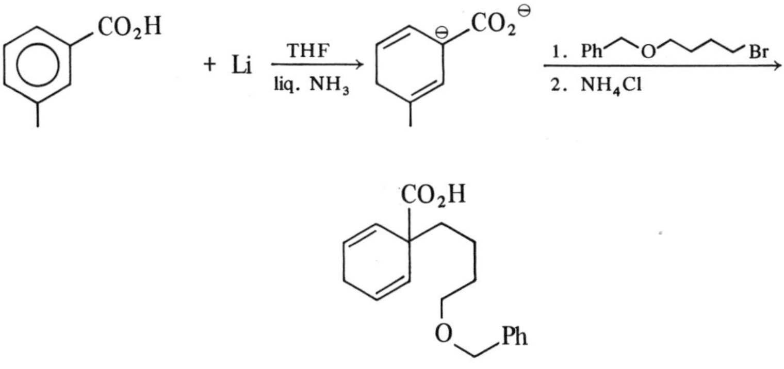

$$+ \text{Li} \xrightarrow[\text{liq. NH}_3]{\text{THF}} \xrightarrow[\text{2. NH}_4\text{Cl}]{1. \text{Ph}\frown\text{O}\frown\frown\text{Br}}$$

(23)

98%

Electrochemical reduction of phenylacetylene in the presence of aklyl halides produced alkylated acetylenes in high yield (Eq. 24) (Tokuda et al., 1976).

$$\text{PhC}\equiv\text{CH} + e^- \xrightarrow[\text{HMPA}]{n\text{-Bu}_4\text{NI}} [\text{PhC}\equiv\text{C}^-] \xrightarrow{\text{EtI}} \text{PhC}\equiv\text{CEt}$$

99% yield
71% conversion

(24)

CHAPTER 2

General Reaction Types

CARBANIONS are very reactive synthetic intermediates and, once prepared, they are promptly treated with another reagent which produces a new bond to the carbon that bore the negative charge. The carbanion may simply act as a base and pick up a proton, or it may behave as a nucleophile and attack a positively polarized carbon atom to generate a new carbon-carbon bond. If the partially positive carbon bears a good leaving group, the result is displacement of that leaving group. If it is π bonded, the carbanion attaches to the positive end of the π bond, leaving a negative charge on the other end of the π bond which may then pick up a proton, giving overall addition. The elaboration of these and other reactions is the subject of this chapter. The number of potential cross combinations between the many kinds of carbanions and the many reactions is enormous. Here we select from these to illustrate the variety of reaction possibilities. In Chapters 3 to 6 the opposite approach is made, showing the applications by type of carbanion, particularly emphasizing those reactions which lead to a new carbon-carbon bond.

2.1 PROTONATION

Addition of a relatively acidic compound, usually water, to a carbanion gives proton transfer, that is, formation of a C-H bond. If the carbanion was originally prepared from an organohalide, the overall process is reduction.

When D_2O or HTO is used, we have the most common way to label compounds isotopically. For instance, 4-bromobiphenyl was converted to the 4-lithio compound by exchange with butyllithium in ether and then quenched with D_2O to give biphenyl-4-d (Streitwieser et al., 1965). Using a relatively weak base and a low concentration of carbanion allows reversible deuterium replacement of all of a set of equivalently exchangeable protons. For high-purity labeling, the D_2O and DHO should be replaced with fresh D_2O several times to wash out all the protons, as in Eq. 1 (Cook, 1976). Many other compounds

$$\text{(1)}$$

>90% D_4 after two treatments

have been deuterated similarly, including for example 3-methylcyclopentanone (Lipnick, 1974), cyclopropyl methyl ketone (Creary, 1976), and fluorene (Streitwieser et al., 1971). Those compounds that are not sufficiently acidic to exchange in aqueous or alcoholic media may be deuterated in dimethyl sulfoxide-d_6 (Eq. 2) (Chen et al., 1971).

$$PhCH_3 + CD_3SOCD_3 \xrightarrow[150° \text{ to } 160°]{CD_3SOCD_2^-} PhCD_3 \tag{2}$$

$$>99\% \text{ } \alpha\text{-exchanged}$$

On the other hand, selective monodeuteration may be accomplished by first using a strong base to convert all of the substrate to the carbanion at once and then adding the deuterated acid. Monodeuteroacetylene was prepared in this manner as shown in Eq. 3 (Lompa-Krzymien et al., 1976). 9-Fluorenyllithium,

$$HC{\equiv}CH + n\text{-BuLi} \xrightarrow[-80°]{THF} HC{\equiv}C^- \text{ } Li^+ \xrightarrow[-80°]{CH_3CO_2D} HC{\equiv}CD \tag{3}$$

$$92\% \text{ singly labeled}$$

prepared from fluorene using butyllithium, gives some dideuteration when quenched directly with D_2O, but the clean monodeutero compound is produced when magnesium bromide is added to the lithium reagent and then the D_2O (Streitwieser et al., 1971).

If the carbanion is reversibly generated from a chiral center or other stereo determinant, stereoisomerization may occur, for example the cis cyclopropane compound in Eq. 4 was converted to the more stable (less crowded) trans epimer via the ketone enolate anion (Welch et al., 1977).

$$\tag{4}$$

When the carbanion is allylic or propargylic, protonation may occur on a carbon other than the one from which an atom was removed, thus leading to structural isomerization. In the allylic cases under equilibrium conditions, the double bond is moved to a lower-energy, more stable position with greater substitution or conjugation as in the examples shown in Eqs. 5 and 6 (Anderson

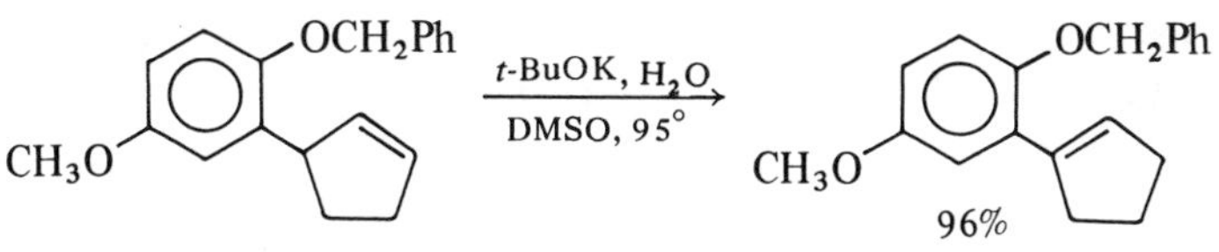

$$\text{(5)}$$

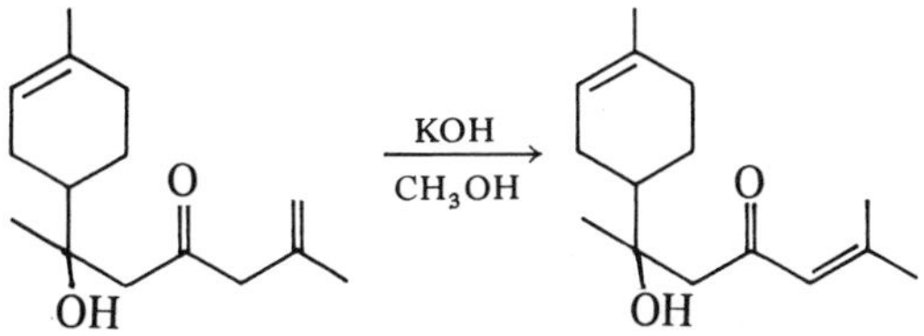

$$\text{(6)}$$

et al., 1977; Kergomard et al., 1976). Allyl ethers may be isomerized to the easily cleaved propenyl ethers by potassium *tert*-butoxide in dry DMSO (Stephenson, 1976). Protonation under kinetically controlled conditions can often be used to isomerize double bonds out of conjugation. For instance verbenone was converted to the β,γ-unsaturated isomer as shown in Eq. 7 (Ohloff et al., 1977). Similar kinetic deconjugation of steroidal α,β-unsaturated ketones

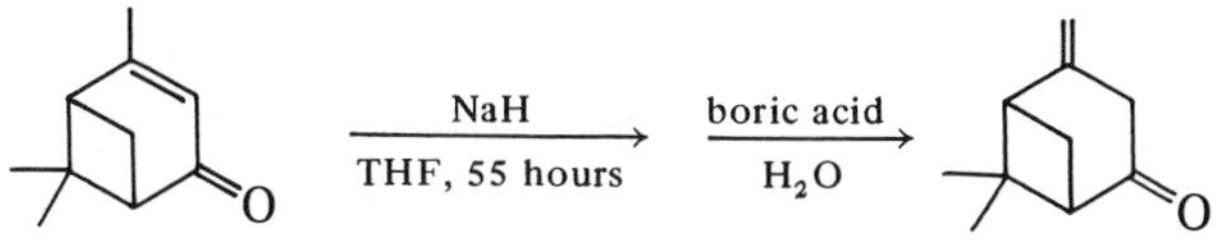

(±)-verbenone   88% yield

$$\text{(7)}$$

(Ringold et al., 1962), an octalone (Armour et al., 1959), and methyl 2-nitro-3-methylbut-2-enoate (Baldwin et al., 1977) are known.

Acetylenes are isomerized by strong bases. 1-Alkynes are converted to 2-alkynes by heating with sodamide in DMSO (Klein et al., 1970) or with lithium acetylide in DMSO or HMPA (Lewis et al., 1967; Tyman, 1975). The reverse change may be brought about by treatment with excess sodamide in hot hydrocarbon solvents (Eq. 8) (Caine et al., 1969; Lewis et al., 1967; Bainvel et al., 1963). This may be a reflection of the insolubility of acetylide salts in nonpolar solvents. More recently, multipositional isomerization of alkynes to the terminal

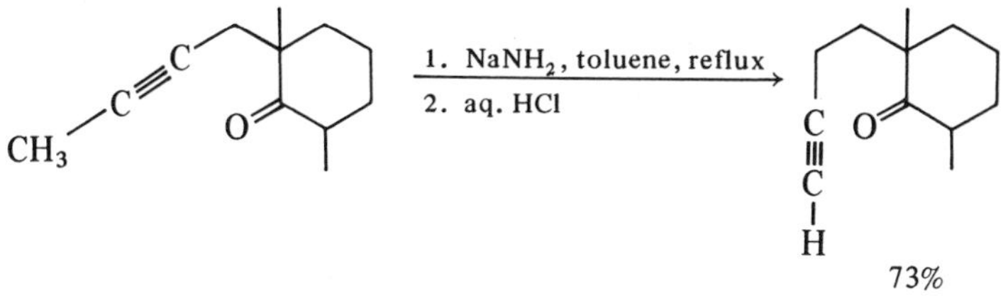

$$\text{(8)}$$

73%

position has been effected under mild conditions using excess potassium 3-aminopropylamide (Eq. 9) (Brown, C. A. et al., 1976).

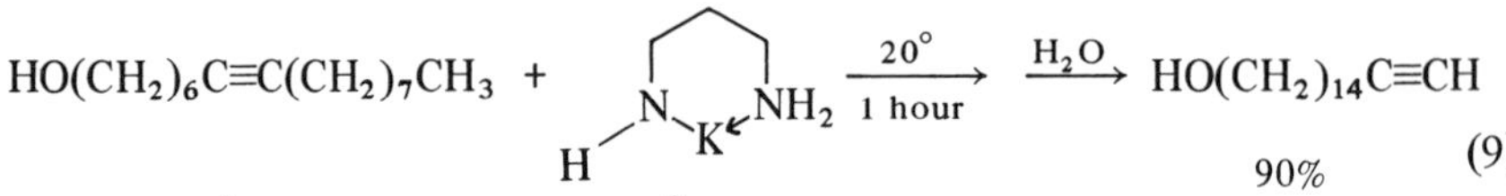

$$HO(CH_2)_6C \equiv C(CH_2)_7CH_3 \ + \quad \underset{\substack{H \\ \ \\ 1}}{\overset{}{\underset{N}{\bigcirc}}}_{K} NH_2 \ \xrightarrow[\text{1 hour}]{20^\circ} \ \xrightarrow{H_2O} \ HO(CH_2)_{14}C \equiv CH$$

$$1 \qquad : \qquad 3 \qquad\qquad\qquad\qquad 90\% \qquad (9)$$

Protonation by atmospheric water is a problem to avoid in most carbanion reactions. The reagents and solvents should be thoroughly dried and the solutions transferred by syringe or cannula. The reactions should be blanketed by an atmosphere of dry nitrogen or other inert gas.

2.2 ALKYLATION OF CARBANIONS

Carbanions are nucleophilic and are frequently used to displace a leaving group from an alkylating agent to form a new carbon-carbon bond. Common alkylating agents include primary and secondary alkyl halides, sulfonates, sulfates, and epoxides. For examples see Eqs. 10 and 11 (Murphy et al., 1973; Kondo

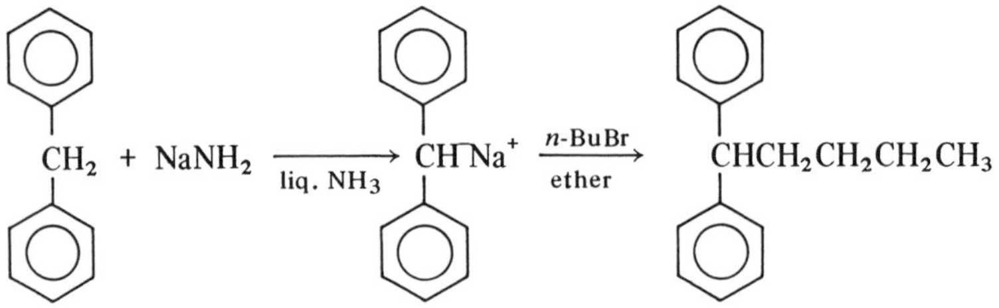

$$\underset{}{\overset{}{Ph_2CH_2}} + NaNH_2 \xrightarrow{\text{liq. NH}_3} Ph_2CH^- Na^+ \xrightarrow[\text{ether}]{n\text{-BuBr}} Ph_2CHCH_2CH_2CH_2CH_3 \quad (10)$$

97%

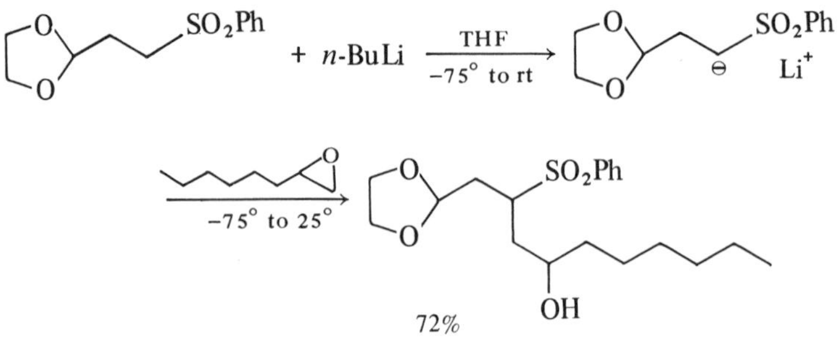

72%   OH

et al. 1975b). Higher yields and faster reactions are found with "active" alkylating agents such as allyl and benzyl bromides, methyl iodide, and ethyl bromoacetate.

The carbanions are often more nucleophilic than the bases used to prepare them, and thus the alkylating agent may be present already when the base is added. In the example shown in Eq. 12, butylamine formation is not a problem even though the concentration of n-butyl bromide is high and the concentration of the dithiocarbanion is low (Schill et al., 1975). In this way alkylations can be carried out with low equilibrium concentrations of carbanion. This is seen again in intramolecular alkylations. In the ring closure shown in Eq. 13, the potassium

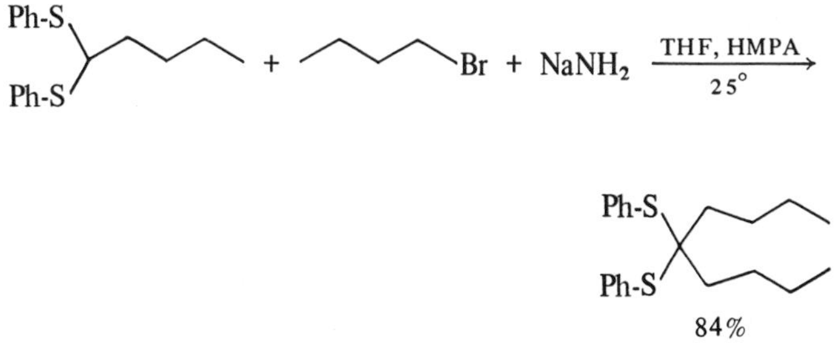

$$(12)$$

amide did not attack the labile epoxide (Stork et al., 1974b).

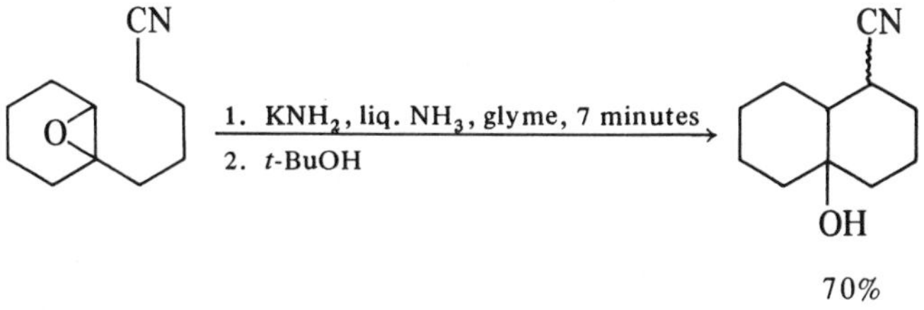

$$(13)$$

Carbanions derived from α,β-unsaturated nitriles or carbonyl compounds are generally alkylated at the α position often leading to β,γ-unsaturated compounds. Examples of this preference for nonconjugated products are shown in Sections 5.1.1, 5.3.1, and 5.6.1. If an α proton remains, the base may cause isomerization to the α-alkylated α,β-unsaturated compound.

2.3 PHENYLATION OF CARBANIONS

Direct phenylation of enolate carbanions with aryl halides does not occur since simple aryl halides are not susceptible to nucleophilic attack. However if the aryl halide is treated with an amide anion to generate transient benzyne, this is subject to attack by enolate anions. Thus the combination of 2-pentanone, bromobenzene, and two equivalents of sodamide gives 3-phenyl-2-pentanone in 65% yield (Leake et al., 1959).

Another method of phenylation is the treatment of enolate anions with diphenyliodonium chloride. The mechanism is complex and involves free radicals. The enolate anion prepared from ethyl phenylacetate using sodamide in liquid ammonia gave ethyl diphenyl acetate in 57% yield when treated with an equimolar amount of diphenyliodonium chloride. Ethyl cyclohexanone-2-carboxylate was similarly phenylated in .60% yield using potassium *tert*-butoxide in *tert*-butyl alcohol to generate the anion (Beringer et al., 1963). γ-Arylation of β-dicarbonyl compounds has been accomplished with this reagent also (Eq. 14) (Review: Harris and Harris, 1969). The high yield in this

(14)

98%

example is based on a 2:1 ratio of dianion to iodonium salt. One mole is arylated and the second mole is then consumed making the dianion of the phenylated diketone since that product is more acidic than the monoanion of the starting diketone (Hampton et al., 1964).

Most alkyllithium reagents give metal halogen exchange with iodobenzene, but methyllithium in ether gives toluene in 91% yield (Whitesides et al., 1969). Lithium dialkylcuprates give some coupling with iodobenzene, but the reaction is again complicated by metal-halogen exchange.

2.4 ACYLATION OF CARBANIONS

Ketones are readily prepared by acylation of carbanions using various carboxylic acid derivatives. The Claisen condensation is an example where a low equilibrium concentration of an ester enolate anion is acylated by more ester. Sulfones, nitriles, ketones, and amides may be acylated similarly. Esters may also be used as acylating agents on oxythioacetal anions as shown in Eq. 15 (Herrmann et al., 1973e).

$$(15)$$

Since the acylation products are usually acidic toward the starting carbanion, they will consume a second mole of that carbanion (or an equivalent of base) to form the enolate anion. In a few cases this is avoided by slowly adding the preformed carbanion to the acylating agent, rather than vice versa (Beck et al., 1977).

Most Grignard or lithium reagents are too reactive to use with esters or acid chlorides under ordinary conditions since a second equivalent will react with the incipient ketone to give tertiary alcohols. The classic alternative has been to convert the Grignard reagents to zinc or cadmium reagents which will react with acid halides but not with ketones (Section 3.2). The lithium reagents may be converted to cuprates which react selectively with acid chlorides at $-78°$ while leaving the keto or a variety of other functional groups intact (Section 3.3.2). The Grignard or lithium reagents may be acylated directly with certain agents that lead to chelates or precipitates of salt derivatives of the ketones. For example anhydrides (at $-70°$) or S-2-pyridyl thioates will acylate Grignard reagents, and lithium carboxylates will do likewise for lithium reagents (Section 3.1.2).

2.5 ADDITION OF CARBANIONS TO CARBONYL GROUPS, NITRILES, AND IMINES

Almost all of the varieties of carbanions discussed in the following chapters will undergo addition to aldehydes and ketones affording alcohols. Illustrative examples are given in Eqs. 16 to 18.

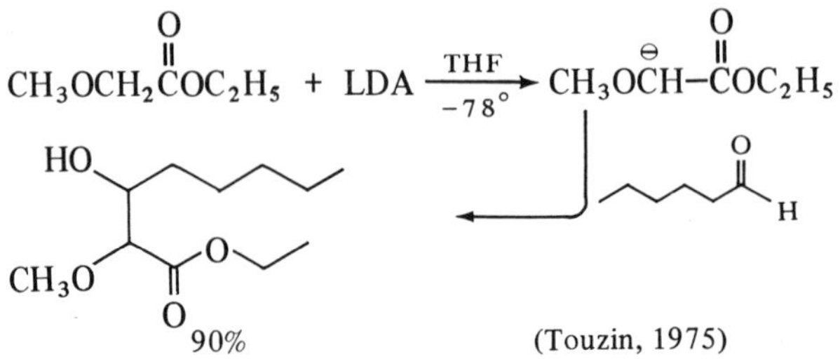

$$CH_3OCH_2\overset{\overset{O}{\|}}{C}OC_2H_5 \;+\; LDA \xrightarrow[-78°]{THF} CH_3O\overset{\ominus}{CH}-\overset{\overset{O}{\|}}{C}OC_2H_5$$

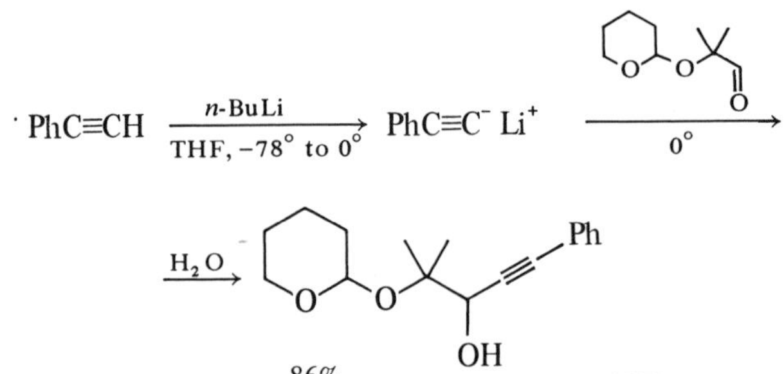

90%   (Touzin, 1975)   (16)

$$PhC{\equiv}CH \xrightarrow[THF,\,-78°\ to\ 0°]{n\text{-}BuLi} PhC{\equiv}C^-\ Li^+$$

86%   (Baldwin et al., 1977b)   (17)

+ CH$_3$MgBr $\xrightarrow{ether}$ $\xrightarrow[H_2O]{NH_4Cl}$ (18)

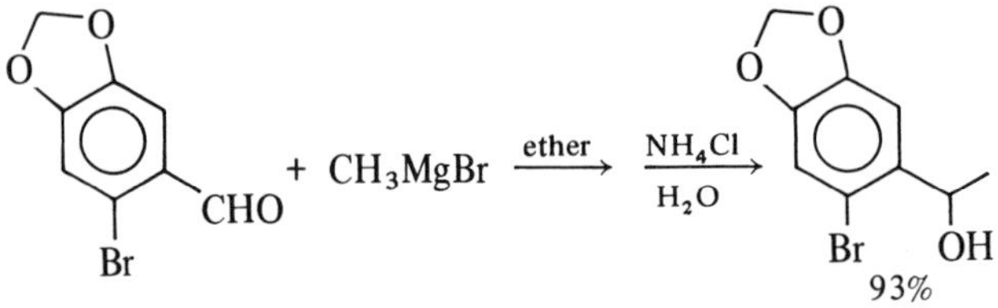

93%

(Ziegler et. al., 1978)

The aldol and the Knoevenagel condensations are examples of addition reactions (often followed by elimination of water) where low concentrations of carbanions are sufficient for high conversions.

The addition reaction involves both the attachment of the carbanion to the carbonyl carbon and also the coordination of the metal cation with the carbonyl oxygen. The attack of resonance-stabilized carbanions may require prior coordination of the metal to increase the degree of carbonium ion character on the carbonyl carbon. This has been demonstrated in the formation of chalcone from o-hydroxyacetophenone and benzaldehyde. Potassium hydroxide will produce the condensation product, but if a crown ether complexing agent is used to coordinate the K$^+$ ions, the reaction fails. Also, comparing potassium, sodium, and lithium hydroxides, the yield is highest with lithium, which is a better coordinator owing to its high charge density (Poonia et al., 1977).

The reagents most reactive toward carbonyl groups such as Grignard, lithium, and sodium reagents will add to carbon dioxide to give salts of carboxylic acids (Wagner et al., 1953). The less reactive enolate carbanions can be carboxylated directly using magnesium methyl carbonate (Eq. 19). In this reaction a methoxide is replaced by the carbanion so it is formally an acylation. The delicate β-ketoacids are stabilized as the intermediate magnesium-dianion chelate (Crombie et al., 1969).

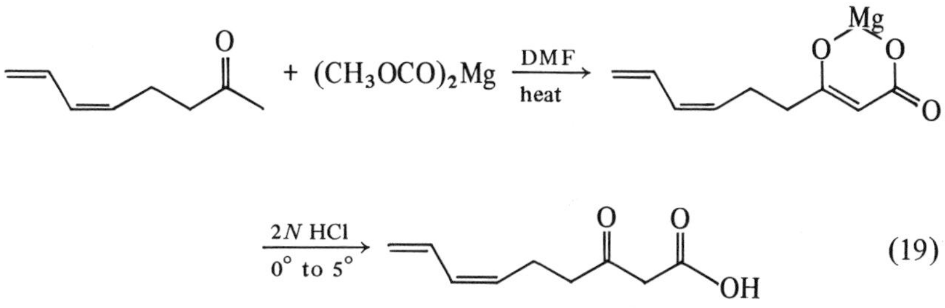

$$+ \ (CH_3OCO)_2Mg \xrightarrow[\text{heat}]{DMF}$$

$$\xrightarrow[0° \text{ to } 5°]{2N \ HCl} \tag{19}$$

100% yield as the potassium salt

Grignard reagents react with nitriles to give intermediate imine salts which are hydrolyzed to ketones (Section 3.1.3). They will also add to imines which do not possess an acidic α proton.

2.6 CONJUGATE ADDITION OF CARBANIONS

α,β-Unsaturated carbonyl and nitrile compounds may attract a carbanion R⁻ to the β position to give a resonance stabilized enolate anion (Eq. 20). This usually

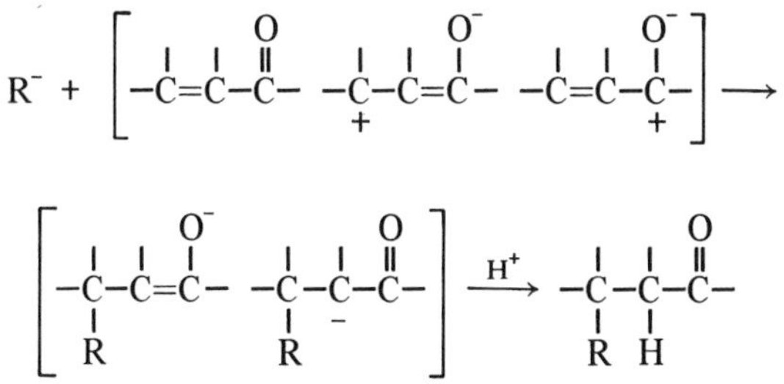

$$\tag{20}$$

gains a proton from another reactant (or from an acidifying workup) to give the overall addition of R-H to the carbon-carbon π bond (Reviews: House, 1972;

Bergmann et al., 1959). The process is frequently termed the Michael reaction. If one considers the metal to be coordinated with the oxygen (or nitrogen) at the intermediate stage, the conjugate addition is called 1,4-addition, while attack directly at the carbonyl function is called 1,2-addition.

Conjugate addition occurs readily if the starting carbanion is highly resonance stabilized, for example see Eq. 21 (Daniewski, 1975). The basic conditions

$$\tag{21}$$

80%

often lead to further reactions when a 1,5-dicarbonyl compound is produced. The combination of conjugate addition of a ketone enolate anion followed by an aldol condensation and dehydration, called the Robinson annulation, has been applied widely to cyclohexenone syntheses (Section 5.1.4).

There is often a competition between direct attack on a ketone function (1,2-addition) and conjugate addition when the substrate is an α,β-unsaturated ketone. The less stable carbanions including lithium reagents, Grignard reagents, allyllithium-TMEDA complexes, and dithianes favor 1,2-addition, while the more resonance stabilized carbanions favor conjugate addition. A likely explanation for this differentiation comes from studies of temperature and solvent effects on the competition.

Under kinetically controlled conditions, attack occurs at the carbonyl group, giving 1,2-addition. If the attacking carbanion is sufficiently resonance stabilized, the initial attack will be reversible owing to the comparable stability of the carbanion and the alkoxide anion from the 1,2-attack. The slower attack at the β position may then accumulate, giving the most stable α-enolate anion as the major final product (Eq. 22).

$$\tag{22}$$

most stable

This process has been demonstrated in the addition of 2-lithio-2-phenyl-1,3-dithiane to 2-cyclohexenone (Eq. 23) (Ostrowski et al., 1977). When the addition was carried out in THF at $-78°$, warmed to $25°$, and then quenched, the conjugate addition product was formed in 93% yield. If the reaction is quenched at $-78°$, conjugate addition was only 35% of the product while the allylic alcohol was 65%. Moreover when a hexane-THF solvent mixture is used and quenched at $-78°$, the allylic alcohol was 95% of the product. Apparently at low temperature and low-polarity solvent, mostly 1,2-addition occurs while at $25°$ this reaction reverses and the slower conjugate addition predominates. 2-Lithio-1,3-dithiane gives only 1,2-addition to cyclohexenone; thus without the stabilization of the phenyl group, reversal of the initial attack is apparently energetically unfavorable.

This same temperature effect on the competition between 1,2- and 1,4-addition was demonstrated with the enolate anions of α-substituted propionates (Section 5.3.4). Some ester enolate anions however do not show reversibility in the 1,2-addition (Neef et al., 1977; Schultz et al., 1976).

The effect of increased resonance stabilization in the attacking carbanion is likewise demonstrated in additions of cyanohydrin acetal anions. Simple cases give mixtures of 1,2- and 1,4-addition to 2-cyclohexenone while the derivative of crotonaldehyde, which has more resonance stabilization, gives high yields of 1,4-addition (Section 5.6.4).

A second factor, steric hindrance, in the attacking carbanion or near the carbonyl group of the substrate can favor the reversal of the 1,2-addition or at least favor the 1,4-addition.

A third factor may operate in the irreversible additions to give some 1,4-addition. A concurrent transfer of the metal cation may be required in aprotic solvents. Where the metal is coordinated with carbon, the distance is short and only the carbonyl is spanned. Where the metal is coordinated at a heteroatom beta to the nucleophilic site in the nucleophile, it can span the distance to the conjugate position. For example, cyanohydrin acetal anions give an appreciable amount of conjugate addition, and the proposed concurrent transfer is illustrated in Eq. 24 (Stork et al., 1974d).

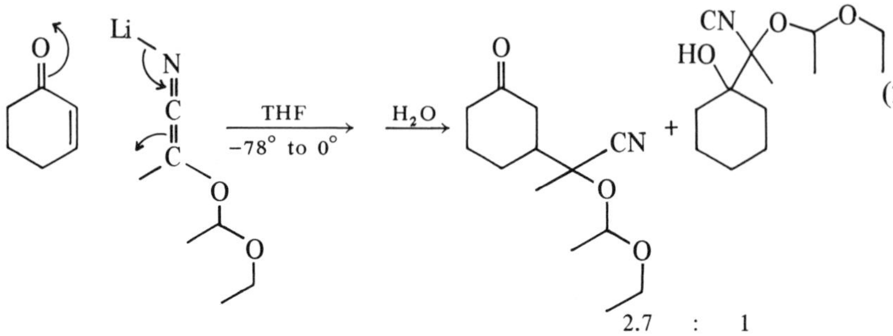

(24)

2.7 : 1

In summary, highly stabilized carbanions give conjugate addition to α,β-unsaturated ketones, intermediate cases give 1,2- or 1,4-addition depending on the temperature and solvent, and carbanions with little or no resonance stabilization give 1,2-addition.

Other α,β-unsaturated substrates may not offer 1,2-addition as a competition to conjugate addition and so will more often undergo conjugate addition. For instance α,β-unsaturated esters will give conjugate addition with some carbanions that give 1,2-addition with α,β-unsaturated ketones (Sections 4.9.4 and 5.3.4). α,β-Unsaturated nitro compounds are susceptible to conjugate addition with enolates and even simple dithianes (Seebach et al., 1975b). Moreover 2-isopropenyldihydrooxazines (Meyers, 1974) and 2-alkenyloxazolines are subject to 1,4-addition by simple Grignard and lithium reagents (Meyers et al., 1975d).

The presence of copper ions in addition reactions produces a dramatic switch from carbonyl addition to conjugate addition, as exemplified by the classic Grignard reaction with isophorone shown in Eq. 25 (Review: Posner, 1972):

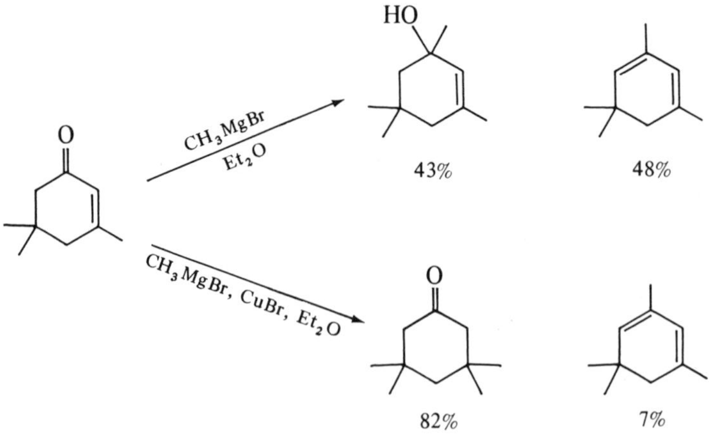

(25)

This selectivity is seen also with lithium diorganocuprates, and so conjugate addition is one of the major uses of these reagents (Section 3.3.3). Equation 26 shows such a reaction applied to the synthesis of prostaglandin E_1 (Sih et al., 1973, 1975).

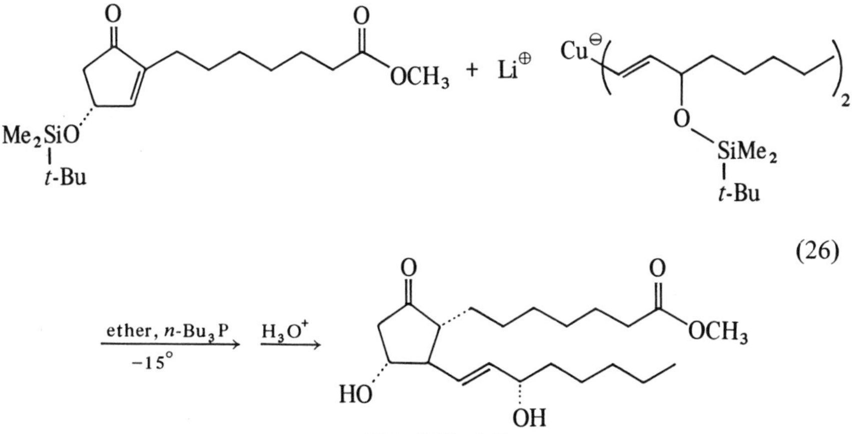

$$(26)$$

65 to 70% yield

2.7 ADDITION TO ACETYLENES AND DIENES

Primary alkylcopper reagents will add to the triple bond of terminal alkynes regio- and stereospecifically in ether in the presence of magnesium bromide or iodide. The resulting vinylcopper reagent can then be protonated, alkylated, halogenated, carbonated, or oxidized to the dimeric diene. An example illustrating the stereochemistry and high yields is the synthesis of nerol from propyne (Eq. 27) (Normant et al., 1974; Levy et al., 1977).

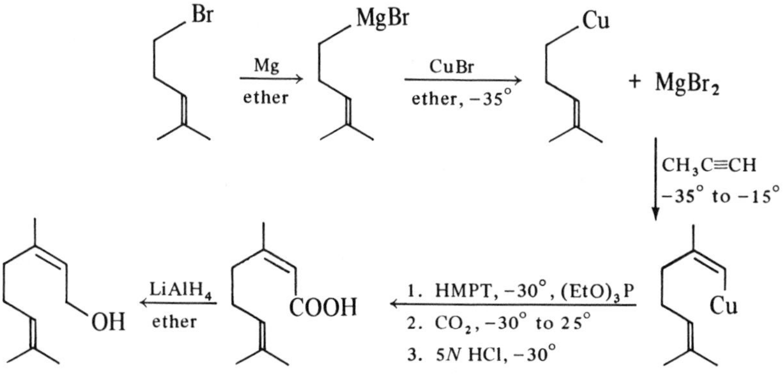

$$(27)$$

90% overall yield

It is proposed that the magnesium bromide serves as a Lewis acid coordinating with and polarizing the triple bond, making it electrophilic and thus attractive to the alkylcopper reagent. Acetylene and phenylacetylene are more acidic than other alkynes and they give an acid-base reaction leading to copper acetylides. This may be avoided by adding pentane to the ether suspension of the organocopper reagent before adding the acetylene.

In the solvent THF, secondary and tertiary as well as primary alkyl copper reagents will add to 1-alkynes. For addition to longer-chain 1-alkynes, the tert-butoxycuprates are superior (Eq. 28) (Westmijze et al., 1976). Addition

$$n\text{-BuCu}-\text{MgCl} + n\text{-C}_6\text{H}_{13}\text{C}{\equiv}\text{CH} \xrightarrow[-10°]{\text{THF}} \xrightarrow{\text{H}_2\text{O}}_{\text{NH}_4\text{Cl}} n\text{-C}_6\text{H}_{13}\overset{\overset{\displaystyle \text{CH}_2}{\|}}{\text{C}}\text{-}n\text{-Bu} \qquad (28)$$
$$\underset{t\text{-BuO}}{|}$$
$$95\%$$

also occurs regio- and stereoselectively to heteroatom-substituted acetylenes, giving new vinylcopper reagents which can then be converted to a wide variety of products. Addition to ynamines gives syn addition with the alkyl group attached to the end remote from the nitrogen, while alkynyl thioethers undergo syn addition with the opposite regioselectivity (Alexakis et al., 1977). Addition to a propargylic carbamate gives an allene since the initial vinyl carbanion has an adjacent leaving group (Pirkle et al., 1978).

Nitrile α-anions will add to unactivated acetylene using benzyltriethylammonium chloride catalyst (Eq. 29) (Makosza et al., 1976b).

$$\underset{\underset{\text{C}_2\text{H}_5}{|}}{\text{Ph}-\text{CHCN}} + \text{HC}{\equiv}\text{CH} \xrightarrow[\text{DMSO, 60° to 70°}]{\text{Et}_3\text{N}^+\text{CH}_2\text{Ph Cl}^-, \text{KOH}} \underset{\underset{\text{C}_2\text{H}_5}{|}}{\overset{\overset{\displaystyle \text{CN}}{|}}{\text{Ph}-\text{C}}-\text{CH}{=}\text{CH}_2} \qquad (29)$$
$$59 \text{ to } 63\%$$

n-Butyllithium adds rapidly to butadiene and the resulting allyllithium compound reacts at least as fast with more diene. This repeating process is anionic polymerization. sec-Butyllithium and tert-butyllithium can be used with equimolar amounts of diene to give good yields of a single addition. Equation 30 shows the preparation of neopentylallyllithium (Glaze et al., 1969, 1972).

$$t\text{-BuLi} + \overset{}{\diagup}\!\!\diagdown\!\!\diagup\!\!\diagdown \xrightarrow[-70° \text{ to rt}]{\text{pentane}} \qquad (30)$$
$$>85\%$$

In subsequent reactions these allyllithium reagents give mixtures of addition products attached at either end of the allylic system, "normal and rearranged products" (Glaze et al., 1977). *tert*-Butyl- and isopropylcopper reagents also add to butadiene at −15°. Subsequent alkylations occur at the primary carbon and additions occur at the secondary carbon (Normant et al., 1975).

Secondary and tertiary lithium reagents add to ethylene in high yield, although the mechanism may not be carbanionic (Bartlett et al., 1969).

2.8 OXIDATION OF CARBANIONS

Many conjugatively stabilized carbanions will absorb molecular oxygen rapidly to give hydroperoxides. If the hydroperoxide is on a tertiary carbon site, it can be reduced to an alcohol, while those on secondary sites usually eliminate water to give the carbonyl compound. Examples of these two kinds of results from the oxidation of ketone enolate anions are given in Eqs. 31 and 32 (Bailey et al., 1962). An extensive list of references is given by Vedejs (1974) (see also Wasserman et al., 1975). The same two kinds of results occur where oxazolines

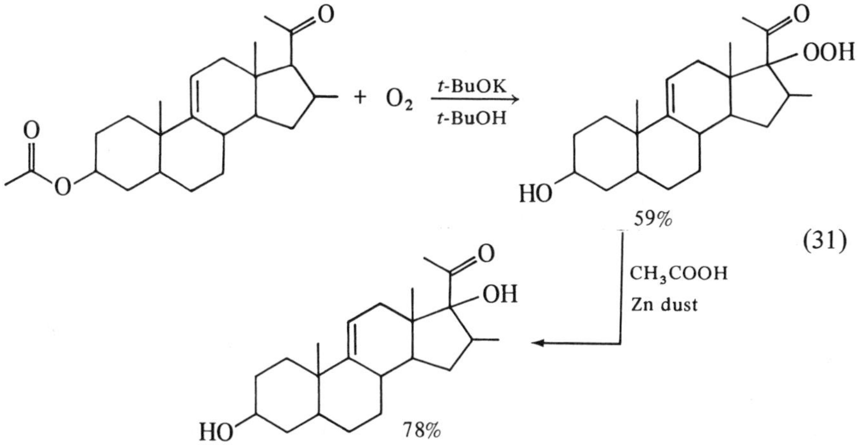

$$(31)$$

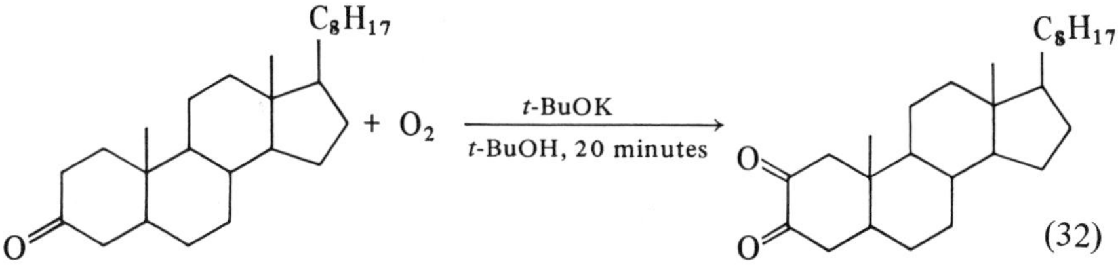

$$(32)$$

cholestanone   as the enols, 87% yield

activate tertiary and secondary sites (Eqs. 33 and 34) (Hansen et al., 1973 and 1976). The carbanion can be generated using potassium hydroxide and phase

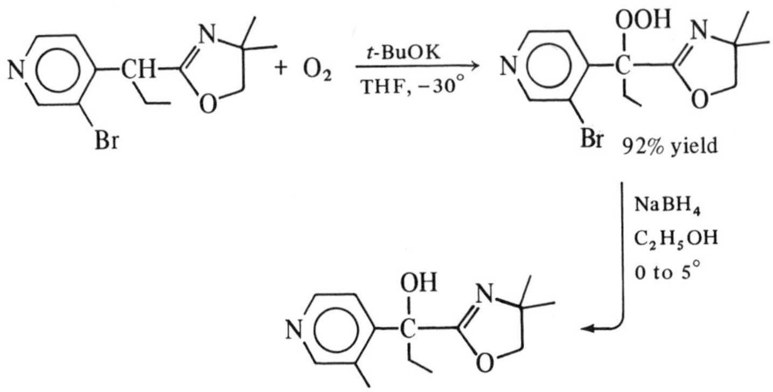

(33)

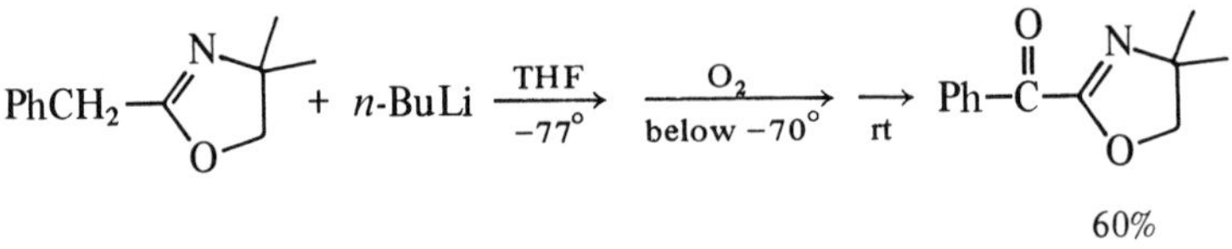

(34)

transfer catalysis, and air can be used as the oxidizing agent in the case of fluorene (Eq. 35) (Gokel et al., 1976b).

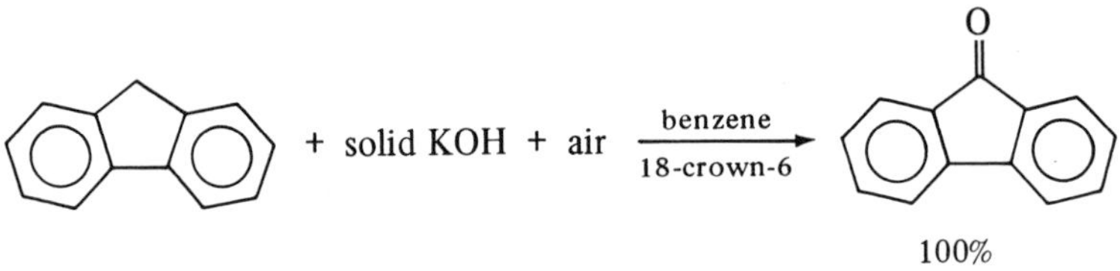

(35)

Nitriles with tertiary α carbons may be oxidized similarly to the hydroperoxide and reduced to the cyanohydrin, which may then be cleaved with base to ketones. If the nitrile was prepared by alkylation at the α position, the overall sequence amounts to a synthesis of ketones where the primary nitrile is an acyl carbanion equivalent (Eq. 36) (Selikson et al., 1975). A similar use has been made of nitriles derived from a Wittig reaction (Raggio et al., 1976).

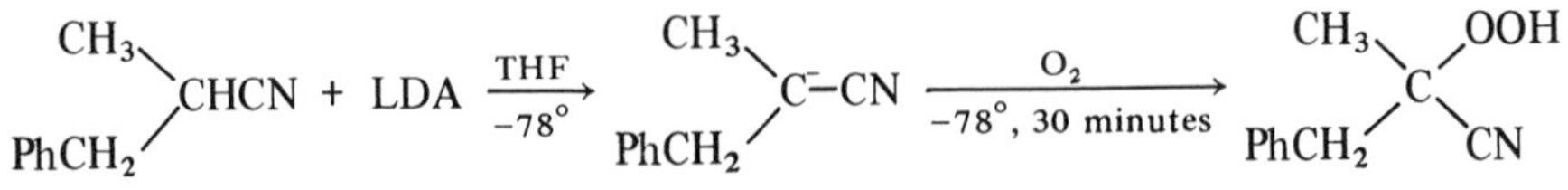

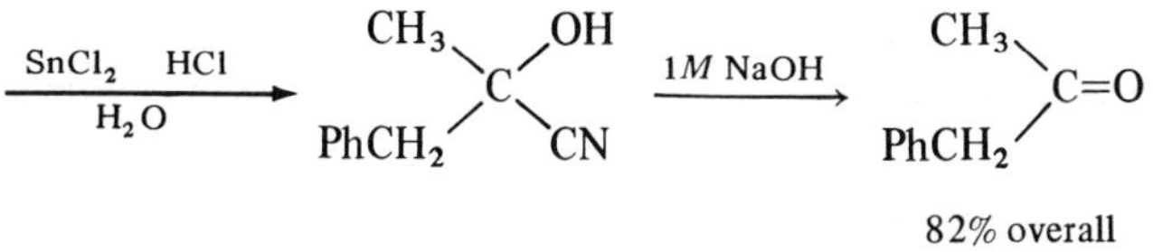

$$\xrightarrow[\text{H}_2\text{O}]{\text{SnCl}_2 \quad \text{HCl}} \quad \underset{\text{PhCH}_2}{\overset{\text{CH}_3}{\diagdown}}\text{C}\underset{\text{CN}}{\overset{\text{OH}}{\diagup}} \quad \xrightarrow{1M \text{ NaOH}} \quad \underset{\text{PhCH}_2}{\overset{\text{CH}_3}{\diagdown}}\text{C}=\text{O} \qquad (36)$$

82% overall

Amides are exceptional in that those with secondary α carbons such as *N,N*-dimethylbutyramide will give the α-hydroxy compound without ketone formation. The intermediate peroxides were reduced with sodium sulfite (Wasserman, 1975). Esters, lactones, ketones, and nitriles with secondary α carbons can all be α-hydroxylated using the peroxide MoO$_5$·pyridine·HMPA. This method does not involve hydroperoxide intermediates so that elimination to the ketone does not occur (Eq. 37) (Vedejs et al., 1974, 1976, and 1978).

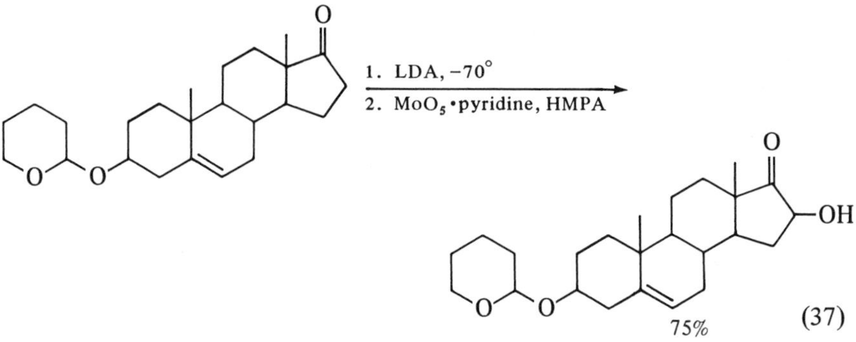

$$\xrightarrow[\text{2. MoO}_5\text{·pyridine, HMPA}]{\text{1. LDA, }-70°}$$

75%   (37)

A complicating side reaction in some of these oxidations is cleavage of the α carbon from the carbonyl carbon. This process has been used to advantage in a synthesis of β-lactams by Wasserman et al. (1976) (Eq. 38).

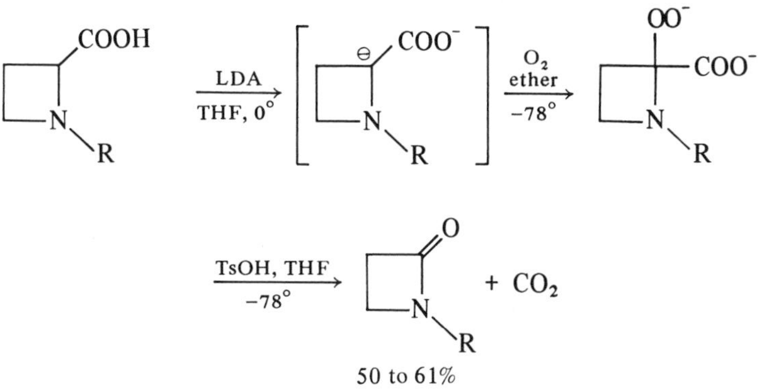

$$\xrightarrow[\text{THF, 0°}]{\text{LDA}} \qquad \xrightarrow[-78°]{\text{O}_2 \atop \text{ether}} \qquad\qquad (38)$$

$$\xrightarrow[-78°]{\text{TsOH, THF}} \qquad + \text{ CO}_2$$

50 to 61%

Unstabilized carbanions such as Grignard and lithium reagents react with oxygen to give alcohols, for example cyclohexylmagnesium chloride gives cyclohexanol in 80% yield (Review: Sosnovsky et al., 1966). This provides a convenient route to cyclopropanols (Eq. 39) (Longone et al., 1969).

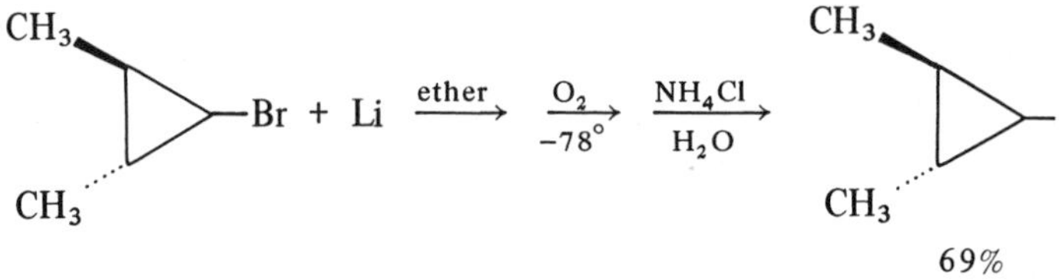

$$\text{(39)}$$

69%

Other oxidizing agents such as iodine or cupric salts will join pairs of carbanions. For example treatment of an ester enolate (Eq. 40) anion with iodine at $-78°$ gives succinates (Brocksom et al., 1975). Methyl and ethyl esters with

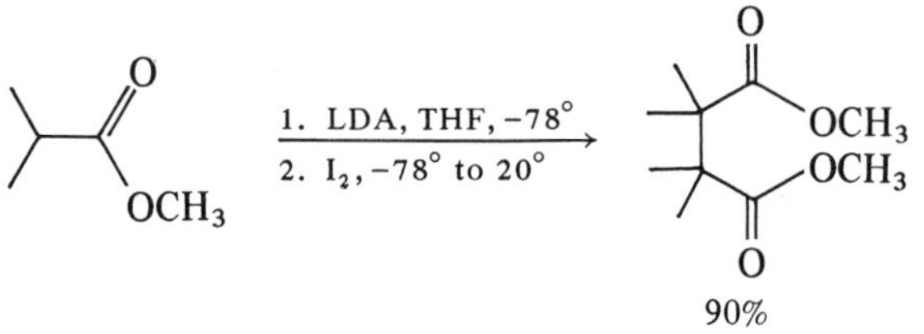

$$\text{(40)}$$

90%

more than one α hydrogen such as propionates give self-condensation, but the corresponding *tert*-butyl esters give the succinates in good yield. The α-anions of carboxylate salts likewise give succinic acids (Ivanov et al., 1975).

Iodine oxidatively joints hydrazone α-anions which can then be oxidized to 1,4-diketones (Eq. 41) (Corey et al., 1976a).

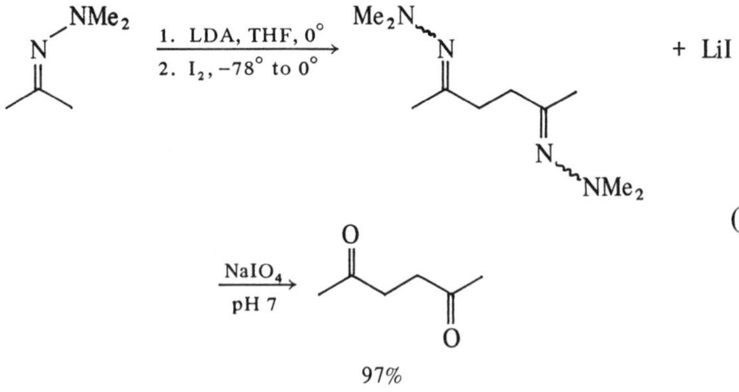

$$\text{(41)}$$

97%

1,3-Dibenzoylpropane was cyclized in high yield to *trans*-1,2-dibenzoylcyclopropane using methanolic sodium hydroxide and iodine (Colon et al., 1972).

Nitronate anions are oxidized by iodine to vicinal dinitro compounds which can be converted to tetrasubstituted olefins by reductive removal of the nitro groups (Kornblum et al., 1977).

Cupric trifluoromethane sulfonate will oxidize ketone enolate anions to give 1,4-diketones (Eq. 42) (Kobayashi et al., 1977). Cupric chloride in DMF also gives 1,4-diketones, particularly from methyl ketones such as β-ionone (Ito et al., 1977).

$$\text{Ph} \overset{\overset{\displaystyle O}{\|}}{\diagup} \xrightarrow[\text{2. Cu(OSO}_2\text{CF}_3)_2,\ \text{Me}_2\text{CHCN}, -78° \text{ to rt}]{\text{1. LDA, THF}, -78°} \text{Ph} \diagup \text{Ph} \qquad (42)$$

80%

2.9   MISCELLANEOUS HETEROATOM ATTACHMENTS

2.9.1   Halogenation. Ketones may be treated with a solution of sodium hydroxide and a halogen to give α-halogenated products. Under these conditions the enolate anion is formed reversibly in low concentration. The α-haloketone is more acidic than the original ketone (if there is another α hydrogen); thus with equilibrium enolate formation polyhalogenation predominates. This is useful for the cleavage of methyl ketones in the familiar haloform reaction, but if monohalogenation is desired, nonequilibrium conditions (or the acid-catalyzed reaction) must be used. If a ketone is first converted completely to the enolate anion with a strong base and then treated with a shortage of bromine, no dihalogenated products are found, for example see Eq. 43 (Anderson et al., 1977). 2-Methylcyclohexanone was selectively brominated in the 2-

$$\xrightarrow[\text{2. Br}_2, \text{CH}_2\text{Cl}_2]{\text{1. LDA, THF}, -78°} \qquad (43)$$

60% conversion
97% yield

position in 92% yield via the enolate anion specifically generated from the enol acetate (Section 5.1.1) with methyllithium. Bromination in the 6-position was carried out selectively in high yield from the other enolate anion (Stotter et al., 1973). These products are useful for elimination to specific α,β-unsaturation. Ester enolate anions may be brominated or iodinated in the same way (Rathke et al., 1971b).

Primary nitroalkanes are selectively monohalogenated when the preformed salt is treated with a halogen at $-78°$. At higher temperatures anion equilibration leads to some dihalo compounds (Erickson et al., 1977).

1-Alkynes are readily halogenated via a low concentration of the anion using aqueous sodium hydroxide-sodium hypobromite in a single operation (Eq. 44) (Pattison et al., 1963). 3-Hydroxy-1-octyne was chlorinated, brominated, and

$$F(CH_2)_4C{\equiv}CH + NaOH + Br_2 \xrightarrow{H_2O,\ ether} F(CH_2)_4C{\equiv}CBr \qquad (44)$$
$$86\%$$

iodinated quantitatively in the same manner (Rose et al., 1977; McConnell et al., 1977). Alternatively all of the alkyne can be converted to the anion using n-butyllithium and then chlorinated with p-toluenesulfonyl chloride or, at low temperature, iodinated with elemental iodine (Pattison et al., 1963; Ravid et al., 1977).

2.9.2 Amination. Enolate anions react with hydroxylamine derivatives to give α-amino compounds in fair yields (Eq. 45) (Scopes et al., 1977).

$$(C_2H_5O)_2\overset{\overset{O}{\uparrow}}{P}-CH_2\overset{\overset{O}{\parallel}}{C}OCH_3 \xrightarrow[\text{2. H}_2\text{NOSO}_2-\langle\bigcirc\rangle-]{\text{1. NaH, DME}} (C_2H_5)_2\overset{\overset{O}{\uparrow}}{P}-\underset{\underset{NH_2}{|}}{C}H\overset{\overset{O}{\parallel}}{C}OCH_3 \quad (45)$$
$$47\%$$

2.9.3 Sulfenylation. Ester enolate anions generated in THF react with dimethyl disulfide to give high yields of the 2-methylthio esters. Ketone enolate anions are less reactive but they are readily sulfenylated with diphenyl disulfide, methylsulfenyl chloride, or phenylsulfenyl chloride (Trost et al., 1976b; Seebach et al., 1973). Specific ketone enolate anions generated by reduction of α,β-unsaturated ketones in liquid ammonia may be sulfenylated with dimethyl disulfide (Gassman et al., 1977a).

Amide α-enolate anions generated with LDA in THF are readily monosulfenylated with dimethyl disulfide. When sodamide in liquid ammonia is used, polysulfenylation occurs; for example N-methyl-N-phenylpropionamide gave N-methyl-N-phenyl-2,2-di(methylthio)propionamide in 60% yield (Gassman et al., 1977b).

Oxidation of thioethers with sodium metaperiodate gives sulfoxides which may be thermally eliminated to afford α,β-unsaturated products overall. For example see Eq. 46 (Trost et al., 1976b). α-Phenyl thioketones are also useful for specific α,α-dialkylation of ketones (Section 5.1.1).

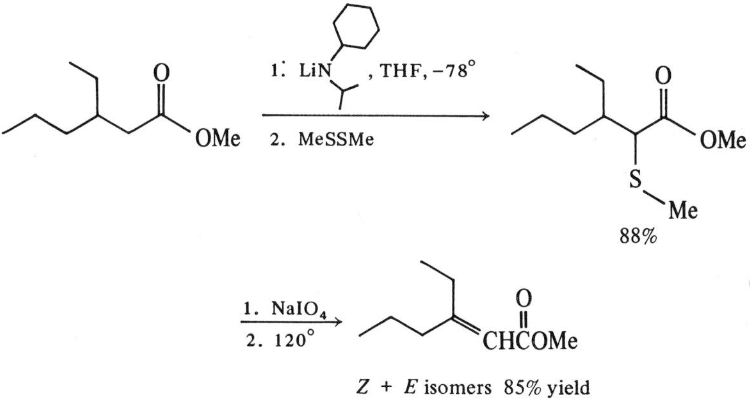

(46)

Z + E isomers 85% yield

The less nucleophilic carbanions from β-dicarbonyl compounds will not react with diphenyl disulfide but the more reactive sulfide sulfone gives thiophenylation (Eq. 47) (Quesada et al., 1978). Likewise, α,β-unsaturated esters can be α-sulfenylated with methyl methanethiosulfonate (Ortiz de Montellano et al., 1976).

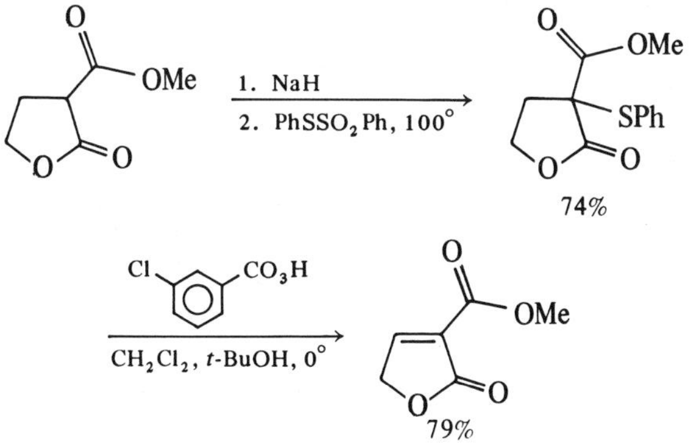

(47)

The anion of 2-substituted 1,3-dithianes may be sulfenylated with dimethyl disulfide which is an effective oxidation from the aldehyde level to the carboxylic acid level (Eq. 48) (Ellison et al., 1972).

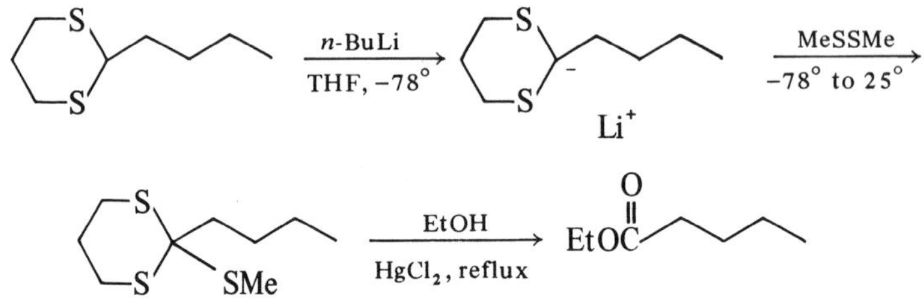

2.9.4 **Selenation.** The enolate anions of ketones and esters react with benzene-selenyl bromide or chloride or diphenyl diselenide at $-78°$ to give the α-phenylseleno compounds (Review: Clive, 1978). Oxidation with hydrogen peroxide or sodium periodate gives the α,β-unsaturated compound without heating since the intermediate selenoxide eliminates below room temperature. The mild conditions allow isolation of sensitive products such as the allenenic ester in Eq. 49 (Kocienski et al., 1977): Where positional selectivity is not

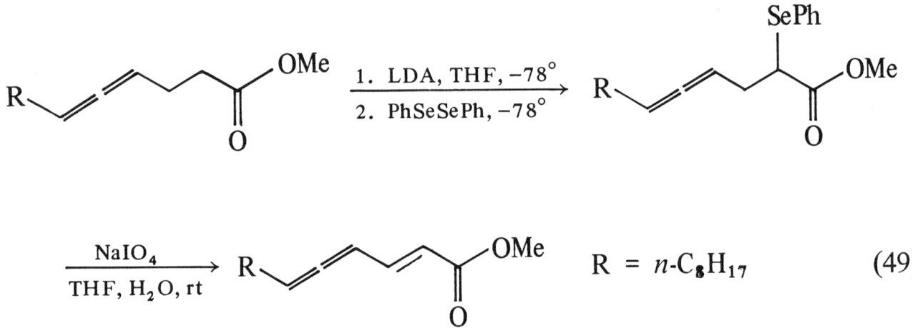

needed, the selenyl compounds can be prepared simply from the neutral carbonyl compound and phenylselenyl chloride in ethyl acetate at room temperature (Reich et al., 1975b; Sharpless et al., 1973).

2.9.5 **Phosphorylation.** Ketone enolate anions and the very reactive diethyl phosphorochloridate combine at oxygen to give the enol phosphate. This is analogous to the room-temperature reaction with acid chlorides (Section 5.1.2). These products are useful for the generation of specific alkenes from ketones, using specific enolate anion formation (Section 5.1.1) and lithium reduction of the phosphate as exemplified in Eq. 50 (Ireland, et al., 1969; Majetich, et al., 1977; Muchmore, 1972).

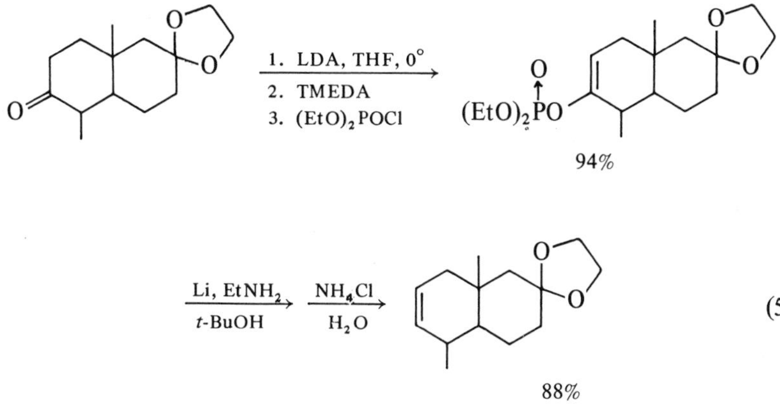

94%

88%

(50)

CHAPTER 3

Carbanions Without Heteroatom Stabilization

SIMPLE ALKYL GROUPS do not stabilize a negative charge; hence an alkyl carbanion associated with a large, very electropositive metal such as sodium is a highly reactive species, in fact too reactive for many synthetic purposes. The ionic nature is shown by the insolubility of, for example, n-butylsodium in hydrocarbon solvents. The concentrated negative charge on one carbon will give strong covalent bonding with the smaller and less electropositive metals so that with lithium we see less reactivity (eg. little coupling with halides) plus high solubility in hydrocarbon solvents. With zinc and cadmium we see even more covalency and reduced reactivity (no reaction with halides or addition to ketones).

When the carbanionic charge is resonance delocalized as in the allylic cases we find generally more ionic character to the compounds whatever metal is associated with them. This is noticed for instance in the low solubility of allyl Grignard reagents in ether solvents. Along with this ionic character we find higher reactivity even though the charge is delocalized. Allylic Grignard and lithium reagents are reactive toward alkyl halides where simple alkyl ones are not, and allylzinc reagents will add to ketones where simple alkyl ones will not. Among the highly resonance-stabilized cases, for example cyclopentadienyl, sodium is a commonly used cation.

Two factors concerned with the carbon atoms besides resonance lead to a range of stabilities. (1) The more s character in the hybridization at the carbanionic site, the more stable the carbanion. This effect manifests itself in the increasing relative basicities in the series acetylide, aryl, and alkyl carbanions (Jones, 1973) and in the direction in which metal-halogen exchange reactions are favorable. (2) High alkyl substitution at the carbanionic site destabilizes (intensifies) the carbanion owing to the electron-donating character of the alkyl groups. This is seen in the relative basicities again; for example sec-butyllithium is a stronger base than n-butyllithium (Section 1.4.6). Also, compare acetone with 3-pentanone in Table I, Section 1.4.1. Furthermore, the equilibration between a primary and a secondary lithium reagent (Eq. 1) gave $> 99\%$ of the more stable primary reagent (Hill et al., 1963).

(1)

These reagents are generally referred to as organometallic compounds and will be represented as RMgX and RLi. There is much evidence that they exist at least partly as aggregate multiples of these formulas in solution and in the crystalline state. The bonding in the aggregates is more complex than that implied by the simple formulas; for example ethyllithium forms a tetramer with alkyl groups bridging pairs of lithium atoms.

Although these carbanions have a fractional negative charge on any one nucleophilic site, they still are generally the most reactive of the synthetically useful carbanions compared to those in Chapters 4 to 6 where the more electronegative heteroatoms bear part of the negative charge.

3.1 ORGANOLITHIUM, MAGNESIUM, AND SODIUM REAGENTS

3.1.1 Alkylation. Epoxides and allylic halides are reactive enough to alkylate alkyllithium and magnesium reagents, but saturated alkyl halides will only alkylate the more reactive alkylsodium, vinyllithium, allyllithium, and allyl-magnesium reagents (Review: Courtois et al., 1974). The alkylation of an organometallic reagent by a halide is generally referred to as a coupling reaction.

Vinyllithium reagents may be alkylated with ethylene oxide or primary halides stereospecifically with greater efficiency than the alkylation of vinylcopper lithium reagents. For example in Eq. 2 an equimolar amount of lithium reagent and alkyl iodide gave a 77% yield of product (Millon et al., 1975;

$$\text{\includegraphics{}} \quad \text{Br} + \text{Li powder (2\% Na)} \xrightarrow[25°]{\text{ether}} \xrightarrow[\text{THF, 0°}]{n\text{-}C_8H_{17}I} \text{\includegraphics{}} \quad (2)$$

77% yield
94% isomeric purity

Cahiez et al., 1976b, 1977a,b). Allyllithium in ether solution, prepared from tetraallyltin and phenyllithium, gave a high yield of alkylation when used in threefold excess over 1-iodopentane (Eq. 3) (Whitesides et al., 1969).

$$\text{\includegraphics{}}\text{Li} + \text{\includegraphics{}}\text{I} \xrightarrow[25°, 1 \text{ hour}]{\text{ether}} \text{\includegraphics{}} \quad (3)$$

99%

Allyllithium reagents may be prepared directly from the hydrocarbon with the strong base combination *n*-BuLi and TMEDA. These complexed reagents give high yields too when used in twofold excess with primary halides (Eq. 4) (Akiyama et al., 1973).

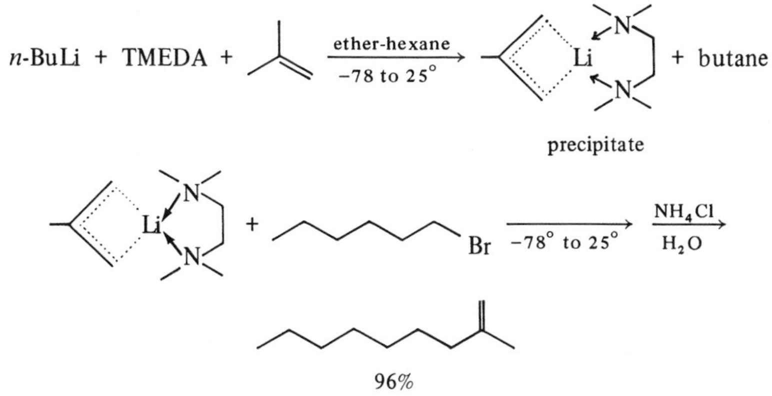

precipitate

$$(4)$$

96%

Allylic Grigard reagents are reactive toward alkyl halides, but they can still be prepared from the halides with only minor amounts of coupling. The crotyl reagent can be prepared from any of three isomers or a mixture of them, and it shows some reactivity at each site. It may be a bridged structure or an equilibrating pair of compounds (Eq. 5) (Review: Benkeser, 1971).

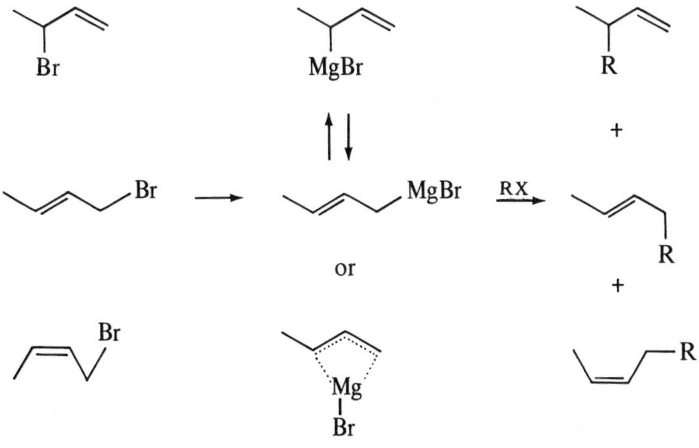

$$(5)$$

Alkyl tosylates are more selective than halides, reacting almost entirely at the more substituted end of the π system (Eq. 6) (Derguini-Boumechal et al., 1977).

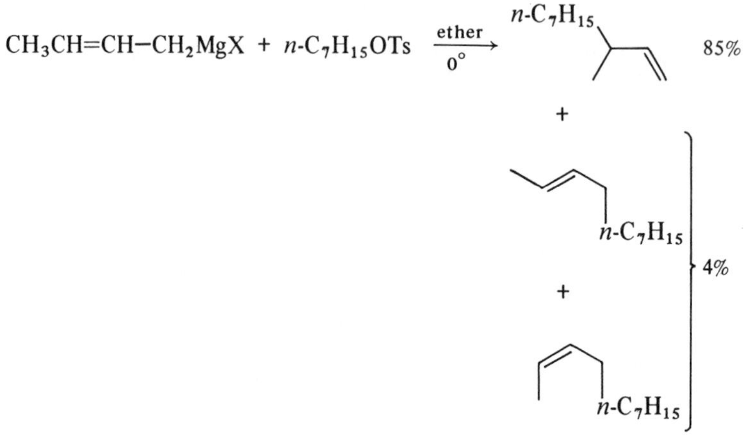
CH3CH=CH–CH2MgX + n-C7H15OTs
ether
0°
n-C7H15
85%
+
n-C7H15
+
n-C7H15
4%

(6)

Allylic chlorides and bromides are readily displaced by simple Grignard reagents, affording allylation in high yields (Eq. 7) (Kajiwara et al., 1975).

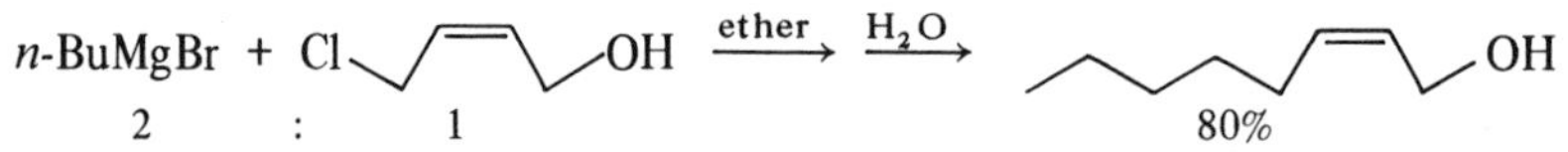
n-BuMgBr + Cl
OH
ether
H2O
OH
2 : 1
80%

(7)

Ethylene oxide is reactive enough to alkylate Grignard and lithium reagents since the strain in the three-membered ring aids in the cleavage of a carbon-oxygen bond. The reaction is not as fast as that with allylic halides, so heating in refluxing benzene is required to decompose the initial complex (Eq. 8)

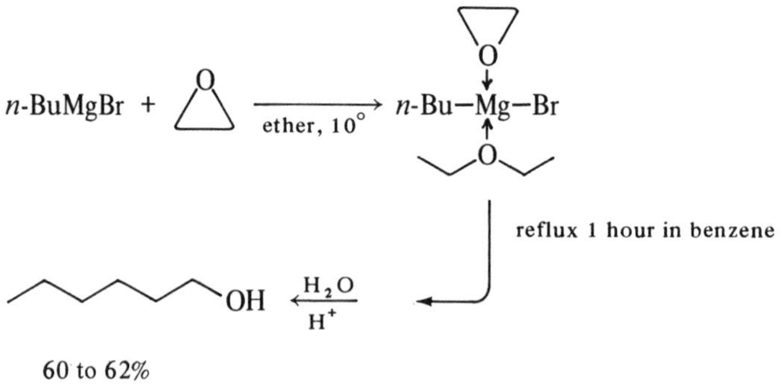
n-BuMgBr +
O
ether, 10°
n-Bu–Mg–Br
O
O
reflux 1 hour in benzene
OH
H2O
H+
60 to 62%

(8)

(Dreger, 1932). With higher homologs such as propylene oxide, the Lewis acid magnesium halide catalyzes rearrangement of the epoxide to a ketone or aldehyde, to which the Grignard then adds. This and halohydrin formation compete to lessen the yield of direct carbanion attack on the epoxide. These interferences may be eliminated by using diorganomagnesium reagents which give predominantly attack of the carbanion at the least hindered ring carbon (Eq. 9) (Schaap et al., 1968).

$$Et_2Mg + \underset{\text{O}}{\triangle} \xrightarrow[\text{0° to reflux}]{\text{ether}} \xrightarrow[\text{H}_2\text{O}]{\text{NH}_4\text{Cl}} \quad + \quad \quad (9)$$

44%   6%

The more resonance-stabilized cyclopentadienide anion is readily alkylated with alkyl halides, tosylates, or epoxides (Ohta et al., 1977). The sodium reagent often gives best results. The initially formed 5-alkylcyclopentadiene rapidly isomerizes at room temperature under the reaction conditions to the 1-alkyl isomer in good yield and isomeric purity (Eq. 10) (Breitholle et al., 1978). The

$$\text{Na}^+ + \text{TsO} \xrightarrow[\text{0° to rt, 4 hours}]{\text{THF}} \quad (10)$$

excess   85%

5-alkyl compounds may be isolated at low temperatures. For example Corey et al. (1969) prepared 5-methoxymethylcyclopentadiene by alkylating the sodium compound at $-55°$ and working up the product below $0°$.

3.1.2 Acylation. Since Grignard and lithium reagents are highly reactive toward ketones, acylation of these reagents with acid chlorides, anhydrides, and esters under ordinary conditions gives substantial amounts of tertiary alcohol as a by-product, see for example Eq. 20. Exceptions are known where the ketone is especially sterically hindered. If free ketone can be avoided in solution, the second attack will not occur. In the case of lithium reagents, reaction with a lithium carboxylate gives an insoluble dilithium salt which is stable in the absence of protic solvents and resists any addition that might lead to tertiary alcohol (Eq. 11) (Bare et al., 1969). This salt must be quenched by pouring it

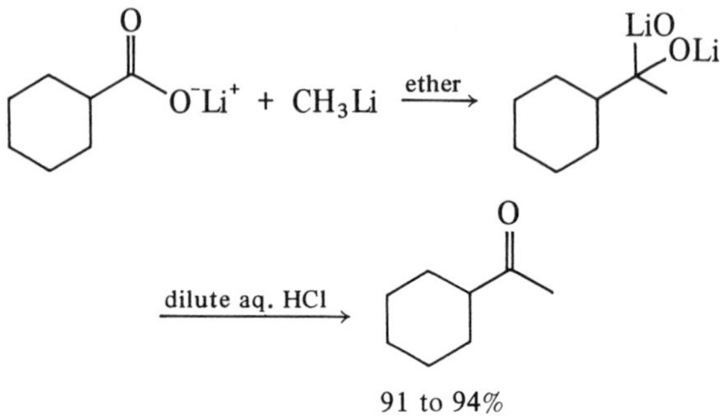

(11)

91 to 94%

into aqueous acid with vigorous stirring to avoid local contact of the rapidly formed ketone with any remaining lithium reagent, which would lead to tertiary alcohol. Vinyllithium is unreactive in ether, but in DME vinyl ketones can be made in high yield (Floyd, 1974).

Grignard reagents may be acylated with anhydrides at low temperatures (Newman, 1948). Here again the intermediate is a stable insoluble precipitate, perhaps the magnesium salt shown in Eq. 12. Acid chlorides were not useful

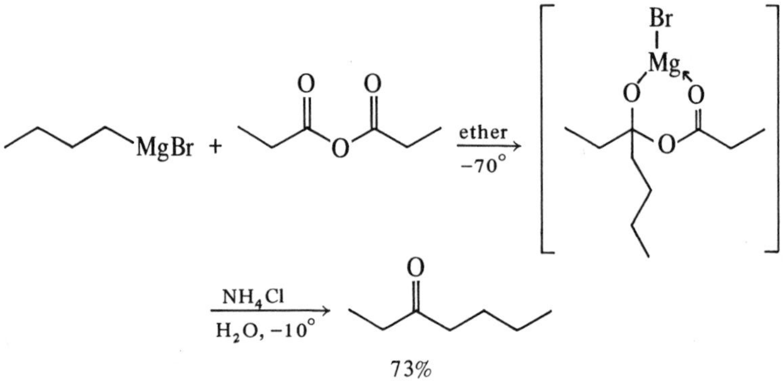

(12)

73%

even at these low temperatures since large amounts of tertiary alcohol result. In the solvent HMPA, Grignard reagents may be acylated with esters. The solvent promotes the acid-base reaction of the ketone giving the enolate which is not attractive toward more Grignard reagent (Eq. 13) (Huet et al., 1973).

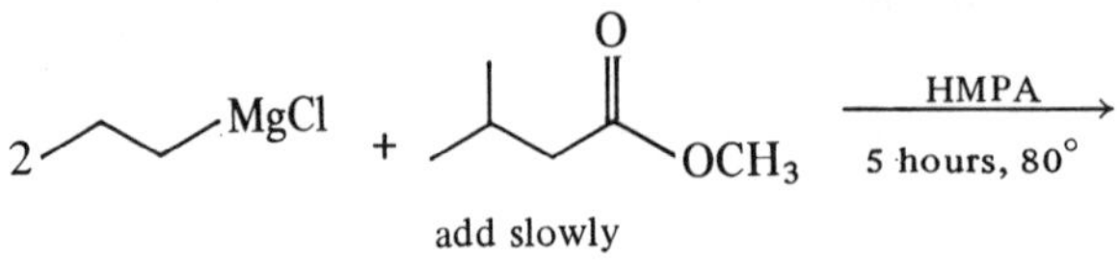

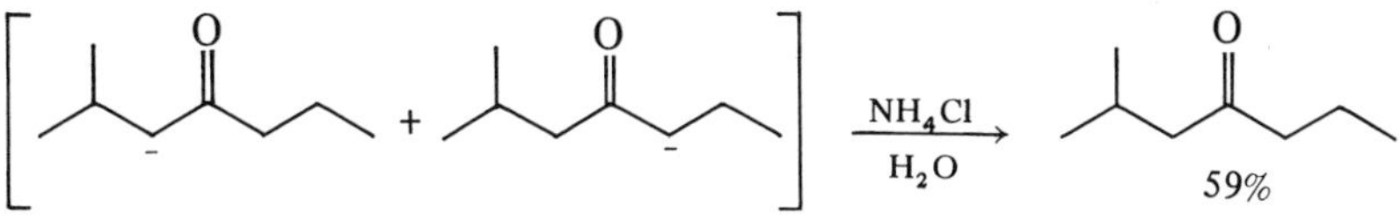

$$(13)$$

S-2-Pyridyl thioates are good acylating agents also. A stable six-membered ring intermediate forms that is unreactive toward carbanions but is readily hydrolyzed (Eq. 14) (Mukaiyama et al., 1973).

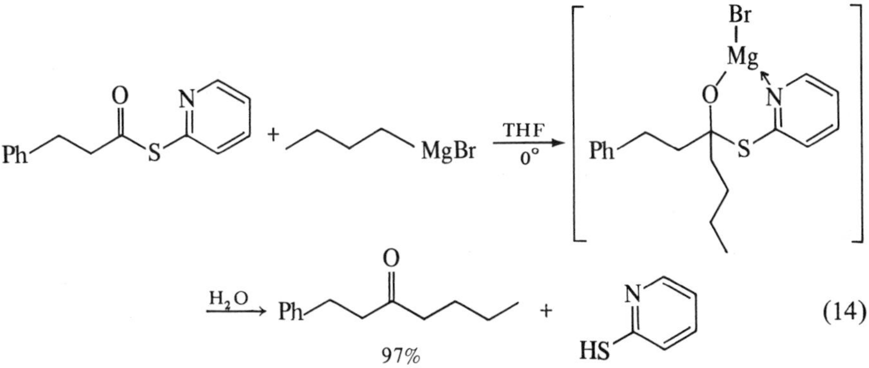

$$(14)$$

Both lithium and Grignard reagents may be acylated in good yield using tertiary amides. For example, isopentyllithium and N,N-dimethylpropionamide in ether at $-20°$ gave 6-methyl-3-heptanone in 75% yield (Izzo et al., 1959; Scilly, 1973). In the synthesis of (R)-marmin a piperidine derivative was treated with excess methylmagnesium iodide and the product converted to the

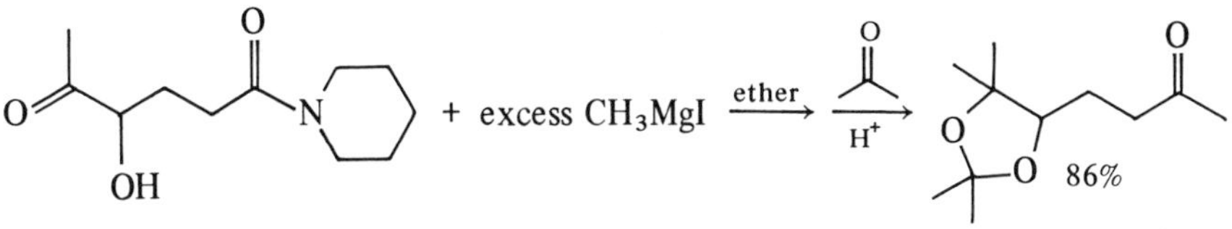

$$(15)$$

acetonide as shown in Eq. 15 (Yamada et al., 1976). Reaction of Grignard reagents with DMF gives the corresponding aldehydes (Eq. 16) (Nelson et al., 1977). The highly reactive N-acylimidazoles likewise give ketones with Grignard reagents (Staab et al., 1968). Aldehydes may also be prepared as the acetals by treating Grignard reagents with phenyldiethyl orthoformate (Stetter et al., 1970).

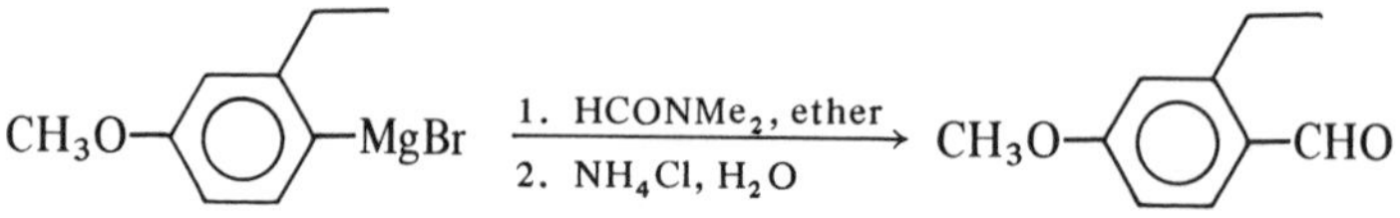

$$CH_3O-\!\!\langle\bigcirc\rangle\!\!-MgBr \xrightarrow[\text{2. } NH_4Cl, H_2O]{\text{1. } HCONMe_2, \text{ ether}} CH_3O-\!\!\langle\bigcirc\rangle\!\!-CHO \quad (16)$$

94%

The highly resonance stabilized carbanions give a variety of results under acylation conditions. Combining equivalent amounts of fluorene, potassium methoxide, and ethyl acetate in ether gave 9-acetylfluorene in 60% yield. Cyclopentadiene and indene gave only tars under these conditions (Von et al., 1944). 9-Fluorenyllithium and acetyl chloride gave only 1,1-bis(9-fluorenyl)-ethanol (Masaki et al., 1961). In contrast to this, indenylsodium and acetyl chloride gave the enol acetate of 1-acetylindene (Riemschneider et al., 1961).

3.1.3 Addition to C–O and C–N π Bonds. The most frequent application of these organometallic reagents is addition to ketones or aldehydes to give alcohols. Of the vast number of examples known, a few are illustrated here by Eqs. 17 and 18 (see also Chapter 1, Eq. 4). When lithium reagents are added to

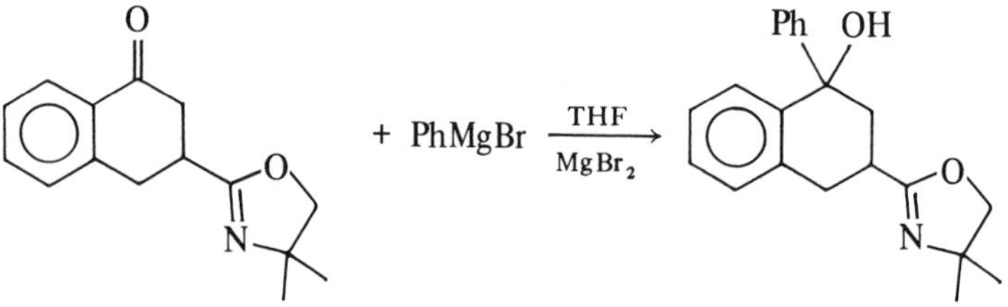

(17)

96%

(Meyers et al., 1970)

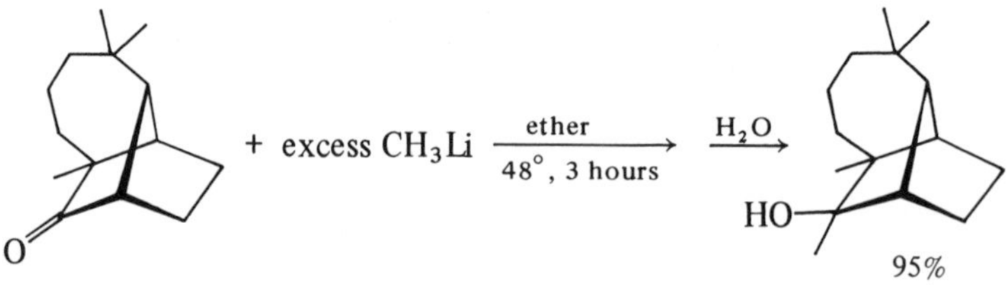

(18)

95%

(McMurry et al., 1972)

enolizable ketones, there is some enolate formed, especially with sterically hindered lithium reagents. The enolate formation is minimized and alcohol formation maximized if the diluted carbonyl compound is added to the lithium reagent at $-78°$ (Eq. 19) (Buhler, 1973). An example of addition to a ketone

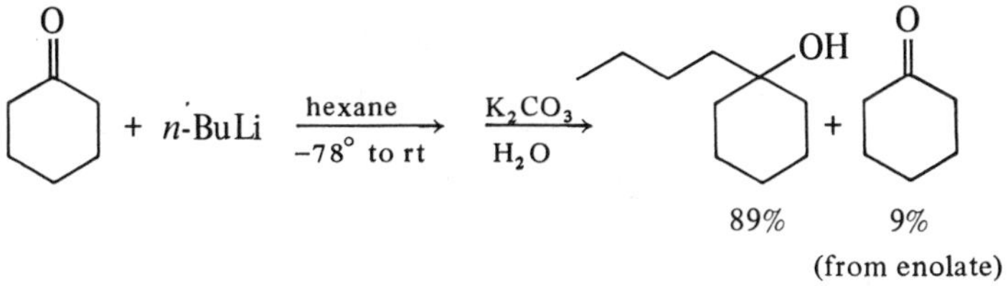

$$89\% \qquad 9\%$$
(from enolate)

formed in situ from an ester is shown in Eq. 20 (Bohlmann et al., 1976).

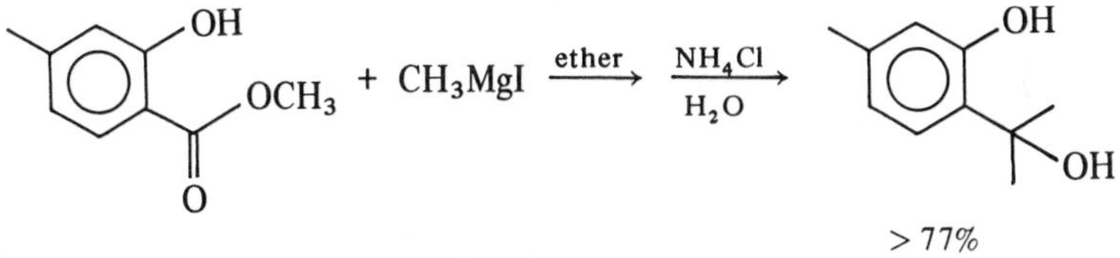

$$> 77\%$$

Vinyllithium reagents, prepared stereospecifically from 1-iodo-1-alkenes by exchange with butyllithium, add readily to ketones, aldehydes, and carbon dioxide in high yield with retention of configuration (Eq. 21). Ether solvent and low temperatures are required to prevent alkylation of the vinyllithium reagent by the butyl iodide produced in the exchange (Cahiez et al., 1976a).

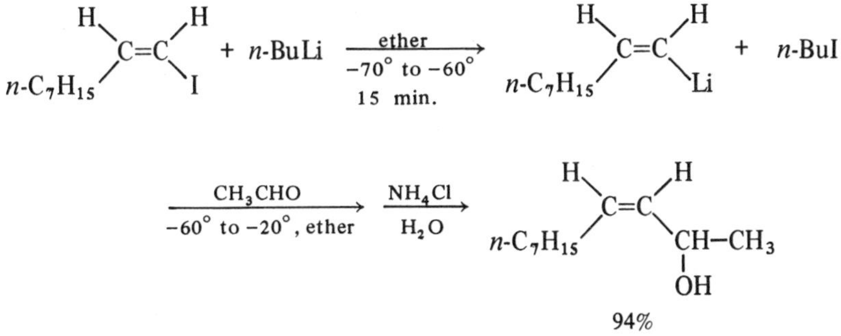

$$94\%$$

A similar addition can be carried out even with a hydroxyl group α to the carbanionic site (Eq. 22) (Corey et al., 1975c). The TMEDA complexed

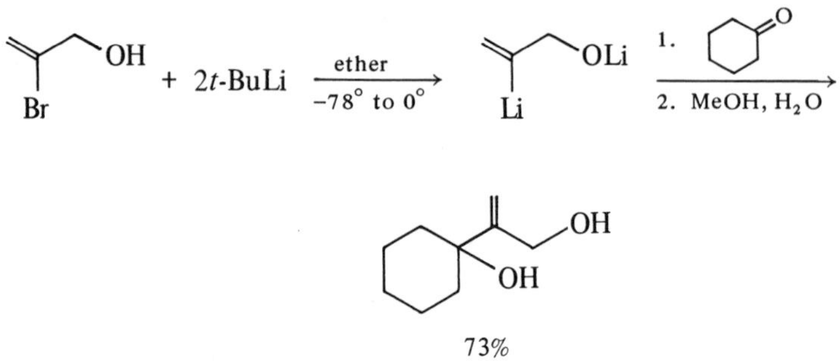

$$(22)$$

73%

allyllithium reagents give addition also as shown in Eq. 23 (Akiyama et al.,

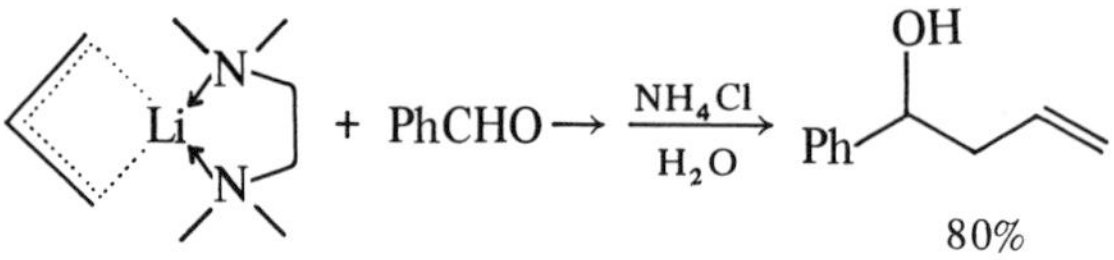

$$(23)$$

80%

1973). Unsymmetrical allyllithium and allyl magnesium reagents usually add to carbonyl compounds or carbon dioxide by attachment to the more substituted end of the π system (Eq. 24) (Reviews: Courtois et al., 1974; Benkeser, 1971; see also Rautenstrauch, 1974).

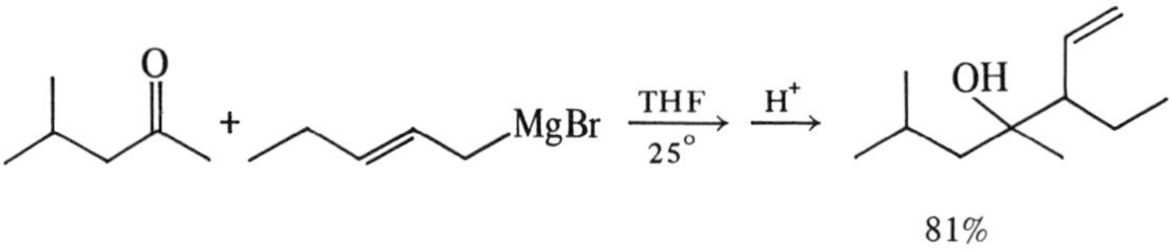

$$(24)$$

81%

Sodium cyclopentadienide gives addition followed by elimination to afford fulvenes (Eq. 24A) (Näf et al., 1977).

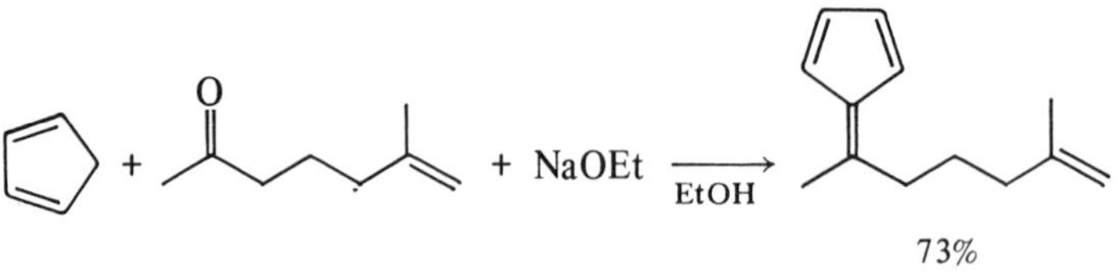

$$(24A)$$

73%

Grignard reagents add to nitriles but higher temperatures are required. The imine salt initially produced may be hydrolyzed to a ketone, which is equivalent to acylation of the Grignard reagent. The imine salt is not attractive toward more Grignard reagent; consequently formation of alcohols is not a problem as it is in acylation with other acid derivatives (Review: Wingler, 1973). This has been applied to the synthesis of racemic solanone by Johnson et al. (1965)

(25)

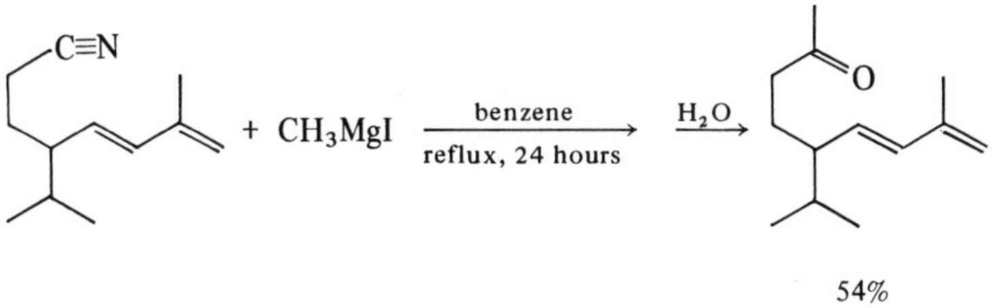

54%

(Eq. 25). The principal reason for the low yields is the action of the Grignard reagent as a base giving the α-anion which then adds to other nitrile groups. This can be suppressed by adding lithium perchlorate, which forms a complex with the Grignard reagent possessing greater reactivity toward addition (Chastrette et al., 1975). Lithium reagents will add similarly except with acetonitrile, phenylacetonitrile, and other more acidic compounds where α proton removal is faster. They are most commonly used with benzonitriles where this problem cannot occur.

Grignard reagents add to benzaldehyde imines to give secondary amines, but aliphatic aldimines and ketimines with α protons give imine enolate anions instead (Section 5.7). The more nucleophilic, less basic allylic reagents (Zn, Mg, and Li) do afford high yields of secondary amines from aliphatic aldimines (Miginiac et al., 1968). Primary amines may be prepared by the addition of organolithium reagents to *N*-alkylidenebenzenesulfenamides (Eq. 26). This is

$$\text{Ph-S-N=C(Me)}_2 \xrightarrow[\text{ether}]{t\text{-BuLi}} \text{Ph-S-N(Li)-C(Me)}_2\text{-}t\text{-Bu} \xrightarrow[\text{3. aq. NaOH}]{\text{1. H}_2\text{O; 2. 6}N\text{ HCl}} \underset{43\%}{\text{H}_2\text{N-C(Me)}_2\text{-}t\text{-Bu}} \quad (26)$$

especially useful for the tertiary alkyl cases where few synthetic alternatives are available. The phenylsulfenyl group serves as an easily cleaved masking group for the imine derivatives of ammonia (Davis et al., 1977).

Carbanion addition to an iminium ion was proposed for the diethylamine-catalyzed condensation of cyclopentadiene with aldehydes (Freisleben, 1963) as exemplified in Eq. 27 (Büchi et al., 1976).

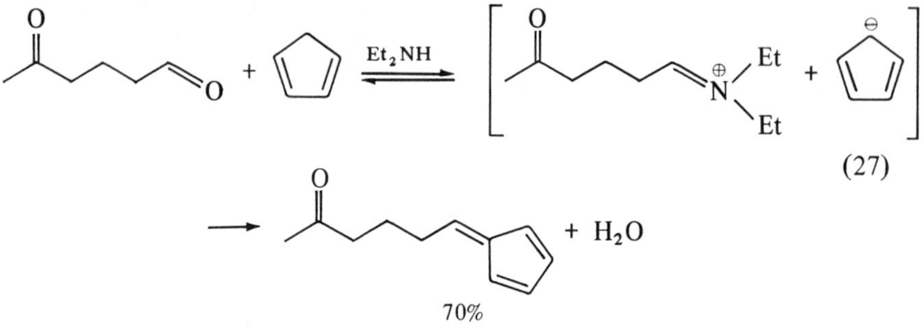

$$(27)$$

70%

Aryl and vinyl (but not the more basic alkyl) Grignard reagents will add to the iminium salt N,4,4-trimethyl-2-oxazolinium iodide. Hydrolysis of the resulting oxazoles affords aldehydes containing one more carbon than the halides used to prepare the Grignard reagents (Brinkmeyer, et al., 1974). A similar process has been used to prepare ketones from oxazinium salts (Meyers, 1974).

Lithium reagents will add to the anions of tosylhydrazones derived from aldehydes. The intermediate dilithio salt eliminates lithium toluenesulfinate and nitrogen to give overall reductive alkylation of the aldehyde (Eq. 28) (Vedejs, et al., 1977).

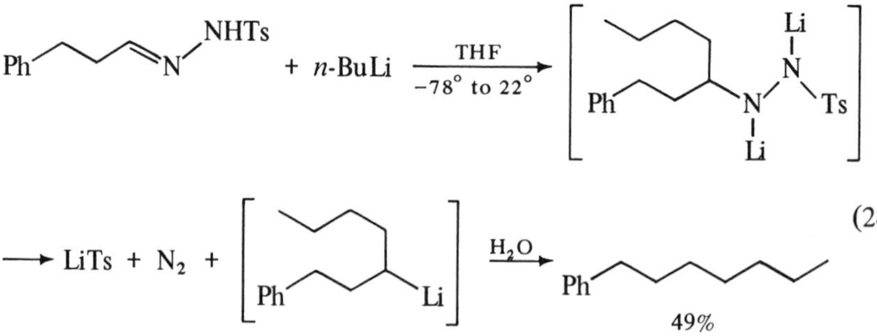

$$(28)$$

49%

3.2 ORGANOCADMIUM, ZINC, RHODIUM, AND MANGANESE REAGENTS

These less electropositive metals are more covalently bonded with carbon and the reagents are therefore less reactive toward electrophiles. They are principally used to prepare ketones from acid chlorides. The reagents may be prepared from the more reactive magnesium and lithium reagents by exchange with the metal halides. Zinc reagents may also be prepared directly from the metal and an organohalide (Reviews: Nützel, 1973c; Furukawa et al., 1974). Typical examples of ketone syntheses are given in Eqs. 29 to 32. In some cases methyl-

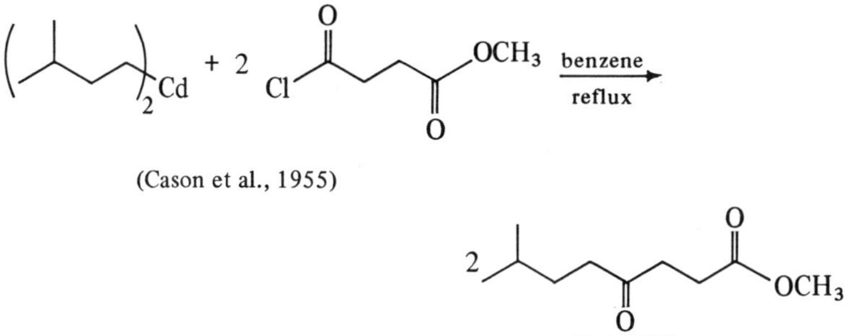

(Cason et al., 1955)

(29)

73 to 75%

cadmium chloride, prepared by combining equimolar amounts of Grignard reagent and cadmium chloride, gave higher yields of cleaner product than dimethylcadmium (Patel et al., 1976). If a dialkylzinc reagent is made from

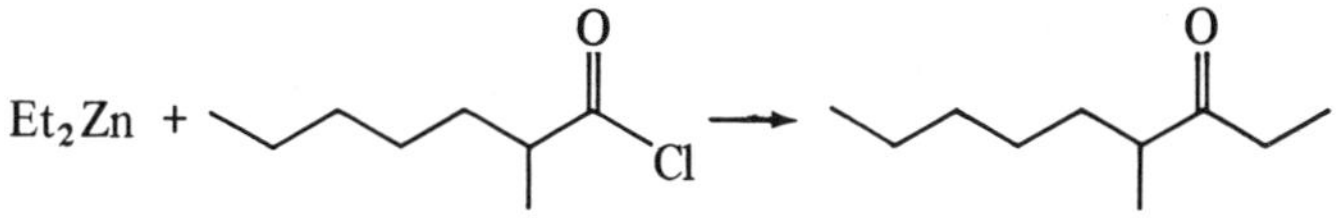

(Karrer et al., 1930)   80%   (30)

the Grignard reagent and used in the presence of the MgX_2, the magnesium salt will catalyze the addition of some of the excess R_2Zn to the ketone product and reduce the yield.

The readily available chloro(carbonyl)bis(triphenylphosphine)rhodium(I) complex exchanges with primary alkyllithium and magnesium reagents to give an alkylrhodium complex that can be acylated. The complex may be recovered ready for reuse after the acylation (Hegedus et al., 1975). This has been demonstrated in the synthesis of the alarm pheromone of the ants of the genus *Atta* (Eq. 31). An analogous polymer-bound complex is even easier to recycle (Pittman, Jr., et al., 1977).

$$RhCl(CO)(PPh_3)_2 + EtLi \longrightarrow LiCl + [Rh(Et)(CO)(PPh_3)_2]$$

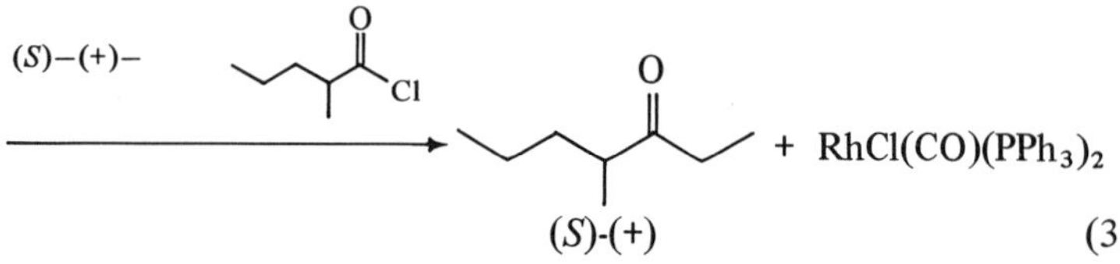

(31)

83% yield

Primary, secondary, or tertiary Grignard reagents can be converted to manganese reagents using manganese iodide. A 1:1 mole ratio of reagents gives high yields in the acylation, although the thermal instability of the secondary and tertiary reagents lowers the yield somewhat (Eq. 32) (Cahiez et al., 1976b).

$$n\text{-BuMgBr} + MnI_2 \xrightarrow[\text{ether}]{} n\text{-BuMnI} + MgBrI$$

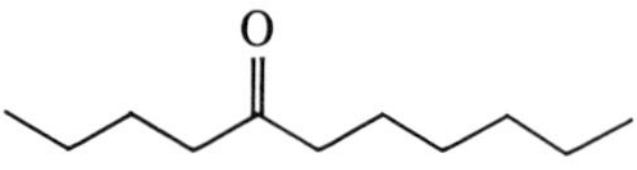

(32)

91%

The high yield of acylation reflects a high reactivity toward the acid halide, but the reagents will also react with ketones and aldehydes to give alcohols. There is good selectivity in the addition also; for example δ-ketohexanal gave the keto alcohol with n-butylmanganous iodide in 78% yield.

Zinc reagents have also found use in addition reactions. Dimethyl- and diethylzinc are sufficiently reactive to add to aldehydes to give secondary alcohols. Higher homologous reagents may reduce the aldehydes to primary alcohols. Magnesium salts may accelerate the addition. The more reactive allylic reagents add readily to ketones and aldehydes in THF even without magnesium halide catalysis. The yields have been improved by passing a solution of allyl bromide and the ketone or aldehyde through a column of hot activated zinc metal (Eq. 33) (Ruppert et al., 1976). As with Grignard and lithium reagents, the

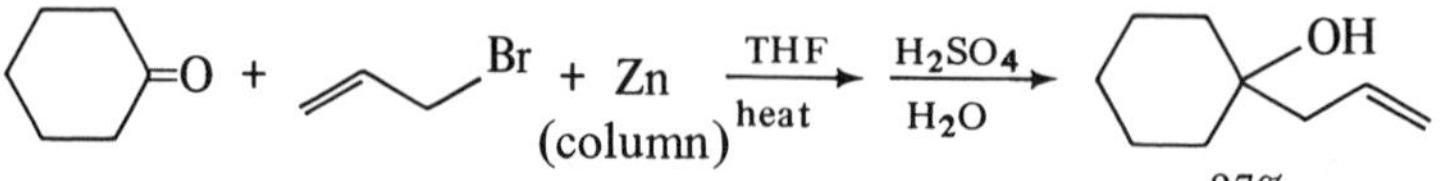

(33)

97%

unsymmetrical allylic reagents add to aldehydes at the more substituted end of the allylic system as shown in a simple preparation of artemesia alcohol (Eq. 34).

$$91\% \qquad (34)$$

3.3 ORGANOCOPPER REAGENTS

The addition of copper salts to lithium and magnesium reagents produces three generally useful changes in reactivity. First, the carbanionic reagents become capable of reaction with alkyl halides to give alkylation. Second, the reagents become more selective and can be acylated with acid chlorides without concomitant attack on ketones, alkyl halides, or other functionality. Third, in reactions with α,β-unsaturated carbonyl compounds, the reagents prefer conjugate (1,4) addition at the β position over addition to the carbonyl group. These characteristics are most frequently demonstrated with the diorganocuprate reagents R_2CuLi. For more efficient use of R groups, mixed cuprates (RR′ CuLi) have been devised. Many similar reactions are also carried out with Grignard reagents using catalytic or molar equivalents of cuprous salts.

3.3.1 Alkylation. The lithium dialkylcuprates are prepared from two moles lithium reagent and one mole cuprous iodide in ether. This gives a homogeneous solution which may be dark colored due to side reactions. The dimethyl compound is more stable thermally than the rest; it can be prepared and used at $0°$ while higher homologs must be kept below $-20°$. Lithium di-n-alkylcuprates will displace primary chlorides, bromides, iodides, tosylates, and epoxides with carbon-carbon bond formation (Review: Posner, 1975). The chlorides and bromides are slow to react in ether but are quite active in THF. The iodides and tosylates are the faster and higher yield reagents in either solvent (Eqs. 35 to 37). Secondary halides and tosylates give low yields in alkylation except with lithium dimethylcuprate. A large excess, typically fivefold, of the cuprate is combined with the alkylating agent in a cold bath and then warmed to room temperature.

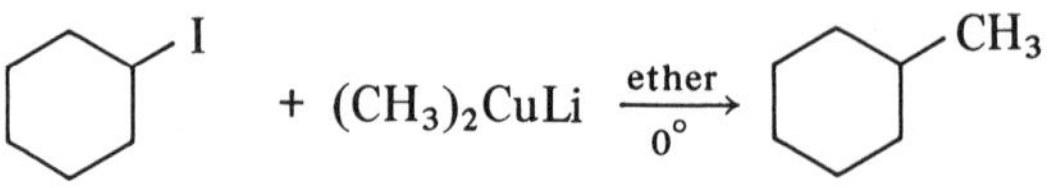

$$\text{(cyclohexyl)}-I + (CH_3)_2CuLi \xrightarrow[0°]{\text{ether}} \text{(cyclohexyl)}-CH_3 \qquad (35)$$

$$1 \quad : \quad 5 \qquad\qquad 75\%$$

(Corey et al., 1967)

$$I(CH_2)_{10}\overset{\overset{\displaystyle O}{\|}}{C}N-Ph + n\text{-}Bu_2CuLi \xrightarrow[-50°]{\text{ether}} CH_3(CH_2)_{13}\overset{\overset{\displaystyle O}{\|}}{C}N-Ph \qquad (36)$$

$$\underset{\text{CH}_3}{} \qquad\qquad\qquad\qquad \underset{\text{CH}_3}{}$$

$$82\%$$

$$Ph\overset{\overset{\displaystyle O}{\|}}{C}CH_2CH_2CH_2OTs + Me_2CuLi \xrightarrow[-78° \text{ to } 0°]{\text{ether-benzene}} Ph\overset{\overset{\displaystyle O}{\|}}{C}(CH_2)_3CH_3 \qquad (37)$$

$$1 \qquad : \qquad 2 \qquad\qquad\qquad 89\%$$

(Johnson et al., 1973b)

The alkylation of lithium diisobutylcuprate with an epoxide gave a short route to a side chain of deoxyharringtonine (Eq. 38) (Auerbach et al., 1973).

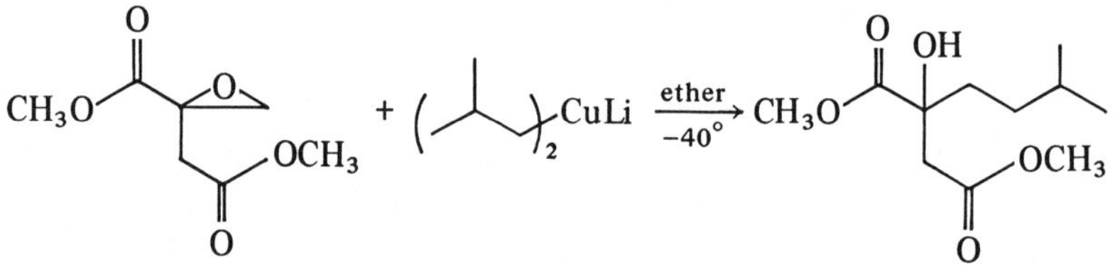

$$(38)$$

60%

The last three examples (Eqs. 36 to 38) demonstrate the selective reactivity wherein alkylation occurs to the exclusion of addition to carbonyl groups.

In some reactions there is a competing halogen-metal exchange to produce a new cuprate derived from the halide which was intended as alkylating agent. This problem is taken care of by subsequently adding alkyl halide to couple with it as exemplified in Eq. 39 (Corey et al., 1968a). In this case the yield was raised to 76% by the addition of n-butyl iodide.

$$I(CH_2)_{10}COOH + n\text{-}Bu_2CuLi \Big\langle \begin{array}{l} CH_3(CH_2)_{13}COO^- \\ \Big\uparrow n\text{-}BuI \\ \Big[n\text{-}BuCu-(CH_2)_{10}COO^- \Big] \\ \qquad\quad |\!|\;\;Li \end{array} \qquad (39)$$

$$\begin{array}{ccc} 1 & : & 5 \end{array}$$

Vinyl halides are normally unreactive toward typical nucleophilic substitution reactions but they are quite easily replaced with the alkyl groups of cuprates. In these reactions the stereochemistry about the double bond is retained (Eqs. 40 and 41).

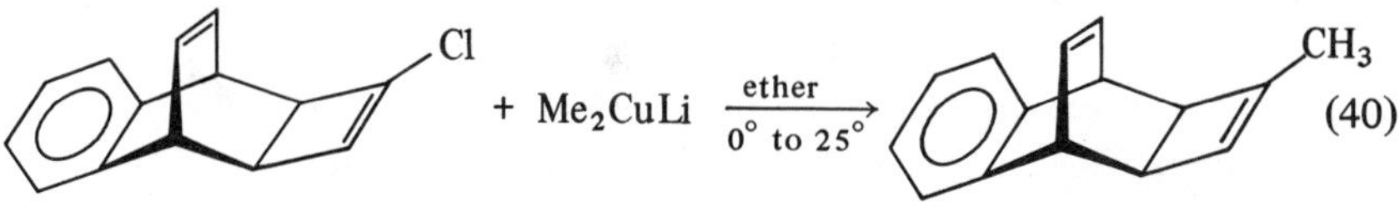

$$\begin{array}{ccccc} & + \text{Me}_2\text{CuLi} & \xrightarrow[0°\ \text{to}\ 25°]{\text{ether}} & & (40) \\ 1 & : & 5 & & 71\% \end{array}$$

(Paquette et al., 1972)

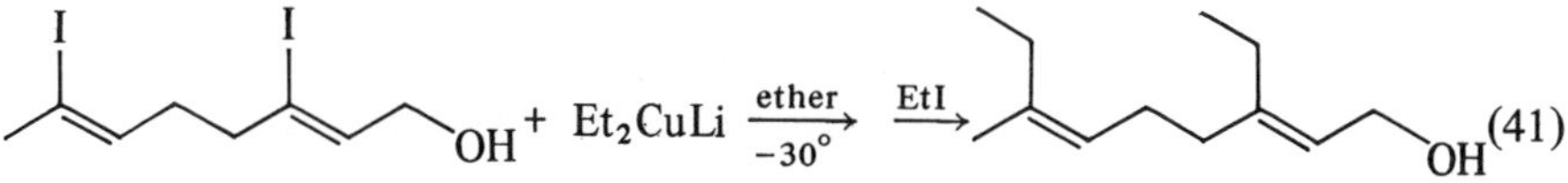

$$\begin{array}{ccccc} & + \text{Et}_2\text{CuLi} & \xrightarrow[-30°]{\text{ether}} & \xrightarrow{\text{EtI}} & (41) \\ 1 & : & 8 & 60\% & \end{array}$$

(Corey et al., 1971b)

Secondary and tertiary alkyl groups may be used in the form of R_2CuLi to replace halogens, but the yields are low. These reagents may be prepared as stabilized complexes with tributylphosphine which give high yields of alkylation with primary halides. For example the *sec*-butyl reagent was alkylated with 1-bromopentane in 94% yield (Eq. 42) and the *tert*-butyl reagent was similarly alkylated in 92% yield (Whitesides et al., 1969). Here again a fivefold excess of cuprate reagent was used:

$$\text{CuI·P-}n\text{-Bu}_3 + sec\text{-BuLi} \xrightarrow[-78°]{\text{ether}} sec\text{-Bu}_2\text{CuLi·}n\text{-Bu}_3\text{P}$$

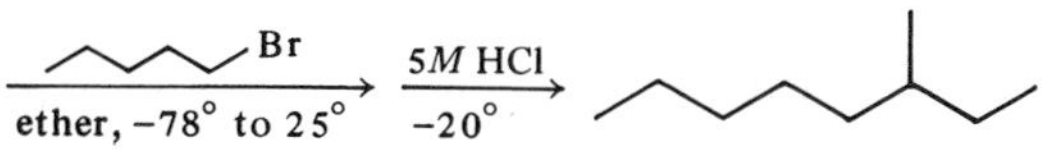

$$\xrightarrow[\text{ether, } -78°\ \text{to}\ 25°]{} \xrightarrow[-20°]{5M\ \text{HCl}} \qquad (42)$$

$$94\%$$

A smaller excess (twofold) may be used if the mixed reagent, lithium phenyl-thio(*tert*-butyl)cuprate is used (Eq. 43). Treatment of the *sec*-butyl compound

$$PhSCu + t\text{-BuLi} \xrightarrow[-78^\circ]{THF} \underset{\underset{Li}{|}}{PhS-Cu-t\text{-Bu}}$$

$$\xrightarrow[THF, -78^\circ \text{ to } 0^\circ]{n\text{-}C_8H_{17}I} \xrightarrow{CH_3OH} \qquad\qquad \tag{43}$$

98%

gave a 67% yield of 3-methylundecane (Posner et al., 1973, 1974). In contrast to Eq. 43, lithium di-*t*-butylcuprate in ether gave no alkylation with 1-iodo-pentane. The mixed organo(cyano)copper lithium reagents are efficient in alkyl-ations with epoxides. They are prepared by combining cuprous cyanide and the lithium reagent in ether at -78°. In this way cyclohexene oxide was converted to 2-*n*-butylcyclohexanol in 66% yield using two equivalents of the mixed reagent (Acker, 1977).

Lithium dialkenylcuprates may be alkylated with primary iodides or tosylates in the presence of HMPA. This is not advantageous over the use of the more stable and efficiently used vinyllithium compounds in simple cases, but the cuprates should be superior where the organohalide or tosylate contains other functional groups such as carbonyls which the organolithium reagents would attack. (*Z*)- and (*E*)-vinylic halides lead to (*Z*)- and (*E*)-alkenes respectively. A high yield of (*E*)-2-undecene was obtained using equimolar amounts of the dialkenylcuprate and *n*-octyl iodide (Eq. 44) (Linstrumelle et al., 1976).

$$\xrightarrow[\text{ether}, -10^\circ]{Li (1\% Na)} \xrightarrow[\text{ether}, -78^\circ \text{ to } -35^\circ]{CuI} \left(\begin{array}{c} CuLi \end{array} \right)_2$$

$$\xrightarrow[\substack{2.\ n\text{-}C_8H_{17}I \\ -35^\circ \text{ to } 0^\circ}]{1.\ HMPA, -35^\circ} \qquad \begin{array}{l} 96\% E \\ 4\% Z \end{array} \tag{44}$$

90 to 93% yield

The advantageous selectivity of the alkenylcuprates is exemplified in Eq. 45, where an ester function survives the alkylation (Garbers et al., 1975).

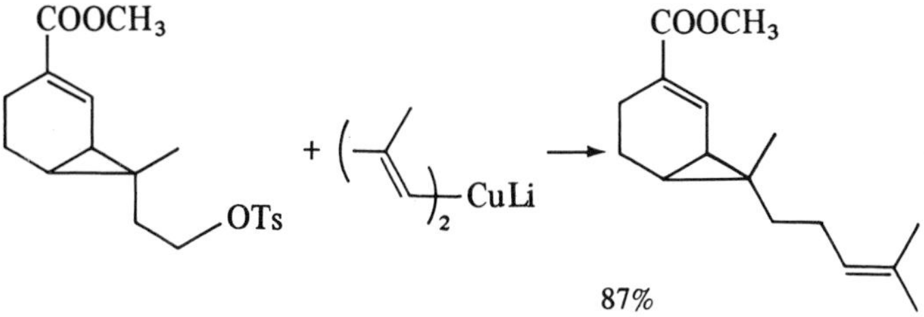

(45)

87%

In light of the excess lithium dialkylcuprates required for most alkylating reactions, it is remarkable that satisfactory yields may be obtained with close to stoichiometric amounts of reactants if the amount of copper salt is cut back to catalytic.

This is especially useful where the carbanionic reagent is more valuable than the ordinary alkyllithiums, and at the same time there is no functionality in the organohalide which would be sensitive to free lithium reagent. This was demonstrated in a three-carbon homologation (Eq. 46) (Anderson, et al., 1975).

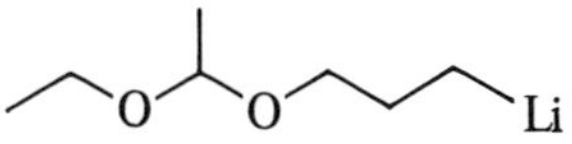

$$\text{1.1} \qquad : \qquad \text{1} \qquad : \qquad \text{0.1}$$

$$\xrightarrow[\text{Cl}_3\text{CCOOH, THF}]{\text{H}_2\text{O}} \quad \text{dodecanol} \tag{46}$$

64%

The copper reactions discussed thus far all began with lithium reagents, but Grignard reagents can also be alkylated using a catalytic amount of copper salt (Eq. 47) (Tamura et al., 1971). Magnesium carboxylate salts are not attacked by

$$n\text{-C}_6\text{H}_{13}\text{Br} + n\text{-BuMgBr} \xrightarrow[\text{THF, 0}^\circ]{\text{Li}_2\text{CuCl}_4} \text{decane} \tag{47}$$

$$\text{0.9} \qquad : \qquad \text{1} \qquad\qquad\qquad \text{78\%}$$

the Grignard reagents at -20° so that at least this functional group is compatible with the alkylation. In Eq. 48 a secondary reagent was used in only 10% excess, which is more efficient than using lithium dialkylcuprates (Baer et al., 1976).

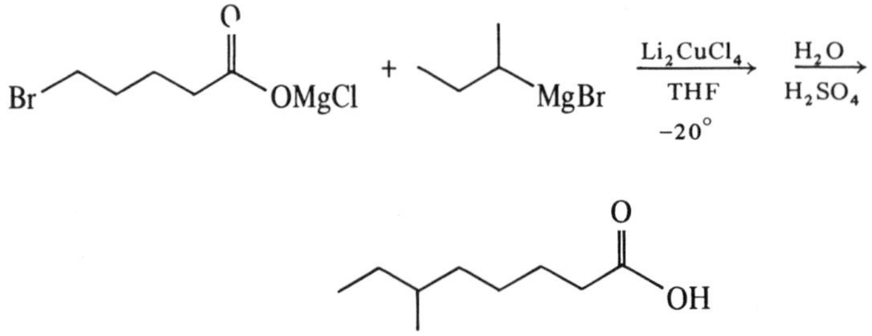

$$\text{(48)}$$

88%

p-Toluenesulfonates give high yields in the Grignard alkylation, and even *tert*-butylmagnesium bromide may be used (Eq. 49) (Fouquet et al., 1974).

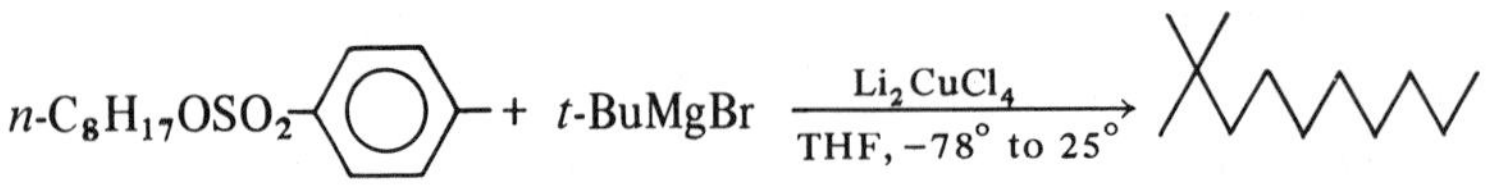

75%
(96% by GLC)   (49)

Vinyl Grignard reagents are alkylated, with retained configuration, by iodides or tosylates at $-30°$ using a catalytic amount of purified cuprous iodide (Derguini-Boumechal et al., 1976). The opposite combination, that is, an alkylmagnesium halide plus a vinyl bromide, likewise gives alkenes with retained configuration, but iron salts are the necessary catalysts (Neumann et al., 1975).

3.3.2 Acylation. Diorganocuprates react rapidly with acid halides at $-78°$ in ether to give ketones (Review: Posner, 1975). At this temperature many other functional groups are untouched including nitriles, ketones, esters, and iodides. The excess cuprate is quenched with methanol while still at $-78°$ to prevent reaction with other groups upon warming (Eq. 50) (Posner et al., 1972).

$$I(CH_2)_{10}\overset{O}{\overset{\|}{C}}Cl + n\text{-}Bu_2CuLi \xrightarrow[-78°]{ether} \xrightarrow[-78°]{CH_3OH} I(CH_2)_{10}\overset{O}{\overset{\|}{C}}\text{-}n\text{-}Bu \quad (50)$$

1   :   3   93%

Lithium phenylthioalkylcuprates may be acylated using only a 10% excess of the cuprate, and moreover secondary and tertiary reagents give high yields as well (Eq. 51) (Posner et al., 1976). Thiophenol esters may be used in place

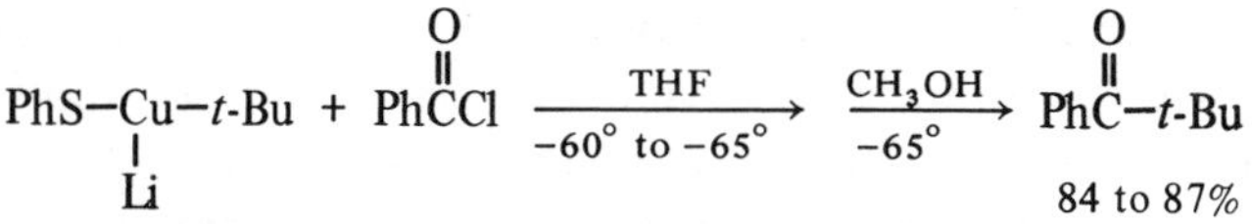

$$\text{PhS-Cu-}t\text{-Bu} + \text{Ph}\overset{\overset{\displaystyle O}{\|}}{C}\text{Cl} \xrightarrow[-60° \text{ to } -65°]{\text{THF}} \xrightarrow[-65°]{\text{CH}_3\text{OH}} \text{Ph}\overset{\overset{\displaystyle O}{\|}}{C}\text{-}t\text{-Bu} \quad (51)$$

$$\underset{\displaystyle \text{Li}}{}$$

84 to 87%

of the acid halides (Rosenblum et al., 1976).

Methylcopper(I) and alkylmagnesium halides form mixed cuprates which give ketones with acid chlorides. The larger alkyl group is transferred more readily than the methyl group (Bergbreiter et al., 1976).

3.3.3 Conjugate Addition. The synthetically valuable addition of alkyl or alkenyl groups to the β position of α,β-unsaturated carbonyl compounds is readily accomplished when copper is present (Review: Posner, 1972). This is frequently carried out with lithium organocuprates or with Grignard reagents and copper catalysis. Examples are shown in Eqs. 52 and 53 here and in Eq. 26

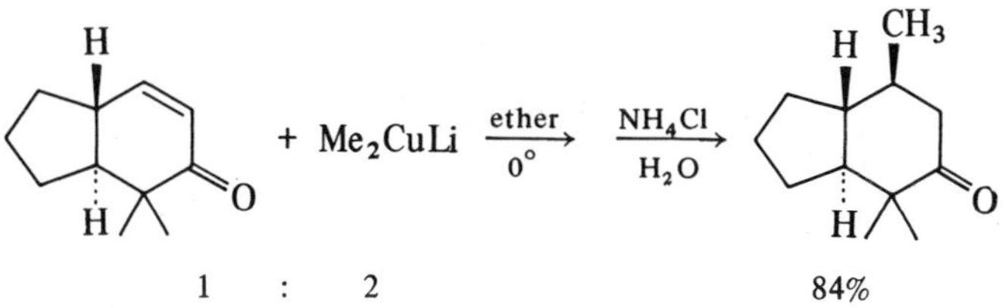

$$+ \text{Me}_2\text{CuLi} \xrightarrow[0°]{\text{ether}} \xrightarrow{\text{NH}_4\text{Cl}}{\text{H}_2\text{O}} \quad (52)$$

1 : 2   84%

(Marshall et al., 1970)

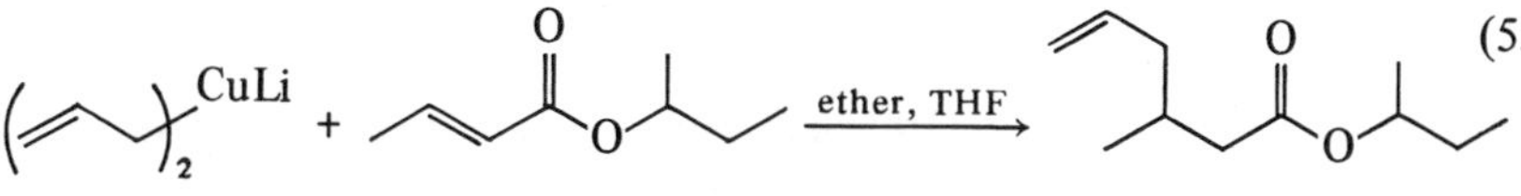

$$(53)$$

$$\text{CuLi} + \xrightarrow{\text{ether, THF}} \quad$$

(Daviaud et al., 1973)   60%

of Chapter 2. As in alkylation and acylation, the tertiary alkyl groups are best transferred using the mixed cuprates (Eq. 54) (Posner et al., 1973). Another

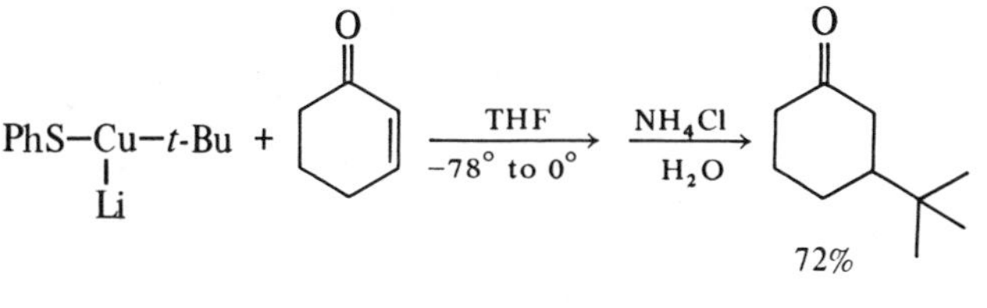

$$\text{PhS-Cu-}t\text{-Bu} + \xrightarrow[-78° \text{ to } 0°]{\text{THF}} \xrightarrow{\text{NH}_4\text{Cl}}{\text{H}_2\text{O}} \quad (54)$$

$$\underset{\displaystyle \text{Li}}{}$$

72%

class of mixed cuprates may be prepared from cuprous acetylides (Corey et al., 1972b). Primary, tertiary, and alkenyllithium reagents combine with the cuprous salt of 1-pentyne (solubilized with the easily separated hexamethylphosphorous triamide) to give mixed cuprates that selectively transfer the alkyl or alkenyl group in conjugate addition. The yields are high using only a 10% excess of the mixed cuprate (Eq. 55) (Corey et al., 1976c).

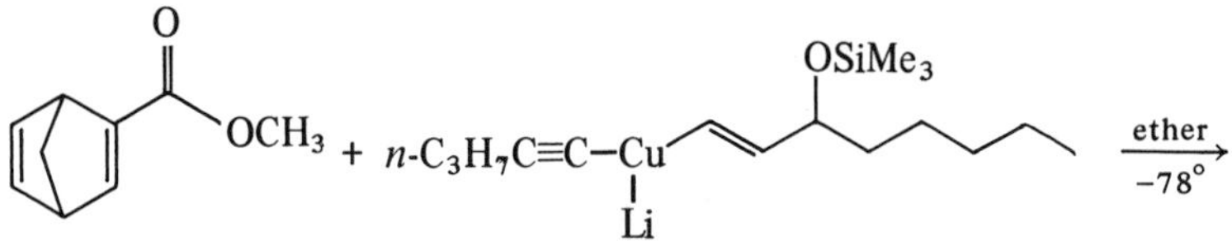

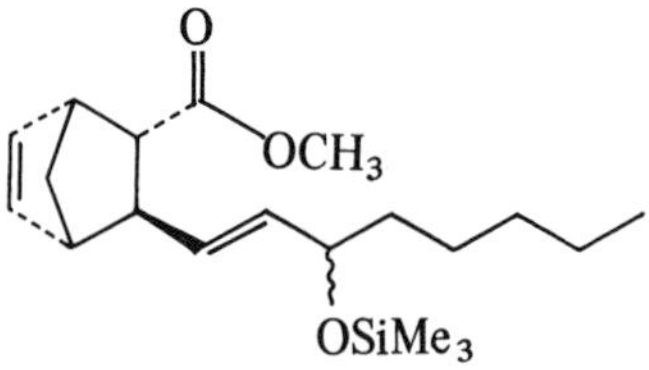

86%

In the above examples, the intermediate enolate anion was protonated with aqueous ammonium chloride to give effective addition to the carbon-carbon π bond. That enolate anion may instead be alkylated, acylated, oxidized, halogenated, O-silylated, or used in addition reactions without equilibration with other enolate sites. Some examples of these uses are given in Section 5.1.

If a halogen, acetoxy, or alkylthio substituent is present on the β position, the transient intermediate enolate anion loses that substituent giving β-alkyl-α,β-unsaturated carbonyl compounds (Eqs. 56 to 58). Mechanistically, this may alternatively be direct substitution for the leaving group (Richards et al., 1976) without appreciable carbanionic charge formation at the α position. It is inter-

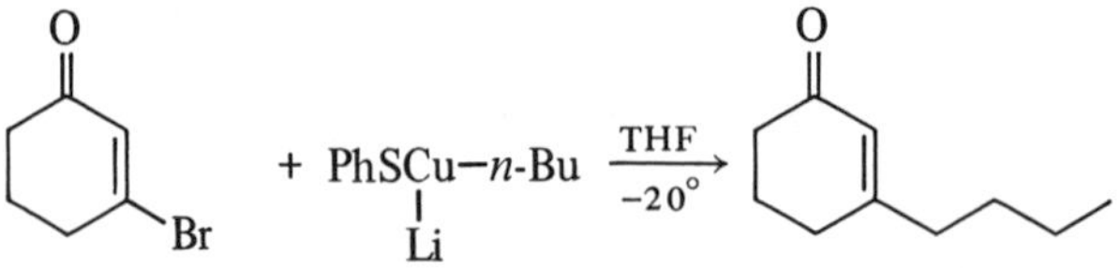

84%

(Piers et al., 1975)

esting to note that the same overall result is obtained with direct addition of
Grignard reagents to the carbonyl of cyclohexane-1,3-dione monoenol ethers

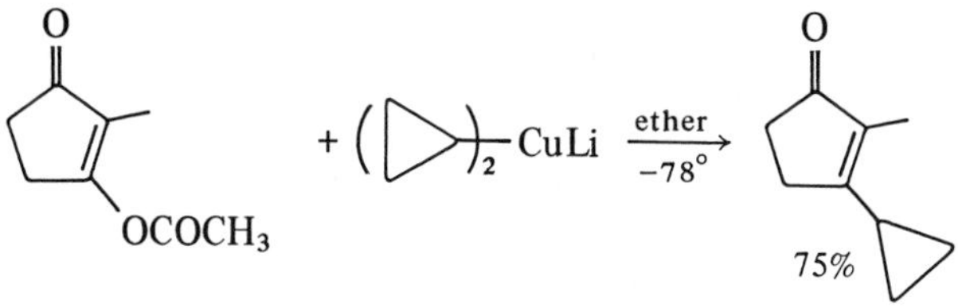

(57)

75%

(Marino et al., 1976)

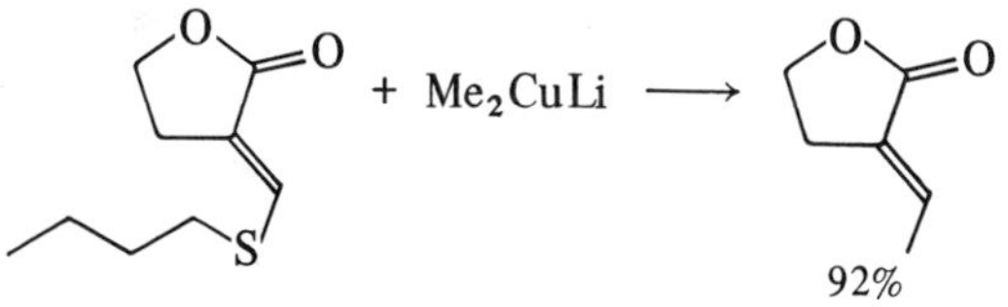

(58)

92%

(Martin et al., 1976)

followed by hydrolysis (for example see Kocienski et al., 1976). When an
excess of lithium dimethylcuprate is added to an n-butylthiomethylene ketone,
a second conjugate addition occurs after the elimination to give an α-isopropyl
ketone (Eq. 59) (Sher et al., 1977).

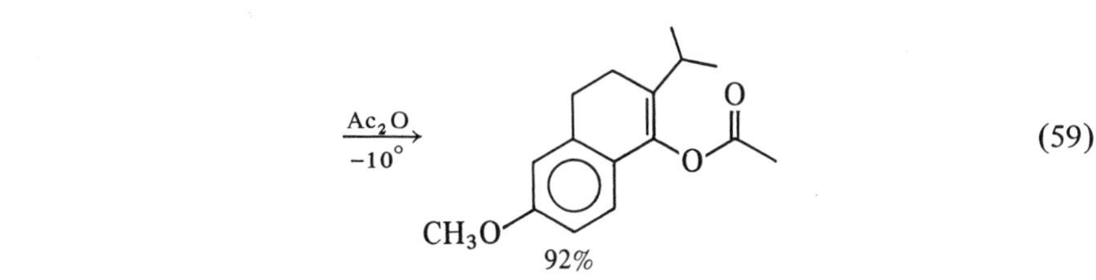

(59)

92%

The addition-elimination sequence may be used to prepare substituted
α,β-unsaturated esters stereospecifically. A β-acetoxy group is replaced with

retention by the dialkylcopper lithium reagent (Eq. 60) (Casey et al., 1974).

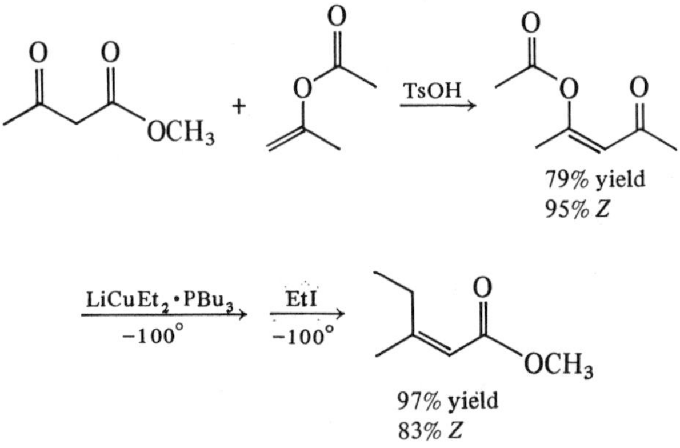

79% yield
95% Z

(60)

97% yield
83% Z

The E isomer is likewise prepared from the E-enol acetate which is available from the keto ester using acetyl chloride and triethylamine in HMPA.

Conjugate addition occurs also on acetylenic esters, and if the resulting ion is kept at low temperature and quenched, the addition is specifically syn (Eq. 61) (Pitzele et al., 1975) (see also Bates et al., 1977).

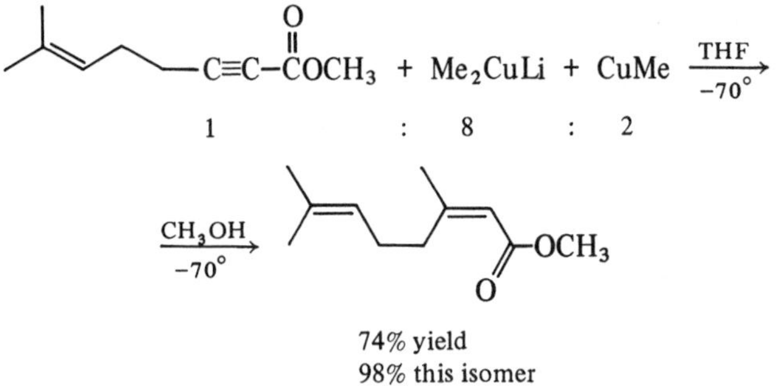

1   :   8   :   2

(61)

74% yield
98% this isomer

α,β-Unsaturated aldehydes accept conjugate addition also (Eq. 62) (Hamon

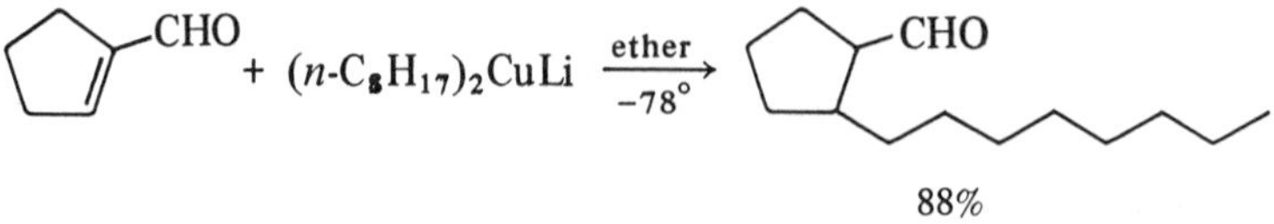

88%

(62)

et al., 1975, 1976). Copper-catalyzed conjugate addition of Grignard reagents may be used as well if the aldehyde is first converted to the diethyl acetal (Eq. 63) (Normant et al., 1975).

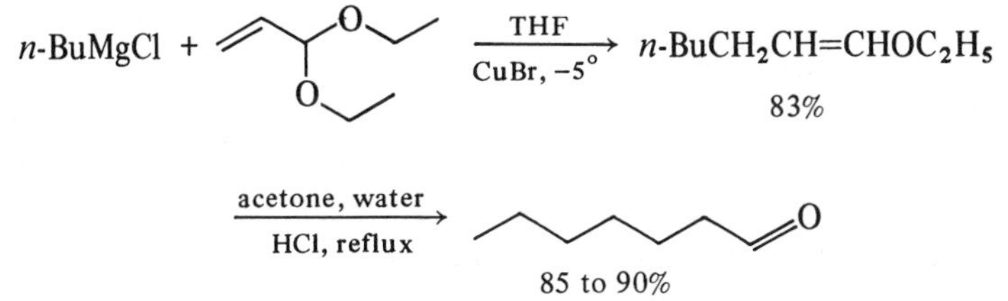
n-BuMgCl +
O
O
THF
CuBr, −5°
n-BuCH$_2$CH=CHOC$_2$H$_5$
83%
acetone, water
HCl, reflux
O
85 to 90%

(63)

Lithium divinylcuprate is best prepared from vinyllithium and the dimethyl sulfide complex of cuprous bromide (House et al., 1975). This same complex may be used catalytically with the more readily available vinylmagnesium bromide if it is kept cold to avoid thermal decomposition of the vinylcopper compound (Eq. 64) (House et al., 1977a). The higher alkenylcuprates retain configuration from the vinyl halide to the conjugate addition product. A

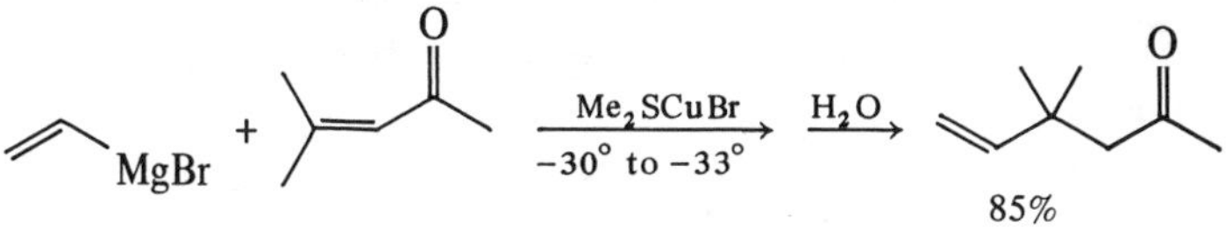
MgBr
+
O
Me$_2$SCuBr
−30° to −33°
H$_2$O
O
85%

(64)

functionally substituted vinyl cuprate was used to conjugatively add the equivalent of acrolein (Eq. 65) (Marino et al., 1975b).

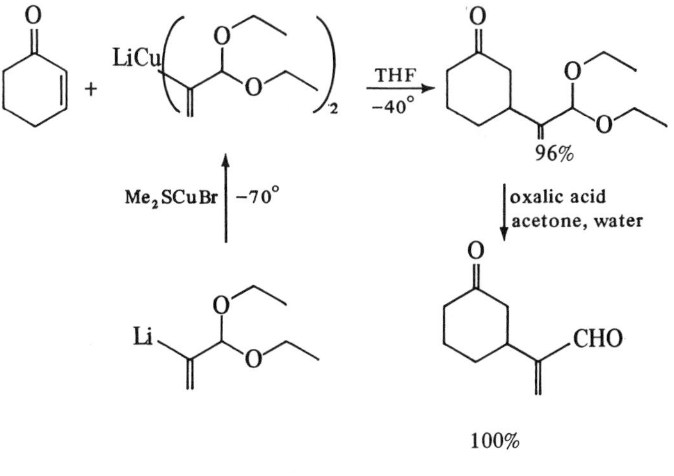
O
+
LiCu
O
O
2
THF
−40°
O
O
O
96%
Me$_2$SCuBr
−70°
oxalic acid
acetone, water
Li
O
O
O
CHO
100%

(65)

What could it be about cuprates that makes them capable of conjugate addition? They are not strong nucleophiles like organolithium and Grignard reagents in that they react very slowly with nonconjugated carbonyl compounds, wherein they resemble zinc and cadmium compounds. However, unlike the latter, they have high reactivity toward α,β-unsaturated carbonyl compounds. It is likely that Grignard and lithium reagents give carbon-carbon bond formation in a single nucleophilic step (Eicher, 1966) while the cuprates react in three steps, the first being transfer of a single electron from copper to the conjugated system (House, 1976). The conjugated systems would be more susceptible to this first step since they have a lower polarographic reduction potential (about -2.25 V and less) than the nonconjugated carbonyl compounds (more negative than -2.9 V). (The E_{red} values are in volts versus saturated calomel electrode in an aprotic solvent.) Copper derivatives are especially amenable to this transfer since copper has a pair of close-lying stable ionization states, unlike the other metals mentioned here.

The proposed mechanism is illustrated in Eq. 66 for the dimethylcuprate case where the reagent is shown as a dimer with bridging methyl groups. The third step is probably an intramolecular rearrangement as shown because, in those cases where the carbon transferred from metal is a chiral center, configuration is retained.

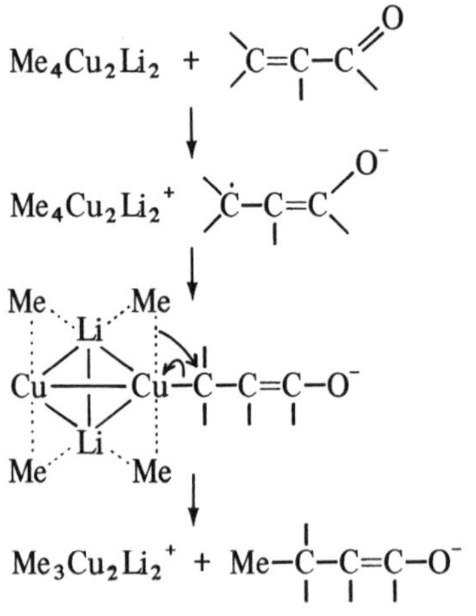

$$\tag{66}$$

The first step in Eq. 66 is the transfer of an electron and only those substrates sufficiently reducible will accept one from the cuprate reagent. The reducibility of these substrates may be determined polarographically and limits defined for the range suitable for conjugate addition. Those α,β-unsaturated carbonyl compounds that will accept one electron at a potential less negative than -2.40 V will undergo conjugate addition with lithium dimethylcuprate. If the reduction potential required is more negative, for example, -2.45 V for 1, no reaction occurs because the copper reagent is not sufficiently reducing to transfer an electron as required for the first step. If the substrate will accept a second electron at a potential less negative than -1.2 V, it forms a dianion which is then diprotonated in the workup to give reduction as for 2. A normal case is 3 which may be reduced at -2.3 V.

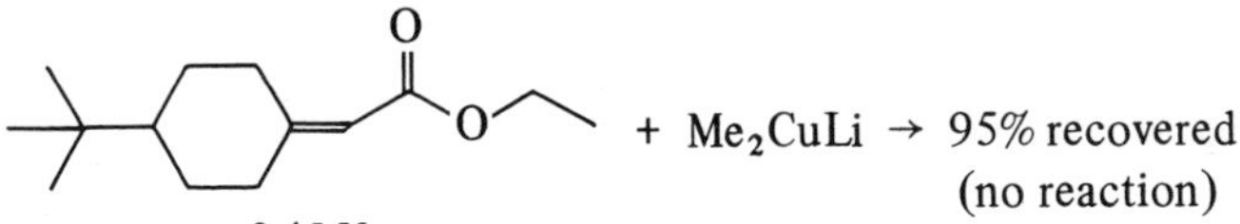

$$-2.45 \text{ V}$$
1

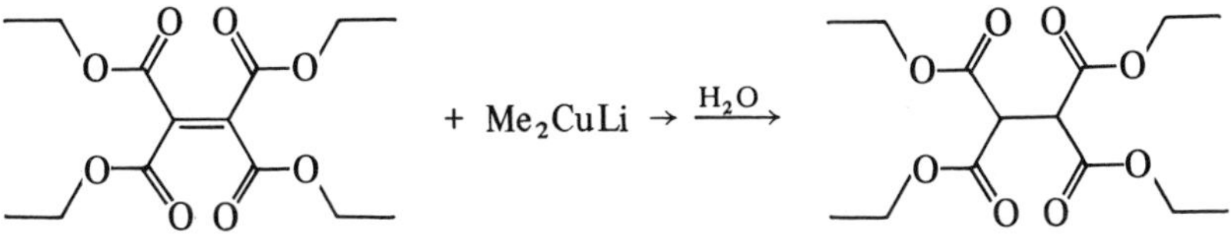

$+ \text{Me}_2\text{CuLi} \rightarrow$ 95% recovered
(no reaction)

-0.97 and -1.10 V
2

$+ \text{Me}_2\text{CuLi} \rightarrow \xrightarrow{\text{H}_2\text{O}}$

76%

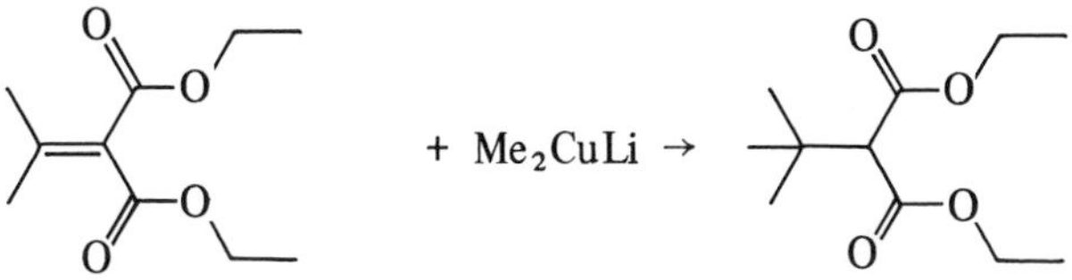

-2.3 V
3

$+ \text{Me}_2\text{CuLi} \rightarrow$

93%

It is interesting that the reduction potential, and therefore a chance for success in the conjugate addition, may be predicted empirically within ± 0.1 V using the values in Scheme 1 (House, 1976). Example calculations are given

SCHEME 1   ESTIMATION OF REDUCTION POTENTIALS FOR
α,β-UNSATURATED CARBONYL COMPOUNDS VERSUS THE SCE
ELECTRODE

	Increments	
	R	R'
alkyl	-0.1	-0.1
first alkoxyl	-0.3	0
first phenyl	+0.4	+0.1

$$\begin{array}{c} R \\ \diagdown \\ \diagup \\ R \end{array} C{=}\overset{R'}{\underset{|}{C}}{-}\overset{O}{\overset{\|}{C}}{-}R$$

base value -1.9 V

additional alkyl groups are additive
additional phenyl groups have no effect

in Scheme 2 along with the measured values. Both of the examples give the normal conjugate addition reaction. Other cuprate reagents have slightly differ-

SCHEME 2   SAMPLE PREDICTIONS OF POLAROGRAPHIC
REDUCTION POTENTIALS

-1.9 base value
+0.4 phenyl
-0.3 alkoxyl

-1.8 V calculated
-1.81 V measured

-1.9 base value
-0.1 alkyl
-0.1 alkyl
-0.1 alkyl

-2.2 V calculated
-2.15 V measured

ent negative limits: $(CH_2{=}CHCH_2)_2CuLi$, -2.1 V; Ph_2CuLi, -2.3 V; and $(CH_2{=}CH)_2CuLi$, about -2.4 V.

3.4   ACETYLIDE ANIONS

The higher apparent electronegativity of carbon in the *sp* hybridization state gives relatively greater acidity to 1-alkynes so that bases such as lithiumamide may be used to generate the carbanion. These anions may be alkylated with primary halides. The earlier work was carried out with sodamide suspension in liquid ammonia and the yields were variable owing to reaction of the halide with the NH_3 and insolubility of some sodium acetylides. A superior procedure is to use THF as solvent and butyllithium as base and to promote the substitution reaction with HMPA. For example see Eq. 67 (Schwarz et al., 1972).

$$\text{[THP]}O(CH_2)_5C\equiv CH \xrightarrow[\substack{2.\ n\text{-}C_7H_{15}X,\ HMPA, \\ 25°,\ 30\ minutes}]{1.\ n\text{-}BuLi,\ THF,\ 10°} \text{[THP]}O(CH_2)_5C\equiv C(CH_2)_6CH_3 \qquad (67)$$

$$92\%$$

Ethylene oxide may also be used as an alkylating agent as shown in Eq. 68 (Borch et al., 1977). The alkyne products are commonly reduced stereoselectively

$$\text{[THP]}OCH_2C\equiv CH \xrightarrow[\substack{2.\ \overset{O}{\triangle},\ 19\ hours \\ 3.\ NH_4Cl}]{1.\ NaNH_2,\ NH_3} \text{[THP]}OCH_2C\equiv CCH_2CH_2OH \qquad (68)$$

$$54\%$$

to the cis or trans alkenes, which are useful particularly in the synthesis of insect sex pheromones.

The alkylation of acetylene itself is complicated by disproportionation of the monoanion, driven by the insolubility of the dianion which precipitates from many organic solvents. The monoanion may be stabilized by liquid ammonia but in this solvent alkylation is again poor. The complex of lithium acetylide and ethylenediamine is stable and soluble in DMSO, which is a good solvent for substitution reactions. This complex may be prepared as a free-flowing powder from N-lithioethylenediamine and acetylene, and it is commercially available (Beumel, Jr., et al., 1963). A typical preparation of a 1-alkyne is illustrated in Eq. 69 (Burgstahler et al., 1977b).

$$LiC\equiv CH \cdot H_2NCH_2CH_2NH_2 + PhCH_2O(CH_2)_4Br \xrightarrow[5°\ to\ 25°]{DMSO}$$

$$PhCH_2O(CH_2)_4C\equiv CH \qquad (69)$$

$$97\%$$

The complex is also useful for addition to carbonyl compounds as exemplified in Eq. 70 (Huffman et al., 1965). The ethylenediamine as well as liquid

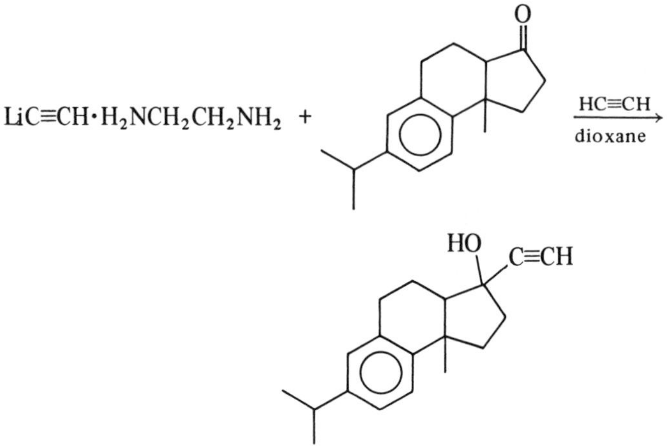

LiC≡CH·H₂NCH₂CH₂NH₂ +

$$LiC{\equiv}CH{\cdot}H_2NCH_2CH_2NH_2 \quad +$$

$$\xrightarrow[\text{dioxane}]{HC{\equiv}CH}$$

(70)

85%

ammonia diminish the reactivity of the lithium acetylide so that for certain reactions under very mild conditions it is still useful to have the uncomplexed reagent in THF. At −78° the disproportionation is slow and does not compete with addition reactions; therefore preparation and use at this temperature gives high yields of adduct (Eq. 71) (Midland, 1975). If the solution of lithium

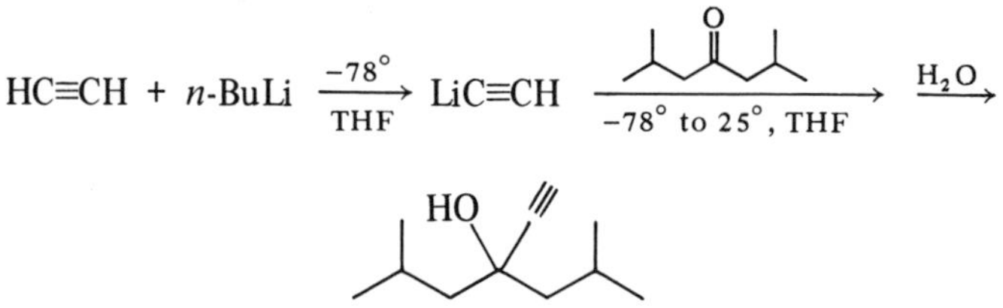

$$HC{\equiv}CH + n\text{-BuLi} \xrightarrow[\text{THF}]{-78^\circ} LiC{\equiv}CH \xrightarrow[-78^\circ \text{ to } 25^\circ, \text{ THF}]{} \xrightarrow{H_2O}$$

(71)

92%

acetylide is warmed or prepared at 0°, a heavy precipitate of Li_2C_2 is formed which gives poor yields of addition reactions.

When only one acidic proton is present as in 1-alkynes, addition reactions are more straightforward (Eqs. 72 and 73):

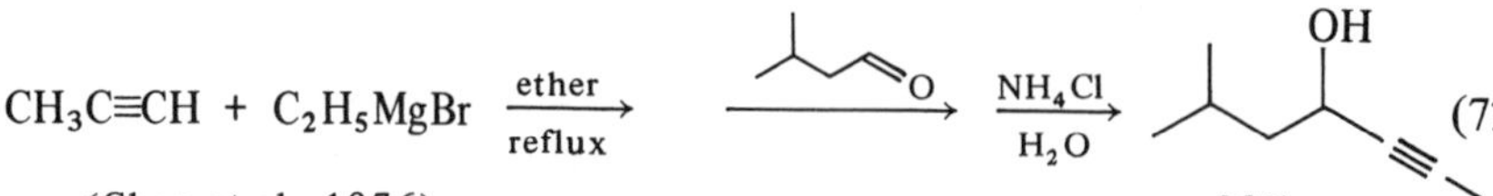

$$CH_3C{\equiv}CH + C_2H_5MgBr \xrightarrow[\text{reflux}]{\text{ether}} \xrightarrow{} \xrightarrow[\text{H}_2\text{O}]{NH_4Cl}$$

(72)

(Chan et al., 1976)

82%

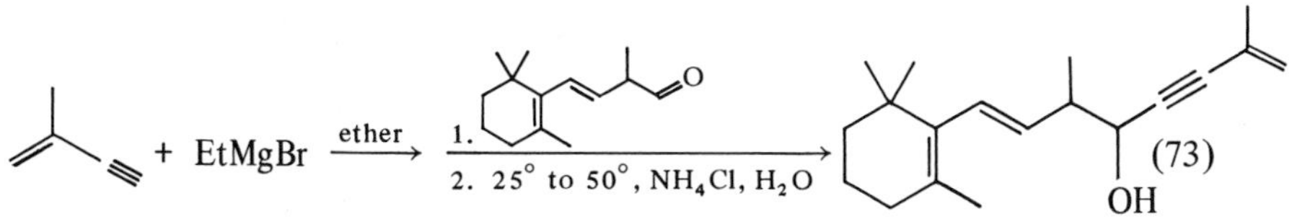

$$(73)$$

(Ramamurthy et al., 1975)   high yield

Propyne can be converted to a soluble dianion using two equivalents of the stronger base combination *n*-butyllithium with TMEDA. This then reacts first at the propargylic site and then at the acetylide end (Eq. 74) (Bhanu et al., 1975).

$$CH_3C{\equiv}CH \ + \ n\text{-BuLi} \ + \ TMEDA \ \xrightarrow[-60^\circ]{ether\text{-}hexane} \ \xrightarrow[2.\ CH_2O]{1.\ n\text{-}BuBr} \ n\text{-}BuCH_2C{\equiv}CCH_2OH$$

$$80\% \qquad (74)$$

Alkynyl ketones are available by acylation of acetylides at low temperature, for example see Eq. 75 (Knox et al., 1977):

$$70\% \qquad (75)$$

3.5 π-ALLYLNICKEL(I)HALIDES

Allyl bromides react with nickel carbonyl (caution: volatile, toxic, and spontaneously flammable) in benzene at 40° to give a complex dimer in high yield (Semmelhack et al., 1972). The complex is generally unreactive in nonpolar solvents, but in polar solvents such as DMF it is selectively alkylated by organohalides (Reviews: Semmelhack, 1972; Baker, 1973). Primary, secondary, and aryl iodides as well as benzyl and vinyl bromides all give high yields, for example see Eqs. 76 and 77. Allyl bromides are reactive, but mixtures of products

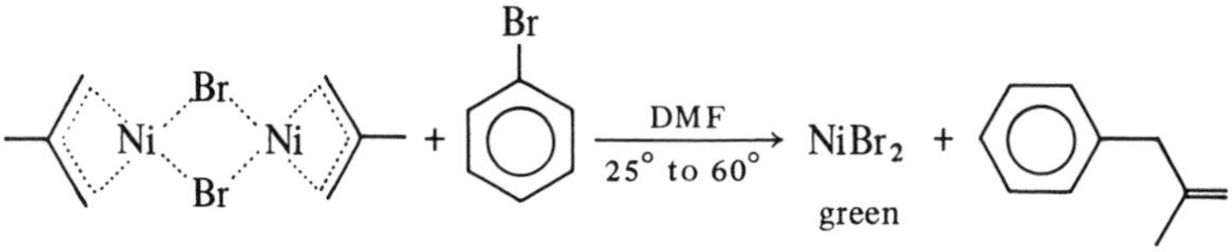

$$(76)$$

red   67 to 72%

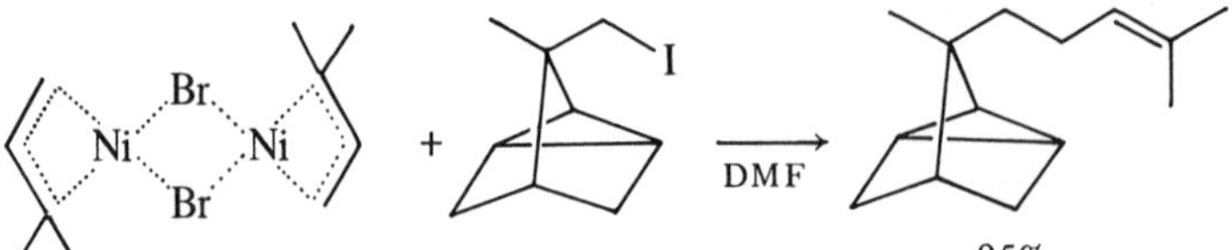

(77)

95%
α-santalene

are often formed owing to metal halogen exchange. Carbonyl or hydroxyl groups are very slow to react with the nickel complexes consequently reaction with bromides or iodides can be carried out in their presence. π-(2-Methoxy-allyl)nickel bromide is useful for overall acetonylation since the intermediate vinyl ether can be hydrolyzed (Eq. 78) (Korte et al., 1977).

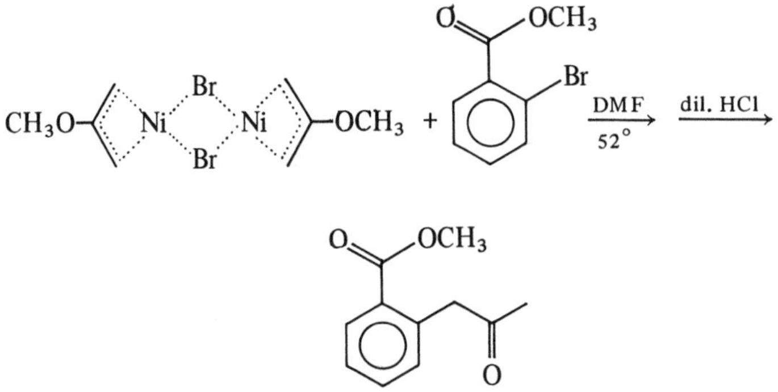

(78)

81%

The coupling of two equal allylic residues may be carried out without isolation of the Ni complex when both steps are done in DMF. This has been applied to the synthesis of macrocyclic 1,5-dienes under high dilution conditions (Eq. 79) (Corey et al., 1975a).

$$BrCH_2CH=CH(CH_2)_{12}CH=CHCH_2Br + Ni(CO)_4 \xrightarrow[50°]{DMF}$$

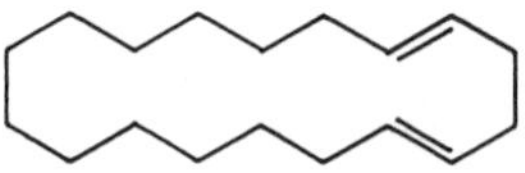

(79)

76 to 84%

CHAPTER 4

Carbanions Stabilized by α-Heteroatoms

ONE OR MORE electronegative atoms directly attached to a carbanionic site stabilize the carbanion by inductively dispersing part of the charge. The more polarizable elements in the third row of the periodic table show a greater effect than those of the second row. This difference is shown in the ready formation of the carbanion from 1,3-dithiane, which cannot be done with 1,3-dioxane in an equally basic solution (Bernardi et al., 1975). The lesser stabilizing effect of oxygen is shown only when the carbanion has some stabilization of another sort as in the α-anions from allyl ethers (conjugation) or in vinyl ethers where the carbon sp^2 hybridization is favorable to carbanion formation.

Positively charged α-heteroatoms can give even more stabilization, as seen in sulfur and phosphorous ylids and related oxides, by favorable electrostatic effects. The charge-contracted $3d$ orbitals may participate in $(p \rightarrow d)$ π bonding to further diminish the negative charge (Peterson, 1967).

4.1 α-HALOCARBANIONS

A halogen atom, especially bromine, gives sufficient stabilization to an attached carbanion to allow preparation by metal halogen exchange from 1,1-dihalo compounds (Review: Köbrich, 1967, 1972). These carbanions decompose rapidly since the halide is a good leaving group, capable of bearing the negative charge better upon itself. A carbene may be the succeeding transient intermediate. Chloromethyllithium can be prepared in low yield at $-110°$, however the preparatively useful reactions involve in situ reactions of the transient carbanions at higher temperatures. The α-halocarbanions may be prepared in the presence of ketones and aldehydes, to which they will add, and then the resulting alkoxy anion displaces the halogen intramolecularly to give epoxides (Eq. 1) (Cainelli et al., 1972). This is analogous to epoxide formation in the well-known Darzens

$$\text{CH}_3\text{CHBr}_2 + n\text{-BuLi} + \text{(ketone)} \xrightarrow[-78° \text{ to } 25°]{\text{THF}} [\text{CH}_3\text{CHBrLi}] \rightarrow$$

$$\left[\text{(alkoxy bromide)} \right] \rightarrow \text{(epoxide)} \qquad (1)$$

95% (cis and trans)

condensation of α-haloesters. The readily available *gem*-dihalocyclopropanes may be converted to the anion at −95° and then alkylated with active alkyl halides as shown in Eq. 2 (Kitatani et al., 1977).

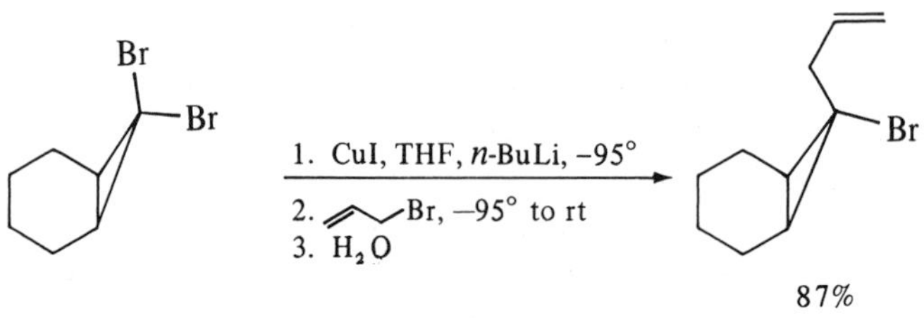
Br
Br
1. CuI, THF, n-BuLi, −95°
2. Br, −95° to rt
3. H₂O
Br
87%

(2)

The α,α-dihalocarbanions may be prepared by removal of a proton from a dihalo compound using lithium dialkylamide or *n*-BuLi-TMEDA bases. This may be done at −78° or higher in the presence of ketones to obtain addition products from the transient carbanions (Eqs. 3 and 4) (Taguchi et al., 1974b, 1976).

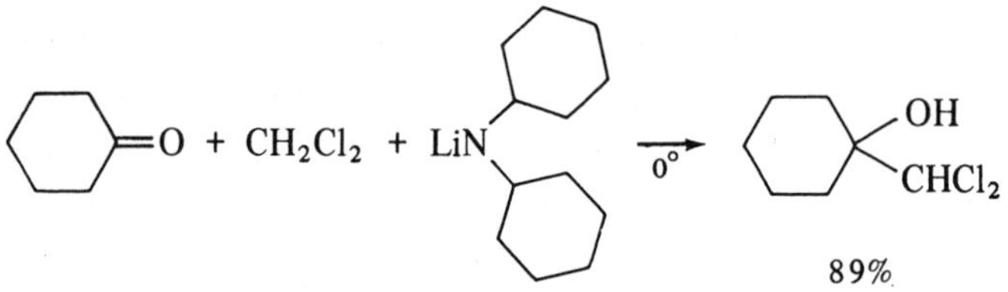
O
+ CH2Cl2
+ LiN
0°
OH
CHCl2
89%

(3)

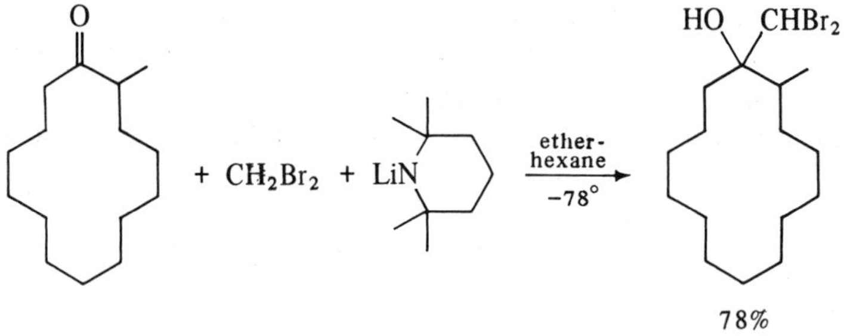
O
+ CH2Br2
+ LiN
ether-hexane
−78°
HO
CHBr2
78%

(4)

Reactions other than addition require the preformed carbanion which may then be treated with CO_2, acid anhydrides, or primary alkyl iodides or bromides to give acids, ketones, or alkylated products respectively. The alkylation requires an equivalent of HMPA to promote the substitution (Eq. 5) (Villieras et al., 1977).

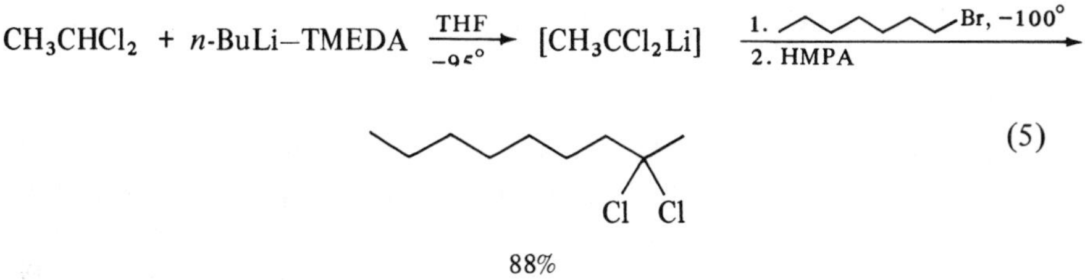

$$CH_3CHCl_2 + n\text{-BuLi--TMEDA} \xrightarrow[-95°]{THF} [CH_3CCl_2Li] \xrightarrow[2.\ HMPA]{1.\ \diagdown\diagup\diagdown\diagup\diagdown Br,\ -100°}$$

$$\text{(5)}$$

88%

The very unstable tribromomethyllithium, as a transient reagent at -78°, gives 1-(tribromomethyl) cyclohexanol in 91% yield (Taguchi et al., 1974b).

A chlorine attached to an sp^2 carbon is not as easily lost as one on an sp^3 carbon, so the α-halovinyl carbanions can be prepared at $-72°$ (Eq. 6) (Köbrich, 1967).

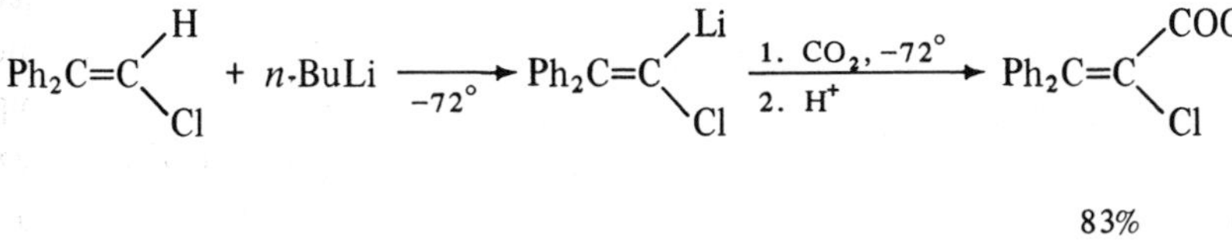

$$\text{83\%} \qquad \text{(6)}$$

The α-chloro-α-trimethylsilyl carbanion is more stable than simple α-halocarbanions and can be warmed to $-40°$ without significant decomposition. It is useful for the simple homologation of aldehydes and ketones as shown in Eq. 7 (Burford et al., 1977; Magnus et al., 1978). The intermediate epoxysilanes may be hydrolyzed to aldehydes or converted to derivatives thereof such as the enol formate shown here.

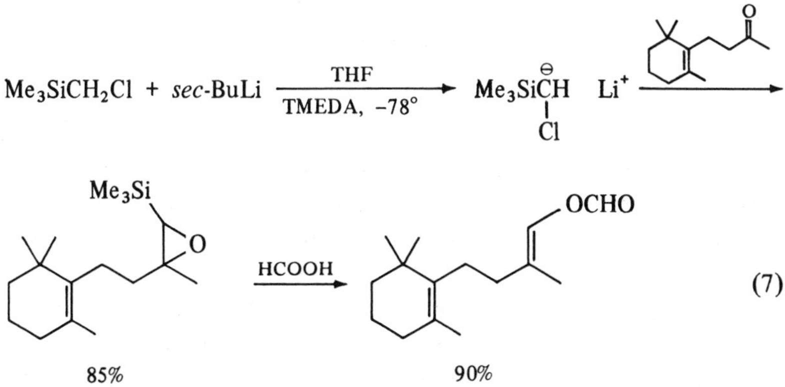

$$\text{(7)}$$

85% 90%

4.2 CARBANIONS BEARING α-OXYGEN

Allylic and vinyl protons adjacent to oxygen are sufficiently acidic to give good yields of carbanion upon treatment with alkyllithium reagents. The early synthetic applications involved 2-furyl carbanions which can be made from furans in refluxing ether (Ramanathan et al., 1962; Review: Meyers, 1974). They will add to ketones and aldehydes to give alcohols in 93 to 98% yield. The carbanions are also readily alkylated by primary bromides as applied by Buchi et al., (1966) in a synthesis of jasmone (Eq. 8). Acid hydrolysis of 2,5-disubsti-

40 to 45% overall

tuted furans leads to 1,4-diketones, so that the 2-furyl carbanions are equivalent to the hypothetical carbanion **1**. Furans with a free 5-position cannot be hydrolyzed directly owing to interfering condensation reactions (Eftax et al., 1965) but they can be opened indirectly via oxidation and reduction reactions (Review: Bayer, 1954; Cavill et al., 1970).

$$\underset{\text{R-C}}{\overset{\text{O}}{\underset{\|}{}}}\text{CH}_2\text{CH}_2\underset{}{\overset{\text{O}}{\underset{\|}{}}}\text{C}^-$$

1

Epoxides serve as alkylating agents also; for example 2-furyllithium and propylene oxide gave 2-(2-hydroxypropyl) furan in 98% yield (Arco et al., 1976).

This process has more recently been extended to the carbanion that may be prepared from methyl vinyl ether using *tert*-BuLi as base. Methoxyvinyllithium may be alkylated with *n*-octyl iodide and the resulting enol ether hydrolyzed to give 2-decanone in 80% yield. Acylation with benzonitrile or benzoic acid gave 1-phenyl-1,2-propanedione (77%). Addition to carbonyl compounds gives the α-hydroxyketone as illustrated in Eq. 9 (Baldwin et al., 1974, 1976).

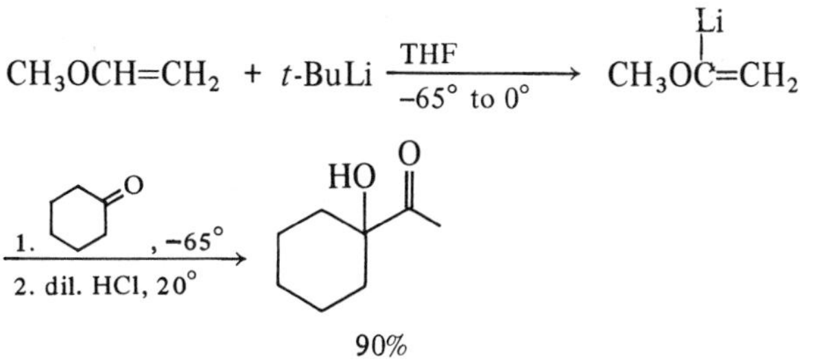

$$CH_3OCH=CH_2 \ + \ t\text{-BuLi} \ \xrightarrow[-65° \text{ to } 0°]{THF} \ CH_3O\overset{\overset{\text{Li}}{|}}{C}=CH_2$$

(9)

90%

With α,β-unsaturated ketones only direct 1,2-addition is found, while the combination with cuprous iodide-dimethyl sulfide gives only 1,4-addition. Hydrolysis of the intermediate with dilute HCl gives the 1,4-diketone in good yield (Chavdarian et al., 1975). In these reactions the methoxyvinyl carbanion is the synthetic equivalent of the hypothetical acetyl carbanion.

Cyclic vinyl ethers may be used similarly, but they are less reactive toward the *tert*-BuLi, requiring a higher temperature where the solvent THF competes for the base. If only 0.5 equivalent of THF is present, the vinyl carbanions can be prepared in good yield. They will then add to ketones or alkylate with reactive halides, but acylating agents gave only tertiary alcohols. The resulting 2-substituted dihydrofurans and pyrans can be hydrolyzed and converted to a variety of products including δ-hydroxyketones, cyclohexenones, and vinylcyclohexenones. For example see Eq. 10 (Boeckman, Jr., et al., 1977). A mixed cuprate

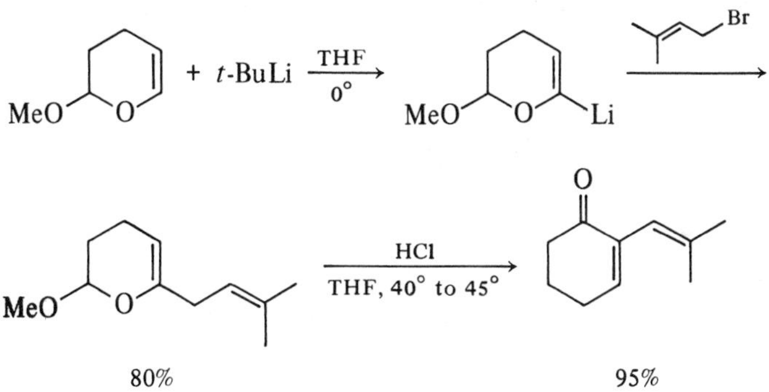

(10)

80%   95%

from dihydropyran and 1-pentyne gave conjugate addition to 2-cyclohexenone in 91% yield.

Allyl ethers can be converted to the corresponding anion, stabilized by the allylic resonance plus the inductive effect of the α-oxygen. The contribution of the oxygen is demonstrated in Eq. 11 where two allylic sites are available in each ether, but carbanion formation occurs only at the one adjacent to oxygen (Evans et al, 1974b). The carbanions may be alkylated with primary halides to

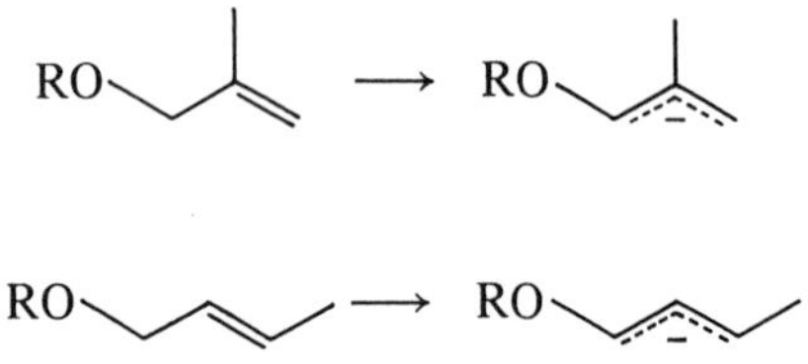

$$RO\diagup\diagdown\diagup \longrightarrow RO\diagup\diagdown\diagup$$

$$RO\diagdown\diagup\diagdown\diagup \longrightarrow RO\diagdown\diagup\diagdown\diagup$$

(11)

give some α and some γ product. Steric hindrance in the R group directs alkylation more toward the γ position, particularly where R is *tert*-Bu or triethylsilyl as exemplified in Eq. 12 (Still et al., 1974). Since the enol ethers from γ attack

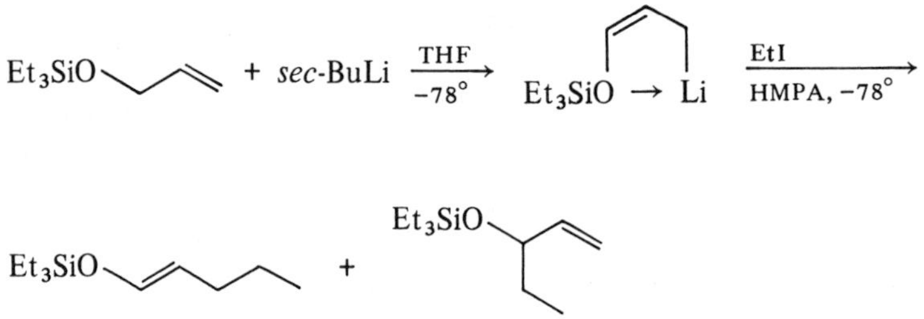

$$Et_3SiO\diagdown\diagup\diagdown \quad + \; sec\text{-BuLi} \; \xrightarrow[-78°]{THF} \quad Et_3SiO \rightarrow Li \; \xrightarrow[HMPA, -78°]{EtI}$$

(12)

$$Et_3SiO\diagdown\diagup\diagdown\diagup\diagdown \quad + \quad Et_3SiO\diagdown\diagup\diagup\diagup$$

83%   17%

can be hydrolyzed to aldehydes, the allyl ether carbanion is equivalent to the hypothetical carbanion 2:

$$\overset{\displaystyle O}{\underset{\displaystyle H\text{-}CCH_2CH_2\,^-}{\|}}$$

2

In the addition to carbonyl groups, steric hindrance has the opposite effect, that is, the larger the R group the more attack there is at the α position. Where R is trimethylsilyl, the attack is > 98% at the α position (Eq. 13) (Still et al.,

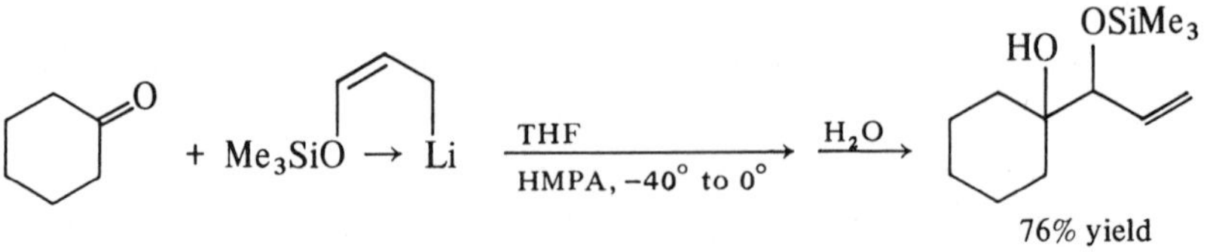

$$\text{cyclohexanone} \quad + \; Me_3SiO \rightarrow Li \; \xrightarrow[HMPA, -40° \text{ to } 0°]{THF} \; \xrightarrow{H_2O} \quad$$

76% yield
> 99% this isomer

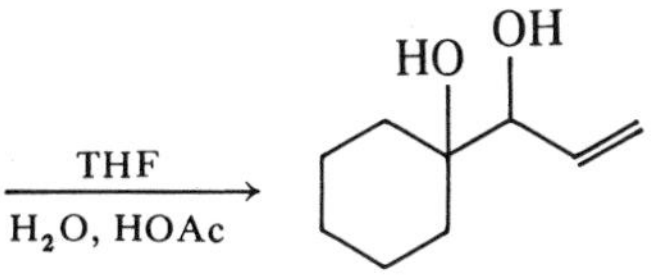

$$\text{(13)}$$

1976) and where R is methyl, 72% of the product is from γ attack. The zinc reagents, prepared from the lithium reagents with $ZnCl_2$ gave exclusively α attack in addition reactions whatever the steric size of the R Group (Evans et al., 1974b).

Since the allylic anions are formed rapidly and alkyllithium reagents are not very reactive toward alkylating agents, the process may be used for cyclizations (Eq. 14) (Still, 1976).

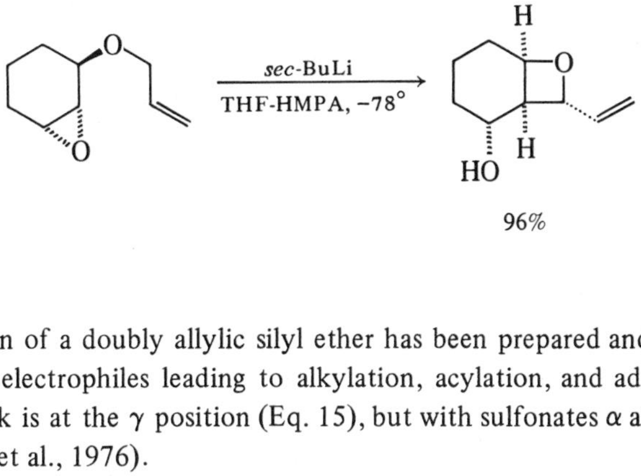

$$\text{(14)}$$

96%

The anion of a doubly allylic silyl ether has been prepared and treated with a variety of electrophiles leading to alkylation, acylation, and addition. In most cases attack is at the γ position (Eq. 15), but with sulfonates α attack is favored (Oppolzer et al., 1976).

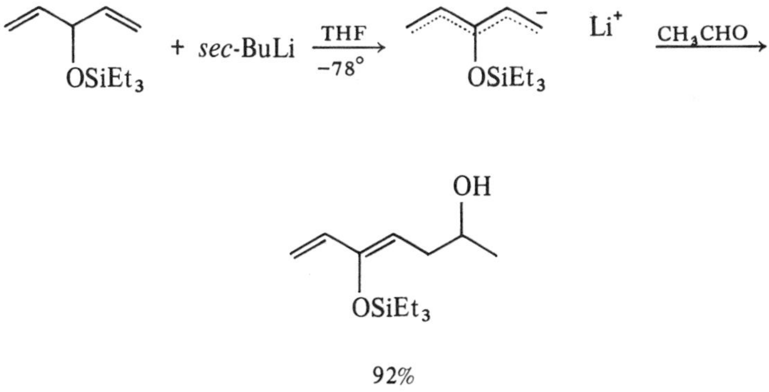

$$\text{(15)}$$

92%

4.3 α-THIOCARBANIONS

One sulfur in the sulfide oxidation state offers sufficient stabilization to an adjacent carbanion to allow preparation from methyl sulfides using the very strong base combination of n-BuLi-TMEDA or DABCO. Thioanisole and dimethyl sulfide have been converted to the anions which were added to ketones and aldehydes to give the corresponding alcohols (Corey et al., 1966; Peterson, 1967; Watanabe et al., 1975).

The addition to ketones may be followed by a reductive elimination step which overall gives methylenation. This is similar to the Wittig reaction but is superior to $Ph_3P=CH_2$ with highly hindered ketones. This also gives less enolate anion formation than the Wittig reagent (Eq. 16) (Sowerby et al., 1972).

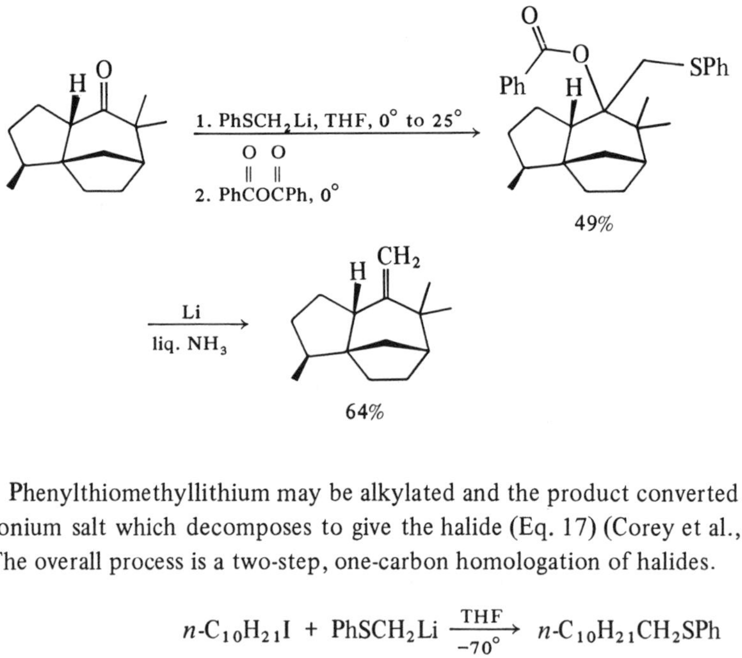

$$(16)$$

Phenylthiomethyllithium may be alkylated and the product converted to a sulfonium salt which decomposes to give the halide (Eq. 17) (Corey et al., 1968b). The overall process is a two-step, one-carbon homologation of halides.

$$n\text{-}C_{10}H_{21}I + PhSCH_2Li \xrightarrow[-70°]{THF} n\text{-}C_{10}H_{21}CH_2SPh$$

$$\xrightarrow[\text{DMF, heat}]{\text{MeI, NaI}} n\text{-}C_{11}H_{23}I$$

93%

$$(17)$$

Higher homologous sulfides give mostly metalation at other sites or elimination reactions, but if $tert$-BuLi-HMPA is used at $-78°$, clean α-carbanion formation is obtained. The carbanion may then be alkylated (Eq. 18) or used in addition reactions (Dolak, et al., 1977). With cyclohexenone, conjugate addition occurs, possibly owing to the ability of the HMPA to form separated ion pairs which may interact with the oxygen and the β position concurrently.

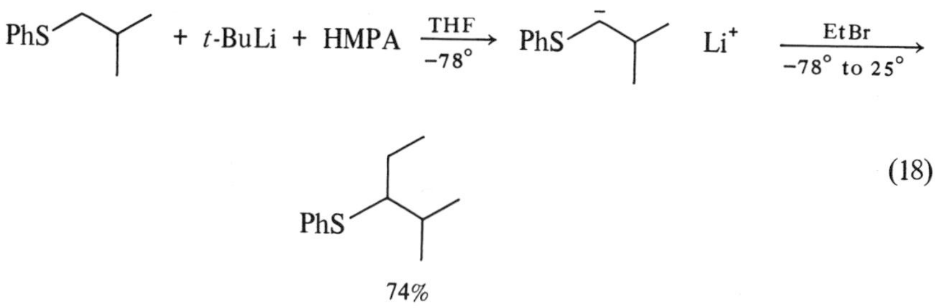

$$(18)$$

74%

Alkyl vinyl thioethers are converted to the 1-(alkylthio)vinyllithium reagents using *sec*-BuLi in HMPA and then may be alkylated with halides and epoxides, or added to aldehydes. The products are still thioenol ethers and therefore may be hydrolyzed, albeit under vigorous conditions, to give the corresponding ketones (Eq. 19) (Oshima et al., 1973a). The more acidic phenyl vinyl thioether is

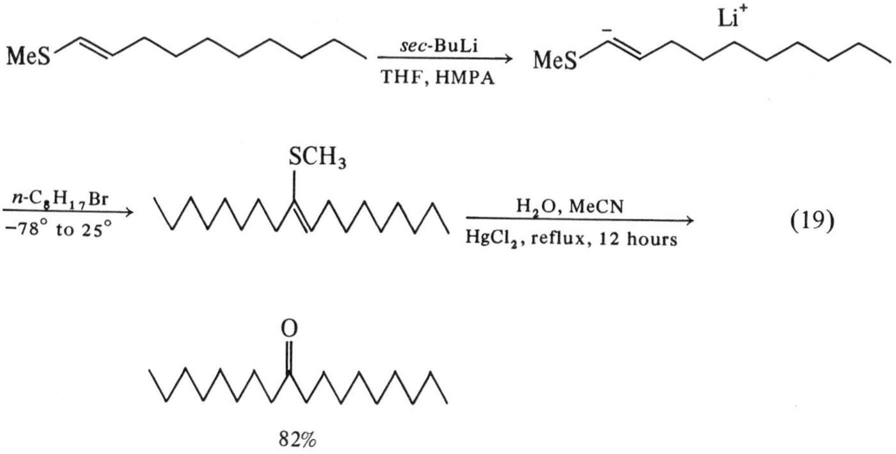

$$(19)$$

82%

converted to the carbanion with LDA-HMPA and then subsequently added to an aldehyde (Eq. 20) (Cookson et al., 1976).

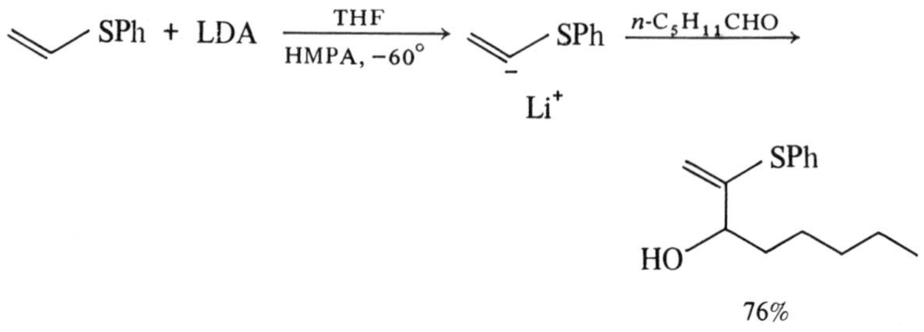

(20)

The superiority of sulfur over oxygen for the stabilization of carbanions is demonstrated in the reactions of 2-ethoxy-1-pentylthioethylene (Eq. 21).

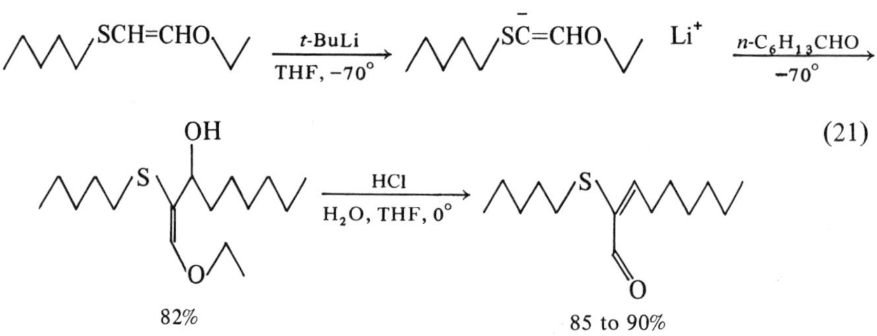

(21)

The carbanion reacts smoothly with ketones and aldehydes and, in the presence of HMPA, may be alkylated with halides and epoxides (Vlattas et al., 1976).

The allylic carbanions stabilized by one sulfur have been alkylated in routes to various natural products. For example an isoprenyl homologation sequence was

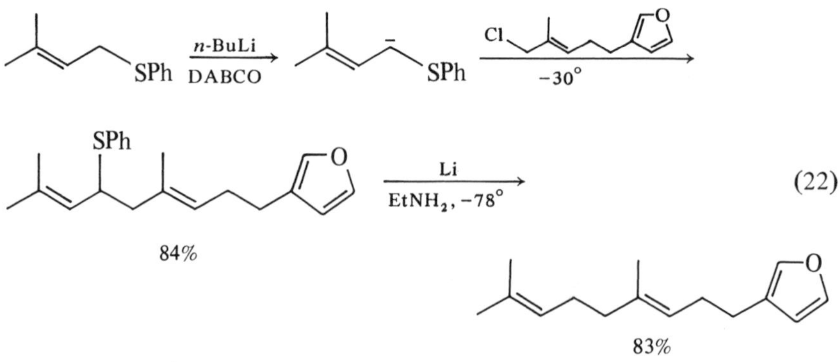

(22)

developed for the preparation of dendrolasin, a sesquiterpene found in a species of ants (Eq. 22) (Kondo et al., 1976; see also Yamada et al., 1976). The phenylthio group, which was temporarily used to activate the allylic position, is removed reductively. The dianion **3** was prepared and used similarly in the synthesis of neotorreyol.

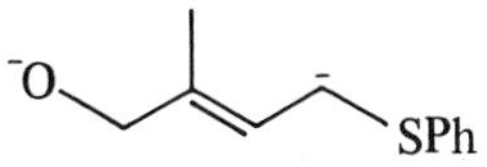

3

Another allylic α-thioanion was alkylated intramolecularly with an epoxide in the preparation of the macrocyclic diterpene nephthenol (Eq. 23) (Kodama et al., 1975).

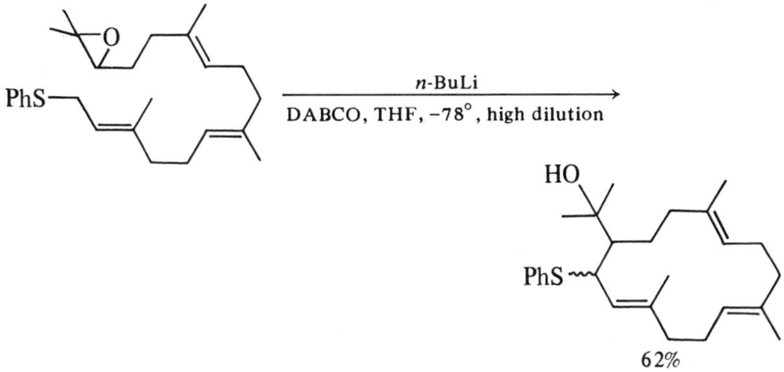

(23)

As with any unsymmetrical allylic carbanion there is a choice of reaction at either end of the π system. When the γ position is highly substituted as in the above cases, the alkylation occurs largely α to the sulfur. Otherwise mixtures appear; for instance alkylation of the anion from phenyl allyl sulfide with 1-iodohexane in THF at −65° gives 75% α- and 25% γ-alkylation. If an N-methylimidazole group is substituted for the phenyl group, the lithium cation is chelated at the α carbon (**4**), and this leads to 99% α-alkylation (Evans et al., 1974a).

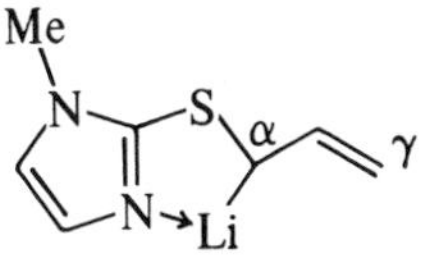

4

Allyl vinyl sulfides are capable of thermal thio-Claisen rearrangement, and the combination of alkylation, rearrangement, and hydrolysis is useful in functional five-carbon homologations (Eq. 24) (Oshima et al., 1973b). Without the ethoxy group the sequence leads to *trans*-γ,δ-unsaturated aldehydes (Oshima et al., 1973c).

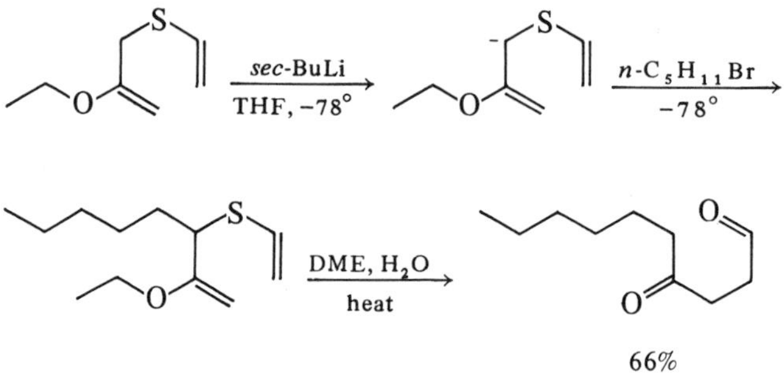

$$(24)$$

4.4 SULFOXIDE α-CARBANIONS

Dimethyl sulfoxide can be converted to the anion by sodium hydride or potassium *tert*-butoxide. This anion has seen frequent use as a strong base (Section 1.4.8) but it also reacts with carbonyl compounds to give acylation or addition. Reaction with esters gives β-keto sulfoxides which may be reduced to methyl ketones with aluminum amalgam (Corey et al., 1964). The intermediate β-keto sulfoxide anions may be alkylated and then reduced to make longer-chain ketones (Gassman et al., 1966b). The methylsulfinyl carbanion will add to non-enolizable ketones and aldehydes to give the salts of β-hydroxysulfoxides which can be converted to alkenes (Eq. 25) (Kuwajima et al., 1972).

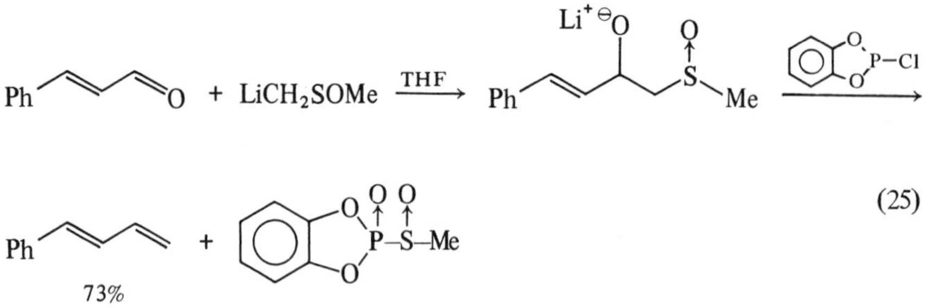

$$(25)$$

The anion of a phenyl sulfoxide gave a high yield when added to isovaleraldehyde (Eq. 26) (Heissler et al., 1976).

$$RCH_2OTs \xrightarrow[DMF]{PhSK} \underset{98\%}{RCH_2SPh} \xrightarrow[CH_2Cl_2]{Cl-\langle\rangle-CO_3H} R CH_2\overset{O}{\underset{\uparrow}{S}}Ph \xrightarrow[2.]{1.\ n\text{-BuLi, THF,}\ -78°}$$

$$\underset{93\%}{\underset{\overset{|}{OH}\ \overset{|}{O}}{\bigvee\!\!\!\bigvee\!\!\!\bigvee}^{R}\!\!SPh} \xrightarrow[AcOH,\ EtOH]{Zn\ dust} \underset{68\%}{\underset{\overset{|}{OH}}{\bigvee\!\!\!\bigvee\!\!\!\bigvee}^{R}} \qquad\qquad (26)$$

$$RCH_2OTs = \text{(bicyclic structure with =CH}_2\text{ and }CH_2OTs)$$

The anions derived from α-halosulfoxides may be prepared and used at −78° but decompose rapidly above −20° as might be expected with the chloride leaving group (Section 4.1). At low temperatures there is no interference from the Ramberg-Backlund rearrangement in sulfoxides with an α' hydrogen (Durst, 1969; Durst et al., 1970; Reutrakul et al., 1977a). They will add to ketones and aldehydes to give sulfoxide alcohols (Eq. 27), which may be caused to eliminate in two different synthetically useful ways.

$$PhS\overset{O}{\underset{\uparrow}{C}}H_2Cl \xrightarrow[\substack{THF \\ -78°}]{n\text{-BuLi}} PhS\overset{O}{\underset{\uparrow}{C}}HCl^{-} \xrightarrow[\substack{-78°\ to\ -20° \\ 2.\ H_2O}]{1.\ n\text{-}C_6H_{13}CHO} PhS\overset{O}{\underset{\uparrow}{C}}H\text{-}\underset{\overset{|}{Cl}}{\overset{\overset{|}{OH}}{C}}H\text{-}n\text{-}C_6H_{13} \qquad (27)$$

$$\underset{70\%}{}$$

First, the products from addition to aldehydes may be heated in xylene to give chloromethyl ketones in high yield (Eq. 28) (Reutrakul et al., 1977a). Second,

$$PhS\text{-}\overset{O}{\underset{\uparrow}{C}}H\underset{\overset{|}{Cl}}{\overset{\overset{|}{OH}}{C}}H\text{-}n\text{-}C_6H_{13} \xrightarrow[160°]{xylene} ClCH_2\overset{O}{\overset{\|}{C}}\text{-}n\text{-}C_6H_{13} \qquad (28)$$

$$\underset{95\%}{}$$

the products from addition to aldehydes or ketones may be cyclized to the epoxysulfoxide with dilute base and then heated in xylene to give an α,β-unsaturated aldehyde with one more carbon than the original carbonyl compound (Eq. 29) (Durst et al., 1970; Reutrakul et al., 1977b).

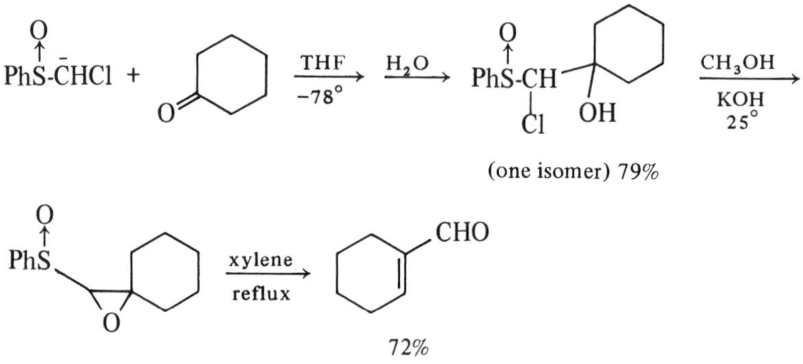

(one isomer) 79%

(29)

72%

The anion derived from optically pure (S)-benzylmethyl sulfoxide will add to acetone, add to CO_2, and even deuterate with a high degree of stereospecificity in generating the new chiral center (Nishihata et al., 1976).

Conjugate addition of sulfoxide α-anions has been applied to the construction of variously substituted naphthalenes as exemplified in Eq. 30 (Hauser et al., 1978). Here the intermediate α-enolate anion is acylated intramolecularly.

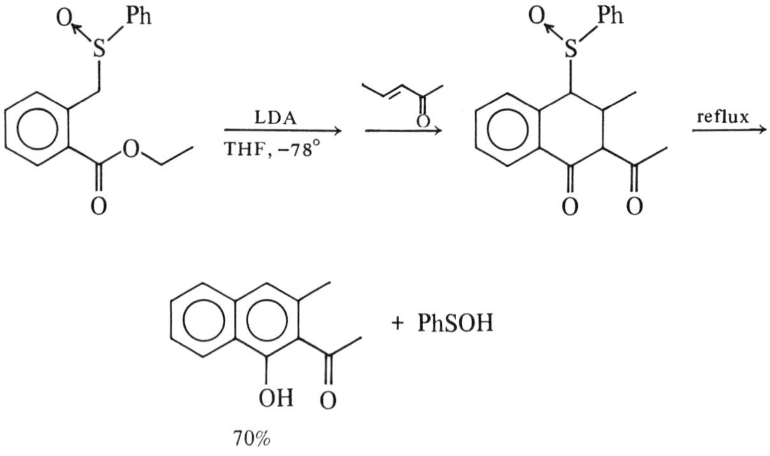

(30)

+ PhSOH

70%

Alkylation of an allylic sulfoxide can be equivalent to alkylation of the hypothetical carbanion HO⌒⌒⁻ owing to the facile rearrangement of allylic sulfoxides to sulfenate esters. This was demonstrated in a synthesis of nuciferal by

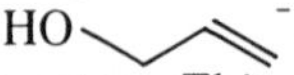

Evans et al., (1973) (Eq. 31). The reversible equilibrium between sulfoxide and sulfenate lies far on the sulfoxide side, but trimethyl phosphite is a strong thiophile and converts the sulfenate to alcohol, thus draining the equilibrium through the sulfenate stage. The alkylation gave 2/3 α product and 1/3 γ product. This selectivity is much improved if alkylation is carried out with the imidazole sulfide and then peracid is used to oxidize the alkylated product to the allylic sulfoxide (Evans et al., 1974a).

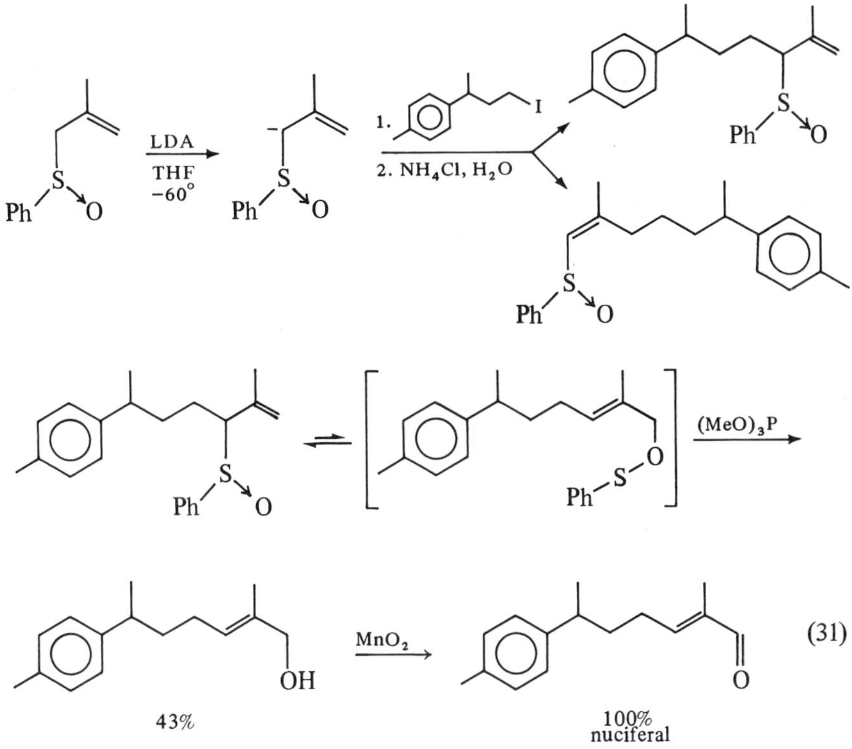

$$
\tag{31}
$$

43%   100%
                                                           nuciferal

4.5 SULFONE α-CARBANIONS

Arylsulfonyl groups are easily eliminated or reductively replaced, and thus they are useful temporary activating groups for alkylation, acylation, and addition reactions (Review: Magnus, 1977). They may be prepared from alkyl halides by substitution with sodium arylsulfinate. A sulfone group is frequently used to activate the β position of protected carbonyl compounds, and for this purpose they are prepared by thiophenoxide displacement of a β chlorine (Eq. 32) (Julia et al., 1975) or by conjugate addition of the sulfinate (Eq. 33) (Fayos et al., 1977; Cooper et al., 1976).

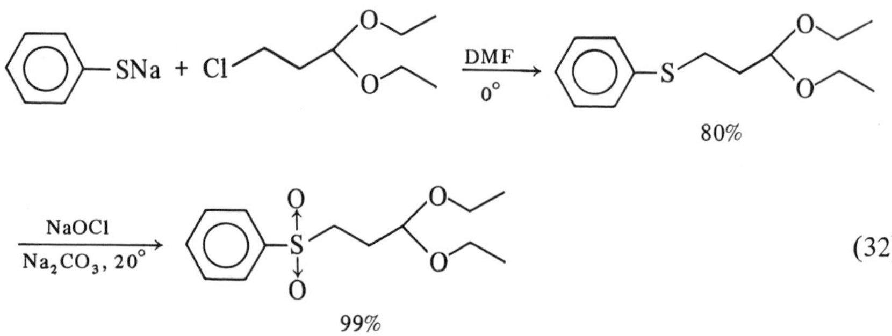

80%

(32)

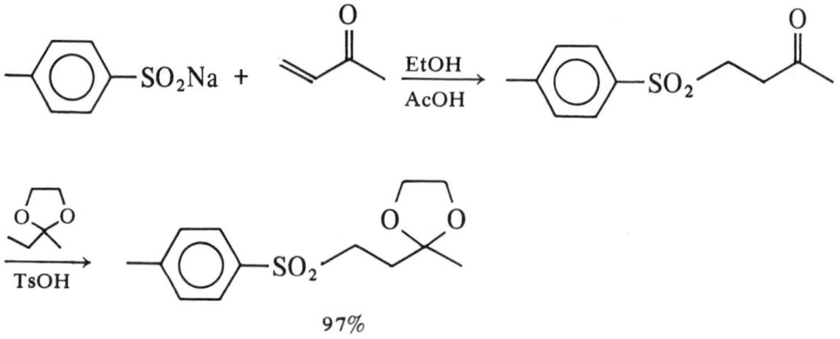

99%

(33)

97%

The anions are prepared from these sulfones using *n*-BuLi and are synthetically equivalent to the hypothetical carbanions **5** and **6** and various methyl-substituted relatives.

$$^-CH=CH-\overset{\overset{\textstyle O}{\|}}{C}-R \qquad\qquad ^-CH_2CH_2\overset{\overset{\textstyle O}{\|}}{C}-R$$

5   **6**

4.5.1 Alkylation. The sulfone α-anions are reactive toward primary bromides, iodides, and epoxides. Equation 34 (Kondo et al., 1975a) shows alkylation followed by elimination of the activating group, and Eq. 35 concludes with reductive removal of the activating group (Kondo et al., 1975b).

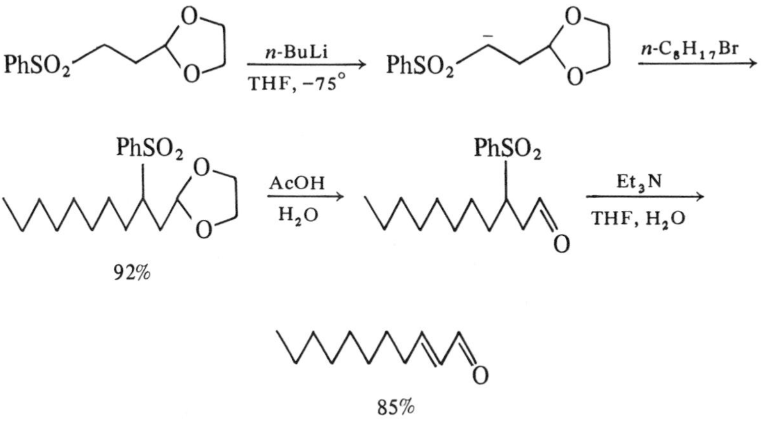

$$(34)$$

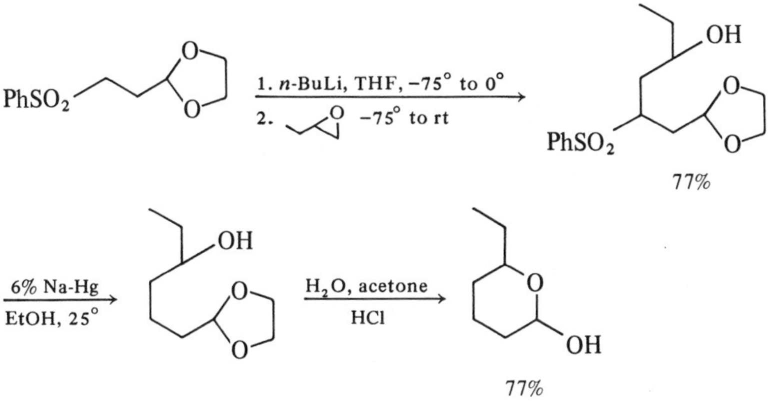

$$(35)$$

Sometimes the low temperatures are unnecessary since Fayos et al. (1977) alkylated a similar sulfone at room termperature to reflux in THF. Trost et al. (1976a) found that control of the pH in the sodium amalgam reduction is important and that disodium hydrogen phosphate buffer in methanol is best.

The coupling of two alkyl halides may be accomplished by first activating one for carbanion formation using sodium toluene- or benzenesulfinate as exemplified in the final steps of a synthesis of α-santalene (Eq. 36) (Julia, et al., 1973a). A similar sequence led to sesquifenchene (Grieco et al., 1975) where lithium in ethylamine was used to reductively remove the sulfone group. Likewise 11-bromoundecanoic acid was converted to myristic acid (Julia et al., 1976).

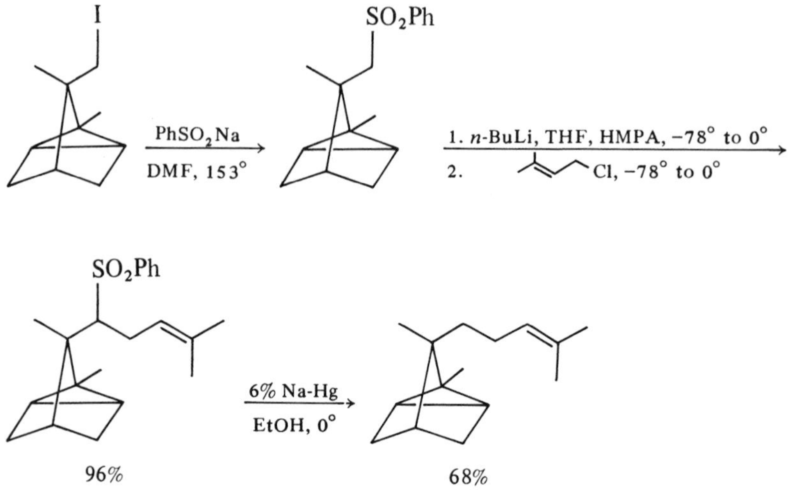

$$
\text{(36)}
$$

The allylation of allylic sulfones followed by reductive cleavage is a convenient route to 1,5-dienes. Elimination of the sulfone leads alternatively to conjugated polyenes. The first has been used to prepare squalene (Eq. 37) (Grieco et al., 1974b) and the second to prepare vitamin A (Eq. 38) (Olson et al., 1976).

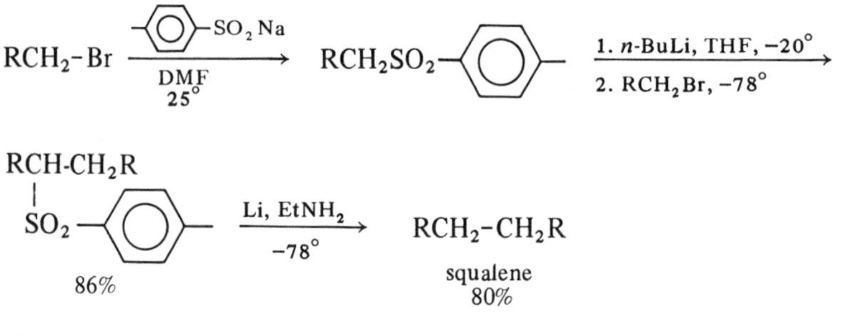

$$
\text{(37)}
$$

RCH$_2$Br = farnesyl bromide

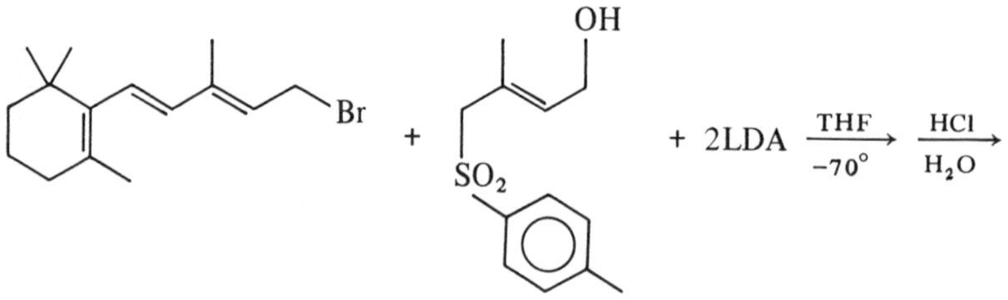

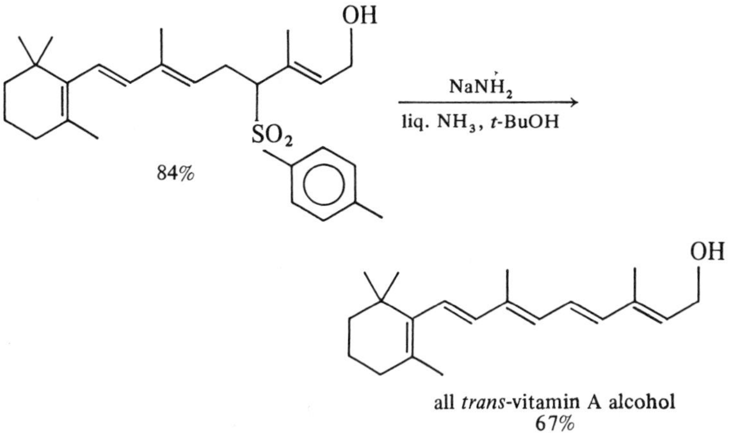

(38)

all *trans*-vitamin A alcohol
67%

Benzylation of an allylic sulfone anion followed by reductive desulfonation was used in the preparation of ferruginol (Torii et al., 1977).

4.5.2 Acylation. Esters readily acylate sulfone anions as shown in a synthesis of 1,4-dicarbonyl compounds (Eq. 39) (Kondo et al., 1975c).

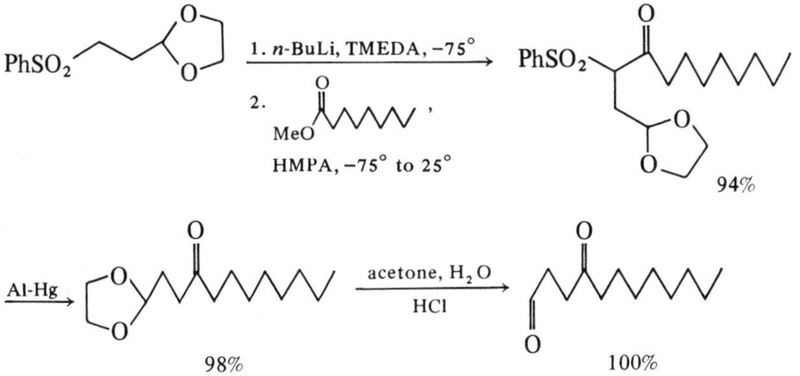

(39)

4.5.3 Addition. Sulfone anions add readily to ketones and aldehydes to give β-hydroxysulfones (Eq. 40) (Cooper et al., 1976). With phase transfer catalysis, the addition to aryl aldehydes and subsequent elimination of water can be carried out in one operation (Eq. 41) (Cardillo et al., 1975).

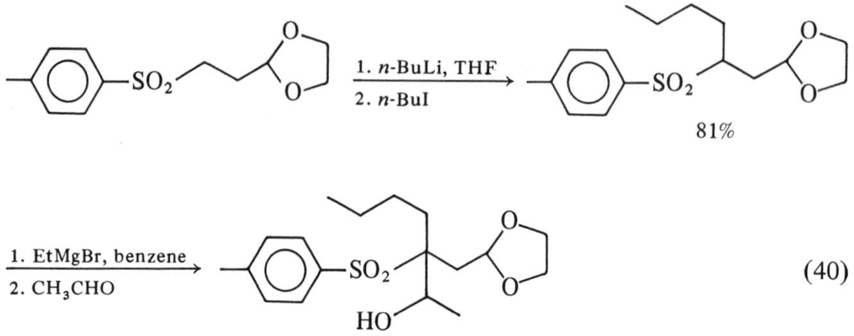

$$\xrightarrow[\text{2. } n\text{-BuI}]{\text{1. } n\text{-BuLi, THF}}$$

81%

$$\xrightarrow[\text{2. CH}_3\text{CHO}]{\text{1. EtMgBr, benzene}}$$

(40)

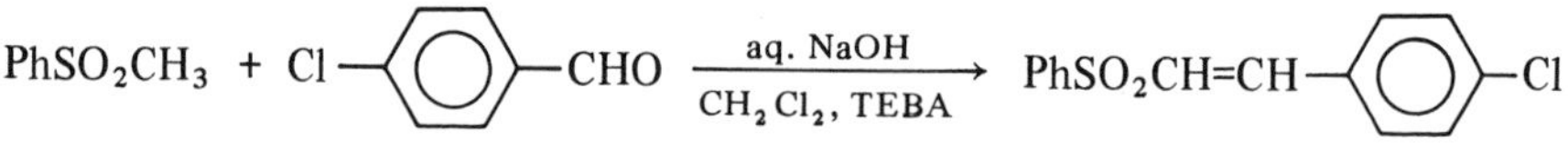

$$\text{PhSO}_2\text{CH}_3 \; + \; \text{Cl}\!\!-\!\!\langle\text{O}\rangle\!\!-\!\!\text{CHO} \xrightarrow[\text{CH}_2\text{Cl}_2,\text{ TEBA}]{\text{aq. NaOH}} \text{PhSO}_2\text{CH=CH}\!\!-\!\!\langle\text{O}\rangle\!\!-\!\!\text{Cl}$$

TEBA = triethylbenzylammonium chloride

98%

(41)

If the simple alkene is the desired end product, the sulfone and hydroxyl groups may be eliminated together by forming the mesylate as shown in Eq. 42 (Julia et al., 1973b). This specific placement of the double bond is an alternative to the Wittig reaction (Section 4.8.1):

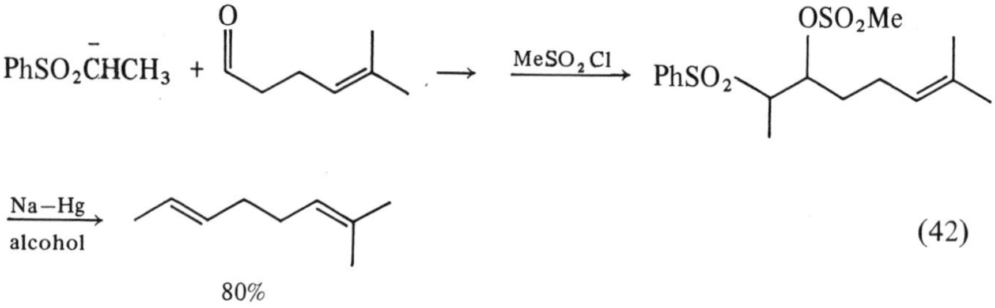

$$\text{PhSO}_2\overset{-}{\text{C}}\text{HCH}_3 \; + \qquad \xrightarrow{\text{MeSO}_2\text{Cl}} \text{PhSO}_2$$

$$\xrightarrow[\text{alcohol}]{\text{Na}-\text{Hg}}$$

(42)

80%

If an α-halosulfone anion is used in the addition, the intermediate alkoxide displaces the halide to give α,β-epoxysulfones (Barone et al., 1978). See also Sections 4.1 and 5.3.3.

4.5.4 Conjugate Addition. Conjugate addition of a sulfone anion leads to an enolate anion which can then displace the sulfinate group to give a cyclopropane ring. This has been applied to the synthesis of methyl chrysanthemate as shown in Eq. 43 (Schatz, 1978).

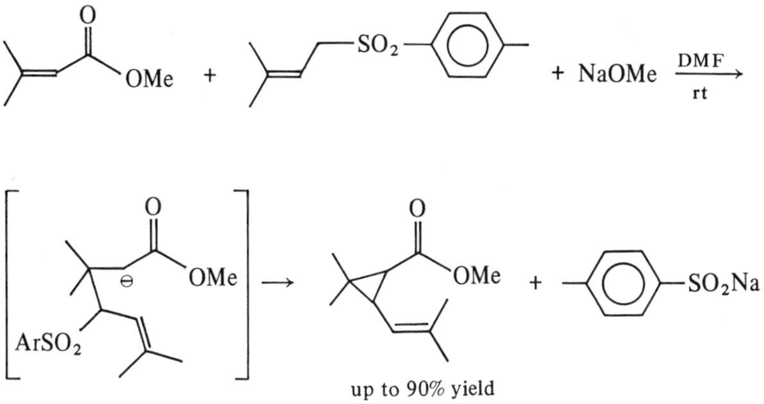

up to 90% yield

4.6 THIOACETAL CARBANIONS, INCLUDING VINYLOGS AND OXIDES

The carbonyl carbon of an aldehyde is partially positive charged and therefore attractive to nucleophiles. It cannot be converted directly to a carbanion for reaction with electrophiles. However, a number of derivatives of aldehydes are sufficiently acidic to allow carbanion formation. If these carbanions are treated with electrophiles and then converted back to the carbonyl compound, we have the equivalent of an acyl carbanion. The temporary reversal of polarity of the carbonyl carbon is referred to by the German word *Umpolung* (Reviews: Seebach et al., 1974b; Lever, Jr., 1976).

Thioacetals, especially the cyclic 1,3-dithianes, are examples of such derivatives. The two sulfur atoms flanking the carbon that was the carbonyl carbon are able to stabilize a carbanion generated by base removal of a proton. This carbanion may then be alkylated, acylated, or added to other carbonyl compounds (Seebach et al., 1975a; Review: Gröbel et al., 1977). Removal of the thioacetal protection then reconstitutes the carbonyl group with electrophilic substitution having been carried out on it. When the starting aldehyde is formaldehyde, two successive substitutions are possible (Eqs. 44 to 46) (Corey et al., 1970b; Seebach et al., 1971). Primary and secondary halides give good yields when the temperature is kept low to suppress competing elimination. Epoxides may be

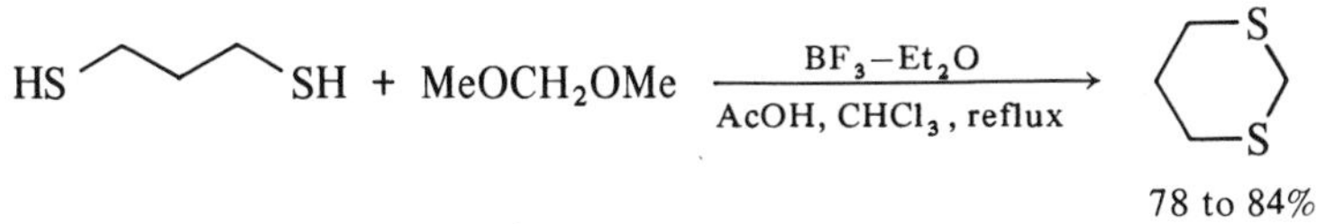

78 to 84%

used similarly to give β-hydroxy compounds, but the reaction is slow even at 0°.

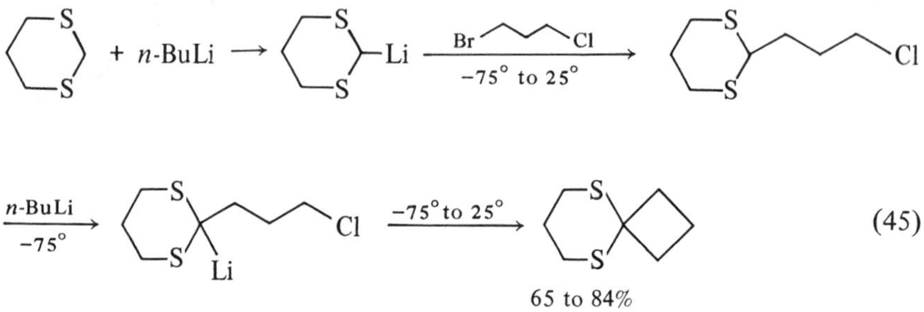

$$(45)$$

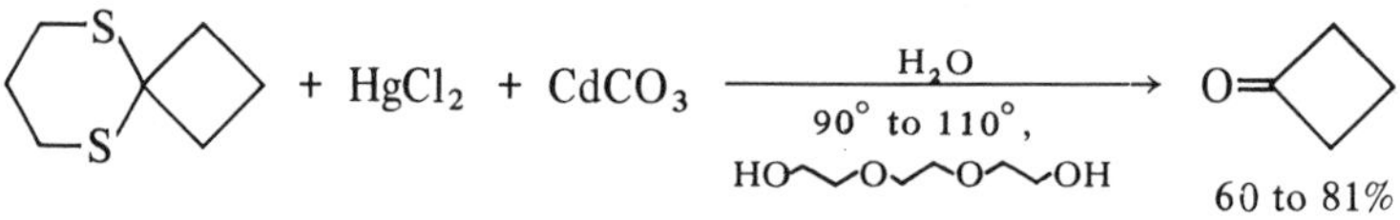

65 to 84%

$$(46)$$

60 to 81%

Addition to aldehydes or ketones gives α-hydroxy compounds. This must be carried out at $-70°$ to $-50°$ to prevent the competing proton transfer which gives the enolate ion of the carbonyl compound (Eq. 47). Addition to α,β-unsaturated ketones gives only 1,2-addition, no 1,4-addition product.

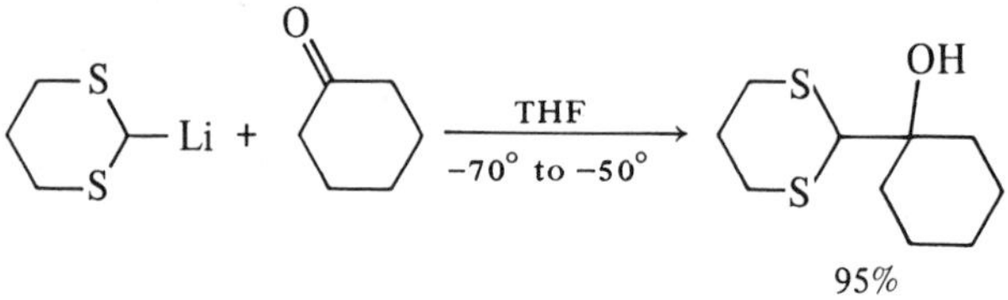

$$(47)$$

95%

Acylation of unsubstituted dithiane occurs with esters or acid chlorides. Since the product is more acidic than dithiane, half of the dithiane carbanion is consumed as a base, forming the anion of the acylated product.

The hydrolysis of the dithioacetals is relatively difficult, requiring heat and complexing of the sulfur by mercury (Eq. 46) or in other cases silver-assisted oxidation with N-chlorosuccinimide (Cory et al., 1971c). These methods give poor results for the preparation of aldehydes, but mercuric oxide plus boron trifluoride etherate serve well (Eq. 48) (Vedejs et al., 1971). A good general procedure for both ketones and aldehydes in the copper-activated process shown in Eq. 49 (Narasaka et al., 1972). Some mild, rapid alternatives which have been demonstrated on a few cases are oxidation of the dithianes with ceric ammonium nitrate or thallium trifluoroacetate (Ho et al., 1972).

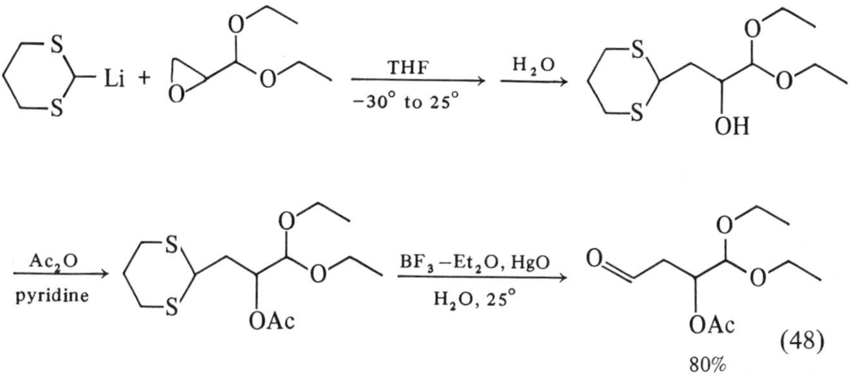

(48)

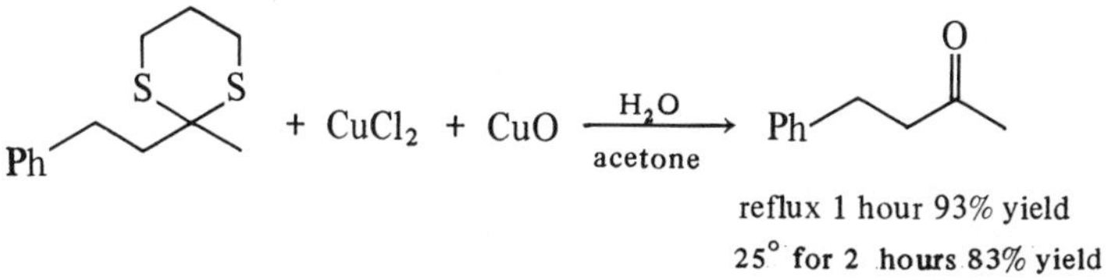

80%

(49)

reflux 1 hour 93% yield

25° for 2 hours 83% yield

Many alternatives to dithiane have been devised, each with some particular advantage. Compound **7** gives odorless crystalline derivatives (Mori et al., 1975); **8** and **9** give products which are more easily hydrolyzed (Hori et al., 1974; Nakai et al., 1974); **10** can be alkylated at room termperature via a low equilibrium concentration of the carbanion generated with sodamide (Chapter 2, Eq. 12) (Schill et al., 1975); **11** gave high yields of addition to lactones, in contrast to **10** which gave mainly enolate anion formation (Trost et al., 1975b). Finally **12** is more reactive than the dithiane anion, and the products can be hydrolyzed under conditions where a dithiane was untouched (Eq. 50) (Balanson, et al., 1977). Ketones cannot be made with **12** however, since the metalation of the 2-alkyldithiazanes was not possible with any of a variety of strong bases.

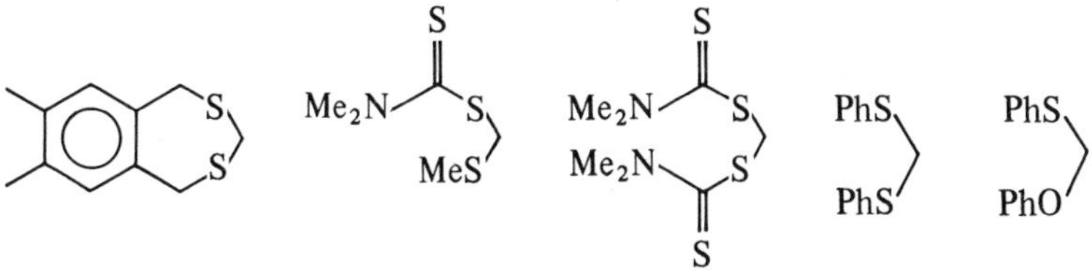

| 7 | 8 | 9 | 10 | 11 |

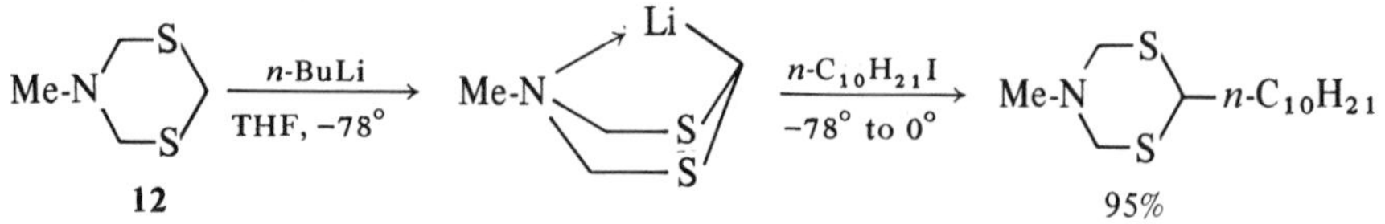

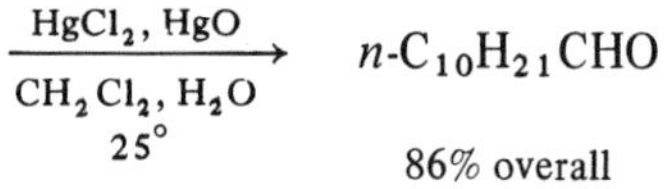

$$n\text{-}C_{10}H_{21}CHO \qquad (50)$$

86% overall

Although nonconjugated dithianes do not undergo conjugate addition, the 2-aryldithiane anions will, as shown with 2-butenolide in Eq. 51 (Ziegler et al., 1975; Damon et al., 1976). See also Chapter 2, Eq. 23. Later in the sequence, the dithiane was removed by Raney nickel desulfurization to give a methylene group rather than the usual hydrolysis to a carbonyl group.

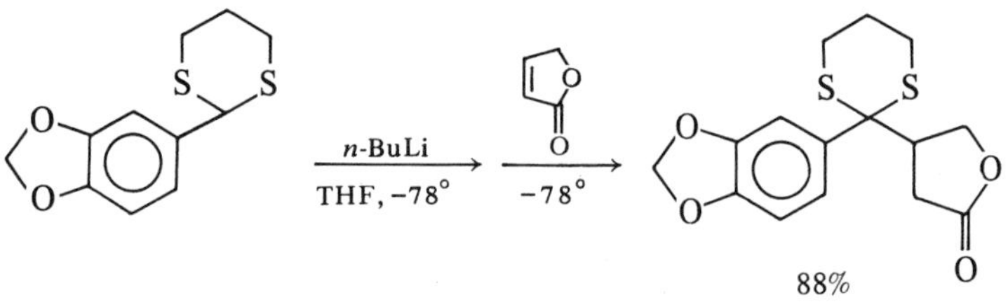

$$(51)$$

88%

If one of the sulfur atoms of a dithioacetal is oxidized ($NaIO_4$) to the sulfoxide level, the positive charge on that sulfur will further stabilize the carbanion. The most generally useful example is the diethyl compound **13**. It can be converted to the carbanion quickly using LDA or n-BuLi at 2°, and that anion is more thermally stable than the dithiane anion. This may be alkylated, realkylated, acylated, and give addition and in some cases conjugate addition reactions in very high yields. Moreover the protected carbonyl group is freed by brief treatment in the cold with a catalytic amount of perchloric acid. Examples of alkylation, addition, and acylation are given in Eqs. 52 to 56 (Richman, et al., 1973; Herrman et al., 1973e).

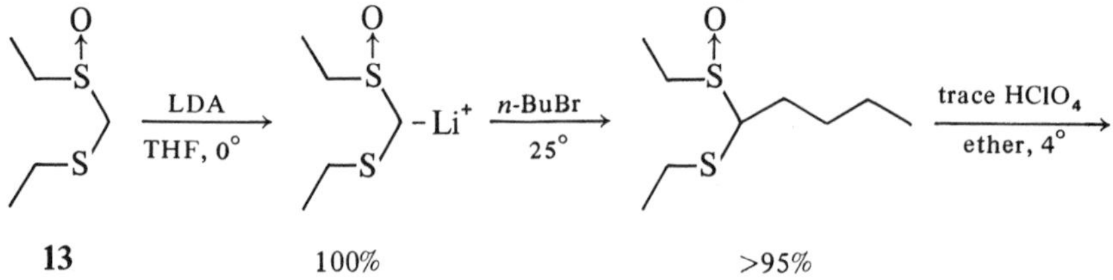

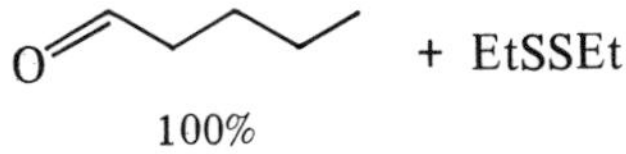

(52)

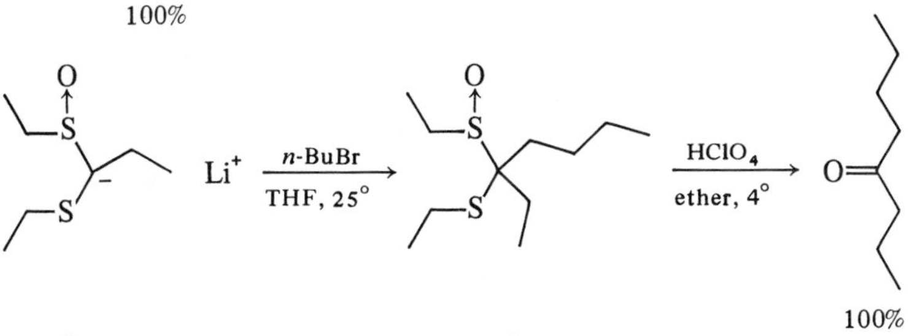

100%

(53)

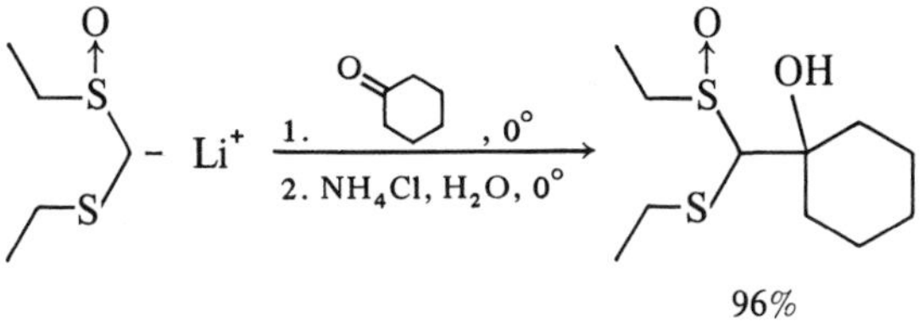

100%

(54)

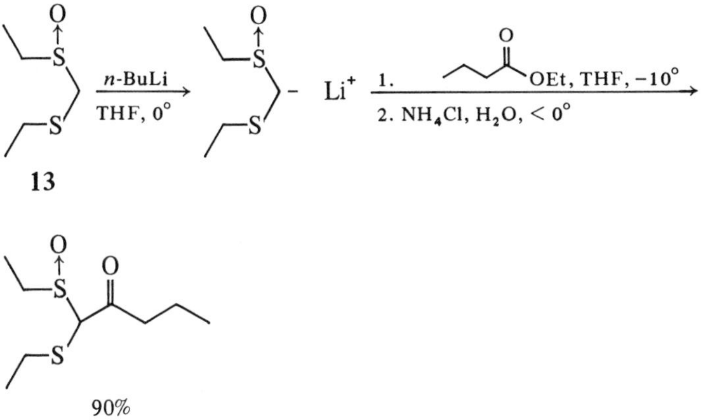

96%

13

(55)

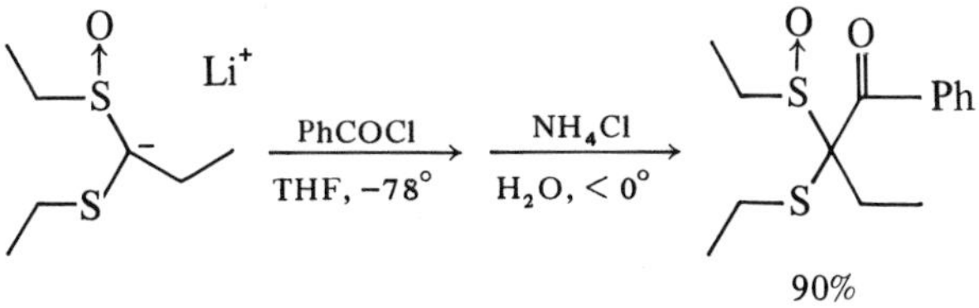

90%

(56)

90%

All of these products can be readily hydrolyzed to the corresponding ketones or aldehydes. The acylation reactions of **13** require two moles of carbanion for each mole of acylating agent since the enolate anion of the product is formed rapidly. The substituted reagents give addition to aldehydes and acylation with acid chlorides (Eq. 56) but are sluggish with ketones and esters.

The substituted and unsubstituted reagents give conjugate addition to α,β-unsaturated esters (Eq. 57) (Herrman et al., 1973d); however the unsubstituted reagent **13** gives 1,2-addition to α,β-unsaturated ketones. Overall, the alkylation followed by conjugate addition is a synthesis of 1,4-dicarbonyl compounds from three parts (Eq. 58).

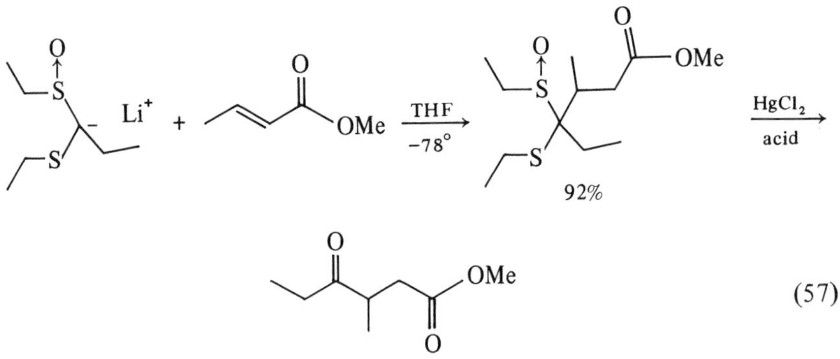

(57)

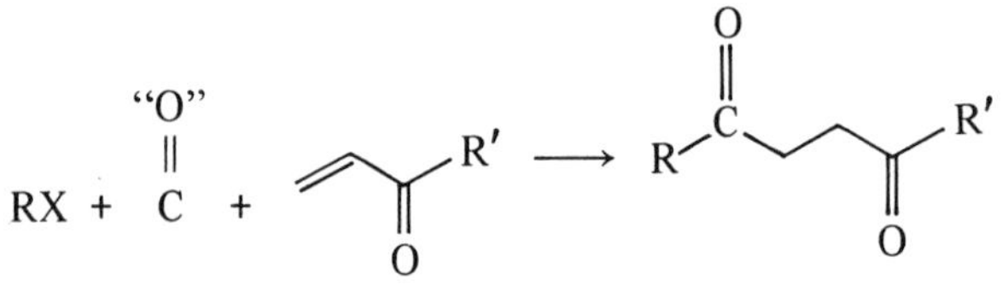

(58)

The similar methylthiomethyl sulfoxide has been alkylated for the synthesis of DL-dopa and of small ring ketones (Ogura et al., 1971, 1976) and also added to aldehydes and ketones (Ogura et al., 1972).

1,3-Bis(methylthio)allyllithium (**14**) is a vinylog of the dithioacetal anions. In use it is the equivalent of the hypothetical carbanion **15**.

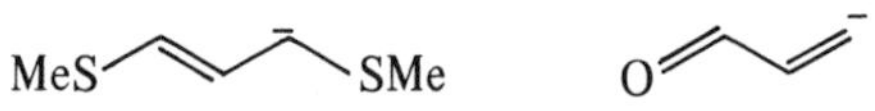

The reagent is generated by treating the methyl ether **16** with two equivalents of LDA causing elimination of methoxide and then formation of the deep-purple carbanion (Eq. 59) (Corey et al., 1971a). This may be alkylated and hydrolyzed to give *trans*-α,β-unsaturated aldehydes. Alkylation with epoxides leads ultimately to δ-acetoxy-α,β-unsaturated aldehydes (Eq. 60). The anion **14** may also be added to ketones and aldehydes to give, after hydrolysis, γ-hydroxy-α,β-unsaturated aldehydes, one of which was used in the synthesis of thromboxane B$_2$ (Corey et al., 1977a).

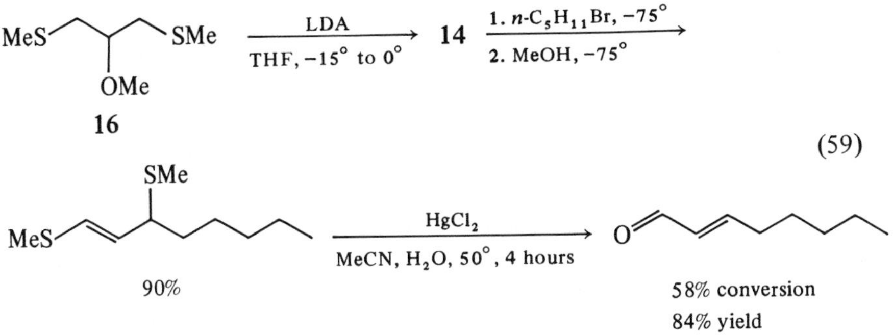

(59)

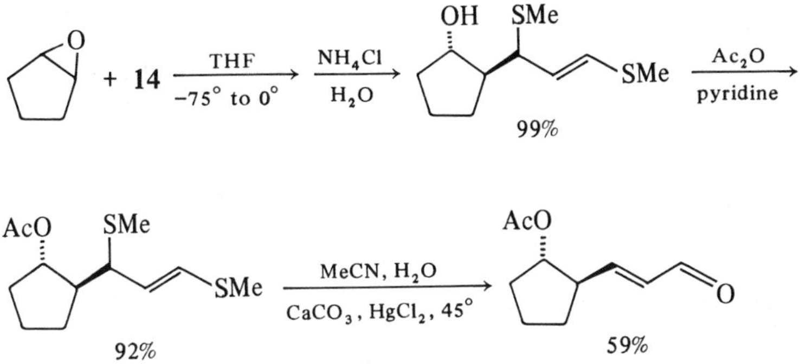

(60)

Reagent **16** is made originally from epichlorohydrin. The comparable phenylthio compounds can be made from the α,β-unsaturated ketones and aldehydes, and then alkylated and hydrolyzed in a manner similar to that used with **16** (Eq. 61) (Cohen et al., 1976). In unsymmetrical cases 1,3-carbonyl transposition is likely.

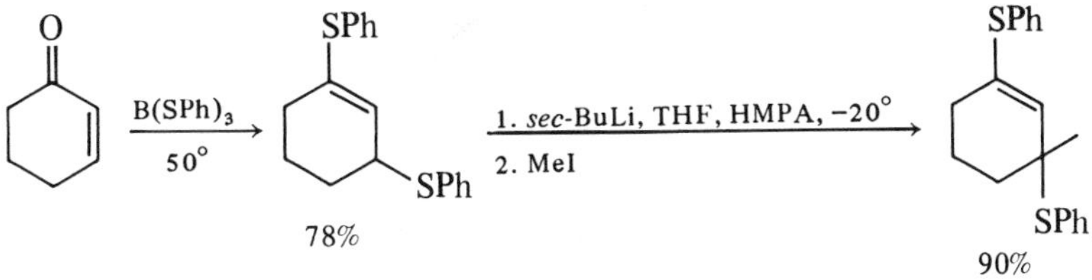

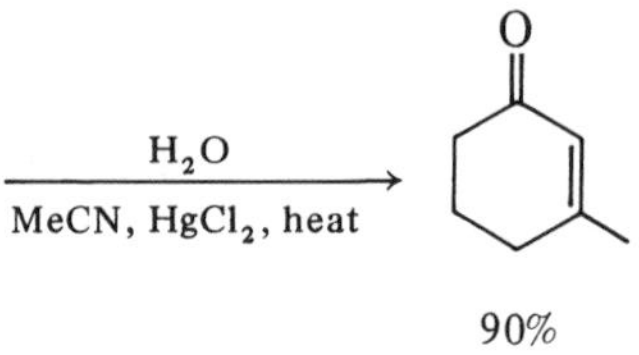

$$\xrightarrow[\text{MeCN, HgCl}_2, \text{ heat}]{\text{H}_2\text{O}}$$

90%

(61)

4.7 SELENIDE AND SELENOXIDE α-ANIONS

The selenoxide, and to a lesser extent the selenide group, will stabilize a carbanionic charge on an adjacent site. This may then be treated with alkyl halides, ketones, aldehydes, esters, and so forth, to generate new carbon-carbon bonds. The selenium is subsequently eliminated to give an alkene. Thus, the selenium-stabilized anions are formally equivalent to vinyl carbanions (Review: Clive, 1978).

Allylic and benzylic selenides (but not simple alkyl) are acidic enough to be deprotonated by LDA (Eq. 62) (Reich et al., 1975a). The simple alkyl selenide anions were prepared by the n-butyllithium cleavage of selenoacetals (Eq. 63) (Van Ende et al., 1975).

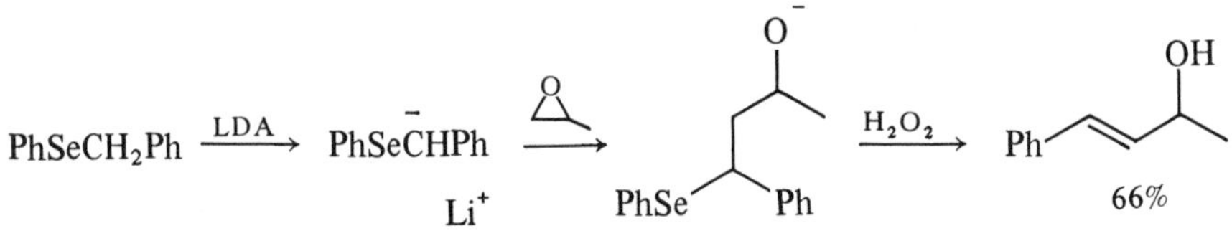

$$\text{PhSeCH}_2\text{Ph} \xrightarrow{\text{LDA}} \text{PhSe}\overset{-}{\text{CHPh}} \quad \text{Li}^+ \longrightarrow \overset{\text{PhSe}}{\underset{\text{Ph}}{\bigvee}} \xrightarrow{\text{H}_2\text{O}_2} \text{Ph} \overset{\text{OH}}{\diagup}$$

66%

(62)

$$\text{CH}_3\text{CHO} + \text{MeSeH} \xrightarrow[0°]{\text{HCl}} \text{CH}_3\text{CH(SeMe)}_2 \xrightarrow[\substack{1.\ n\text{-BuLi, }-78°,\text{ hexane, THF} \\ 2.}]{} $$

80%

, −78° to 25°

1. MeI, 25°
2. KO-t-Bu, DMSO, 25°

90%   97%

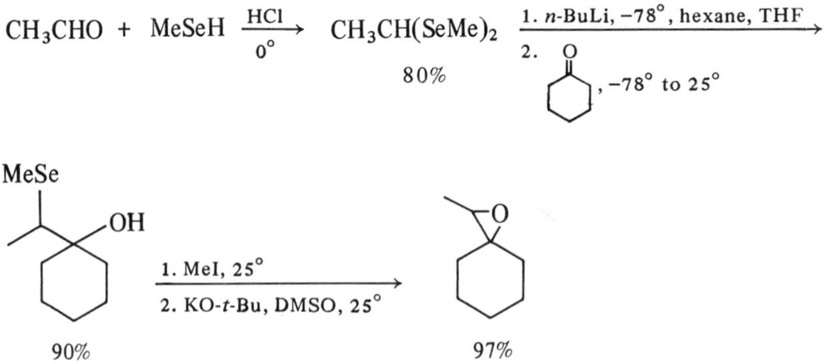

(63)

The more stabilizing selenoxide group allows preparation of a variety of alkyl carbanions using LDA (Eqs. 64 and 65) (Reich et al., 1975a). Selenoxides such as those below which have β-hydrogens must be handled below 0° to prevent premature elimination.

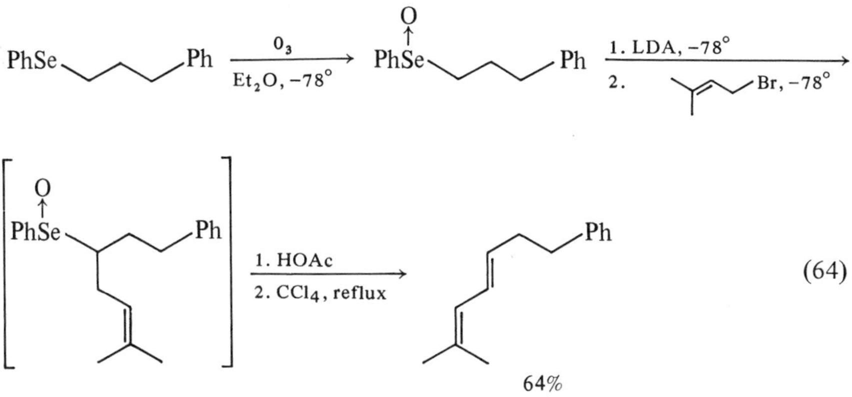

$$(64)$$

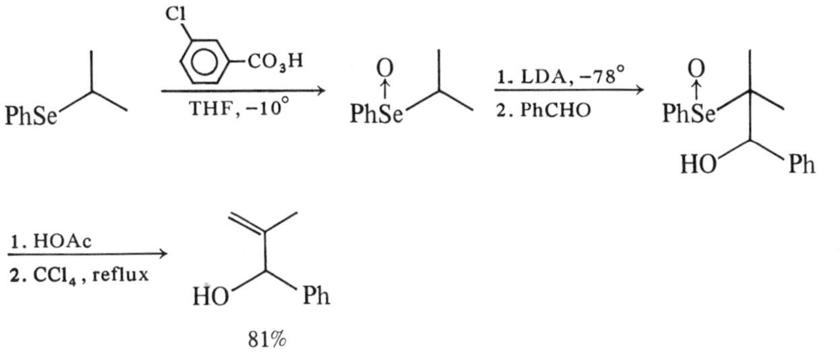

64%

$$(65)$$

81%

4.8 SULFUR YLIDS (Reviews: Trost et al., 1975c; Johnson, 1966)

A sulfonium ylid may be generally represented by two resonance forms, 17 and 18 (more resonance forms contribute when π bonding is in conjugation with the carbanionic site). Because there is no net negative charge on an ylid molecule, it differs from the rest of the carbanions, but there is a considerable negative charge on the nucleophilic carbon in the ylid. Also the basicity, reactions, and methods of formation of these ylids closely resemble those of net carbanions.

$$\overset{\oplus}{\underset{/}{S}}-\overset{\ominus}{\underset{\backslash}{C}} \qquad \underset{/}{S}=\underset{\backslash}{C}$$

$$\textbf{17} \qquad\qquad \textbf{18}$$

Treatment of a sulfonium salt with a base removes a proton, leaving a negative charge on carbon. This charge is diminished (stabilized) by electrostatic interaction with the adjacent positive charge and by π bonding overlap of the carbon $2p$ orbital with a vacant, charge-contracted $3d$ orbital on the sulfur.

In the sulfoxide oxidation state, sulfur is effectively more electronegative, and the corresponding sulfoxonium ylids are more thermally stable. The electrostatic and conjugative stabilization may be represented with resonance forms **19** through **21**.

$$\overset{O^-}{\underset{|}{\overset{|}{\underset{++}{S}}}}-\overset{-}{\underset{\backslash}{C}}\overset{/}{} \qquad \overset{O}{\underset{|}{\overset{\|}{\underset{+}{S}}}}-\overset{-}{\underset{\backslash}{C}}\overset{/}{} \qquad \overset{O^-}{\underset{|}{\overset{|}{\underset{+}{S}}}}=\underset{\backslash}{C}\overset{/}{}$$

$$\textbf{19} \qquad\qquad \textbf{20} \qquad\qquad \textbf{21}$$

The major synthetic uses of the sulfur ylids are the formation of epoxides from ketones and aldehydes and the cyclopropanation of α,β-unsaturated compounds (Corey et al., 1965). The most commonly used reagents are dimethylsulfonium methylid, which must be used directly after preparing, and the less reactive, more stable dimethylsulfoxonium methylid, which may be stored for weeks at $0°$. The experimental details for their preparation are given by Trost et al. 1975c) and Corey et al. (1973) (Eqs. 66 and 67). The sulfonium ylid

$$\text{Me}_3\text{S}^{\oplus}\text{I}^{\ominus} + \text{Me}\overset{O}{\overset{\uparrow}{\text{S}}}\text{CH}_2^{\ominus}\ \text{Na}^{\oplus} \xrightarrow[\text{THF, 0° to 5°}]{\text{DMSO}} \overset{\ominus}{\text{CH}_2}-\overset{\oplus}{\underset{|}{\text{S}}}-\text{Me} \qquad (66)$$
$$\qquad\qquad\qquad\qquad\qquad\qquad\qquad\qquad\qquad\qquad \underset{\text{Me}}{|}$$

$$\text{DMSO} \xrightarrow[\text{reflux}]{\text{MeI}} \text{Me}_3\overset{O}{\overset{\uparrow}{\text{S}}}{}^{\oplus}\ \text{I}^{\ominus} \xrightarrow[\text{DMSO or THF, rt}]{\text{Me}\overset{O}{\overset{\uparrow\ominus}{\text{S}}}\text{CH}_2\ \text{Na}^{\oplus}} \overset{\ominus}{\text{CH}_2}-\overset{O}{\overset{\oplus\uparrow}{\underset{|}{\text{S}}}}-\text{Me} \qquad (67)$$
$$\qquad\qquad\qquad\qquad\qquad\qquad\qquad\qquad\qquad\qquad\qquad\qquad\qquad \underset{\text{Me}}{|}$$

nucleophilically attacks the least hindered face of a ketone (axial in simple cyclic cases) irreversibly, and the intermediate oxy anion intramolecularly displaces the plus charged sulfur leaving group (Bessière-Chrétien et al., 1972) (Eq. 68).

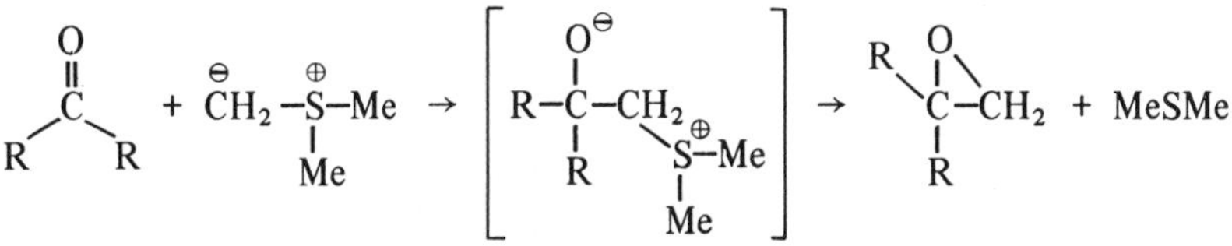

$$\underset{R}{\overset{O}{\underset{\overset{\|}{C}}{\overset{\|}{C}}}}\overset{R} + \overset{\ominus}{CH_2}-\overset{\oplus}{\underset{Me}{S}}-Me \rightarrow \left[R-\overset{O^{\ominus}}{\underset{R}{\underset{\overset{|}{C}}{\overset{|}{C}}}}-CH_2\overset{\oplus}{\underset{\underset{Me}{\overset{|}{S}}-Me}{}} \right] \rightarrow \underset{R}{\overset{R}{\underset{\overset{|}{C}}{\overset{O}{C}}}}-CH_2 + MeSMe \quad (68)$$

The oxosulfonium ylids appear to attack reversibly, giving eventually some of each stereoisomer, often more the one derived from equatorial attack, which reflects the greater thermodynamic stability of the intermediate where the Me_2SOCH_2 group is in the equatorial position. This difference makes the two reagents synthetically complementary (Eqs. 69 and 70). When the hindrance is great on one face of the carbonyl, both ylids will give attack from the least hindered side, the more reactive one giving the higher yield (Eq. 71).

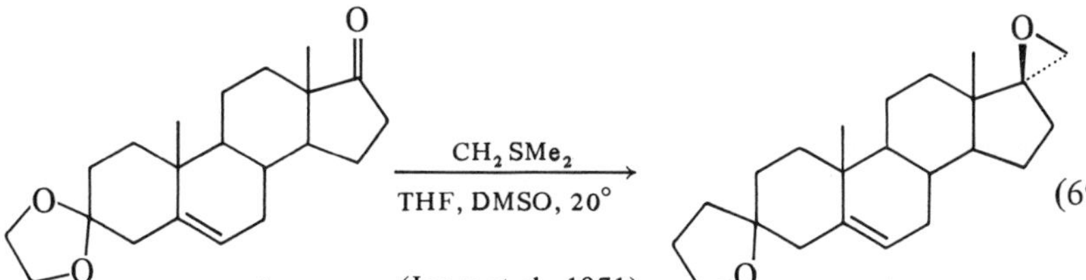

$$\xrightarrow[\text{THF, DMSO, 20}^\circ]{CH_2SMe_2} \quad (69)$$

(Jones et al., 1971) 97% yield

$$\xrightarrow[\text{DMSO, 20}^\circ \text{ to 50}^\circ]{CH_2SOMe_2} \quad (70)$$

2 : 1

96% total yield

(Jones et al., 1972)

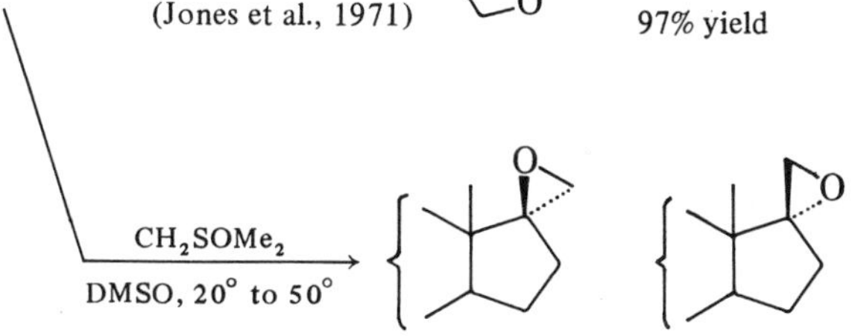

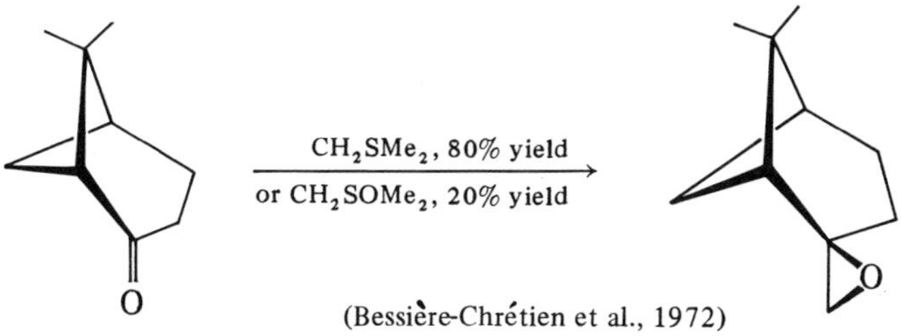

$$\xrightarrow[\text{or } CH_2SOMe_2, \text{ 20\% yield}]{CH_2SMe_2, \text{ 80\% yield}} \quad (71)$$

(Bessière-Chrétien et al., 1972)

Aqueous sodium hydroxide can be used as the base with phase transfer catalysis, but only aldehydes give good yields (Eq. 72) (Merz et al., 1973).

$$Ph\diagdown\diagup\diagdown_O + Me_3S^+\ I^- + NaOH \xrightarrow[H_2O,\ n\text{-}Bu_4NOH,\ 50°]{CH_2Cl_2} Ph\diagdown\diagup\diagdown\triangle$$

$$85\%$$

$$(72)$$

For higher homologs than methylids, the other two groups on sulfur should not
have acidic α-hydrogens, so that only one ylid may result from a given salt. For
the alkyl cases a common choice is two phenyl groups, but diphenyl sulfide is
not nucleophilic enough to react with alkyl halides. However silver fluoroborate
will promote the formation of alkylsulfonium salts. The ylids from these salts
are thermally unstable and so are used in situ at low temperature, or generated
with KOH at 25° in low concentration in the presence of the ketone substrate.
Among the oxosulfonium ylids, the ethyl and isopropylid may be prepared from
(dimethylamino)ethyl-p-tolyloxosulfonium fluoroborate (Johnson et al., 1973a).

An example using an isopropylid is shown in Eqs. 73 and 74 (van Tamelen et
al., 1970). The diphenyl sulfide was ethylated with the strong alkylating agent
triethyloxonium fluoroborate, and the ethylid was alkylated with methyl iodide.

$$Ph_2S + Et_3O^+\ BF_4^- \rightarrow Ph_2\overset{+}{S}Et\ BF_4^- \xrightarrow[DME,\ CH_2Cl_2,\ -70°]{LDA} Ph_2\overset{+}{S}\text{-}\overset{-}{C}HCH_3.$$

$$\xrightarrow[-70°\ to\ -50°]{MeI} Ph_2\overset{+}{S}\text{-}\underset{I^-}{\overset{\overset{\displaystyle CH_3}{|}}{C}H}\text{-}CH_3 \xrightarrow[-70°]{LDA} Ph_2\overset{\oplus}{S}\text{-}\overset{\ominus}{C}\diagup^{CH_3}_{\diagdown CH_3} \qquad (73)$$

$$+ \ Ph_2\overset{\oplus}{S}\text{-}\overset{\ominus}{C}\diagup^{CH_3}_{\diagdown CH_3} \xrightarrow{-60°\ to\ -70°}$$

$$+ \ Ph_2S \qquad\qquad (74)$$

$$75\%$$

Conjugate addition to α,β-unsaturated ketones or esters results in cyclopropanation. The oxosulfonium ylids give high yields in this process (Eq. 75) (Corey et al., 1965), but the unstabilized sulfonium ylids give only epoxides with the same substrates.

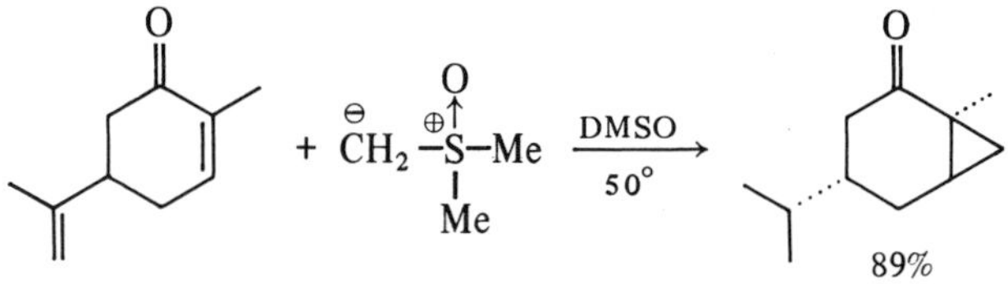

(75)

Sulfonium ylids which are stabilized by conjugation of the carbanionic site with electronegative atoms give cyclopropanations under ordinary conditions (Eq. 76).

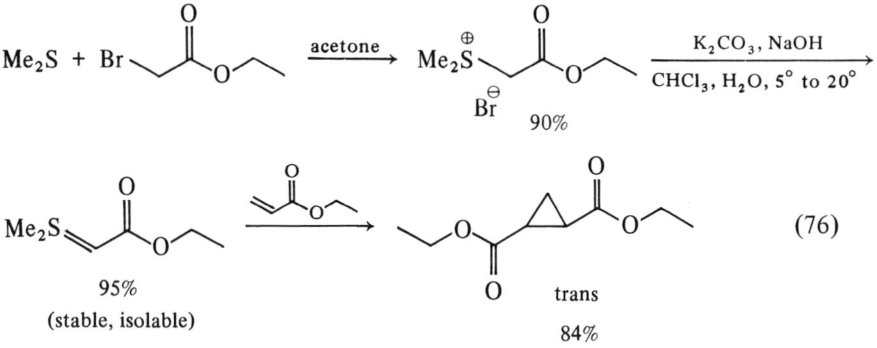

(76)

A great many examples of these cyclopropanations have been carried out (Trost et al., 1975c).

4.9 PHOSPHORUS YLIDS (WITTIG REAGENTS) (Reviews: Maercker, 1965; Johnson, 1966; Bestmann, 1968)

Phosphorus ylids, like the sulfur ylids (Section 4.8), have no net negative charge on the molecule but the nucleophilic carbon bears a partial negative charge as shown by the resonance forms of the simple example triphenylphosphinemethylene (**22**). In view of the partial $(p \to d)\pi$ bonding, these ylids are often called alkylidenephosphoranes.

$$\overset{+}{Ph_3P} - \overset{-}{CH_2} \qquad\qquad Ph_3P = CH_2$$

22

The highly colored ylids are prepared by treatment of phosphonium salts with bases such as PhLi, BuLi, *tert*-BuOK, or sodium methylsulfinylmethide. They are not isolated but used directly after preparation because they are sensitive to oxygen and moisture. The requisite phosphonium salts are available in great variety from the reaction of triphenylphosphine and primary or secondary alkyl halides, usually bromides.

4.9.1 Addition-Elimination. The phosphorus ylids attack ketones and aldehydes nucleophilically to give a zwitterionic intermediate; but unlike the sulfur ylids, the high affinity of phosphorus for oxygen leads to elimination of triphenylphosphine oxide (Eq. 77). Thus an alkene is generated with a controlled,

$$\underset{C}{\overset{O}{\|}} + \overset{\ominus}{CH_2} - \overset{\oplus}{PPh_3} \rightarrow \left[\begin{matrix} O^{\ominus} \overset{\oplus}{PPh_3} \\ | \quad | \\ -C-CH_2 \\ | \end{matrix} \rightarrow \begin{matrix} O-PPh_3 \\ | \quad | \\ -C-CH_2 \\ | \end{matrix} \right] \rightarrow \begin{matrix} O{\leftarrow}PPh_3 \\ + \\ C=CH_2 \end{matrix} \qquad (77)$$

specific location for the double bond (Wittig reaction). It appears only between the carbons that bore the phosphorus and oxygen. The preparation of methylenecyclohexane is outlined in Eq. 78 (Wittig et al., 1973). The special value of this method is shown by the comparison with the elimination of the tertiary alcohol

$$Ph_3P + CH_3Br \xrightarrow[25°]{\text{benzene}} Ph_3\overset{+}{P}CH_3 \ Br^- \xrightarrow[\text{DMSO}]{\overset{O}{\underset{\uparrow}{}}\ MeSCH_2 \ \overset{\ominus}{}\ Na^{\oplus}}$$

99%

$$\overset{\oplus}{Ph_3P} - \overset{\ominus}{CH_2} \longrightarrow \quad + \ Ph_3PO \qquad (78)$$

(dark red)

60 to 78%

from cyclohexanone and methylmagnesium bromide, where the choice of β-hydrogens gives the thermodynamically stable endocyclic alkene. An application in steroid synthesis is shown in Eq. 79 (McMorris et al., 1976). The ylid was prepared from the phosphonium bromide using potassium *tert*-amyloxide.

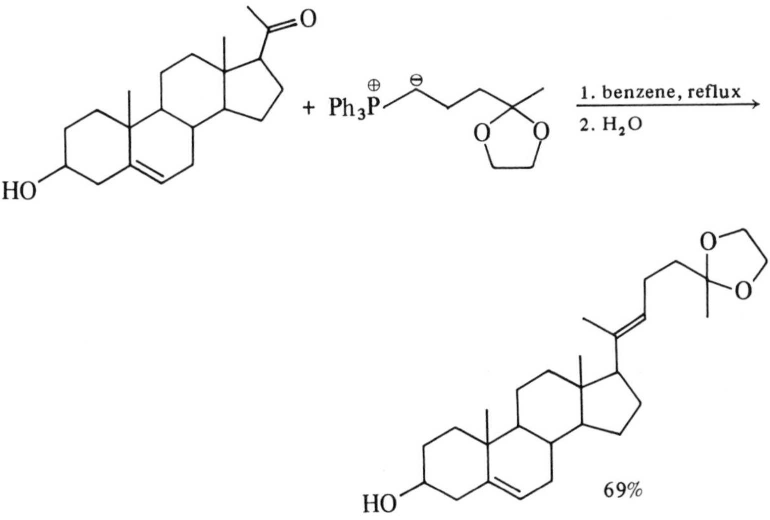

(79)

69%

The methoxymethylid finds frequent use for the one-carbon homologation of ketones and aldehydes where the vinyl ether is hydrolyzed to a new aldehyde group (Eq. 80) (Danishefsky et al., 1975).

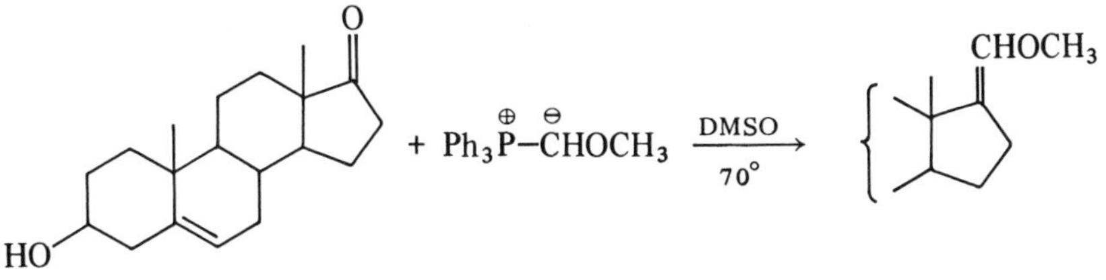

androstenolone

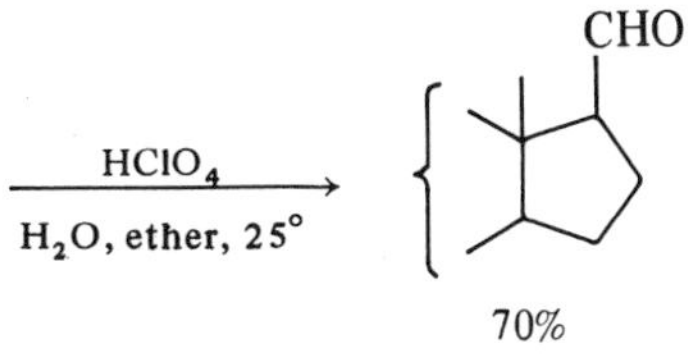

(80)

70%

Aldehydes will undergo the Wittig reaction under phase transfer conditions with an aqueous solution of NaOH as the base, under a benzene solution of the aldehyde (Tagaki et al., 1974). With crown ether catalysis, even potassium carbonate will generate ethylidenetriphenylphosphorane in methylene chloride (Boden, 1975).

When a lithium reagent is used as the ylid-forming base, the zwitterionic intermediate from carbonyl addition forms a stable complex with the lithium bromide present. This retards olefin formation, but the addition of HMPA precipitates the LiBr from ether solution, giving the alkene rapidly and in higher yield (Magnusson, 1977).

The reaction of aldehydes with primary Wittig reagents gives in many cases high cis stereoselectivity (>95%). This can be obtained using bases such as potassium *tert*-butoxide or sodium hexamethyldisilazide (Bestmann et al., 1976) to form the ylid; for example see Eq. 81 (Anderson et al., 1975). This is especially valuable for synthesis of the many insect pheromones and fatty acid derivatives which contain cis double bonds (Review: Henrick, 1977; Rossi, 1977). Soluble lithium salts in the ylid solution will interfere with the cis stereoselectivity; so if lithium reagents are used as bases to form the ylid in THF, a substantial amount of trans isomer will be present with the cis product. However polar solvents such as DMF or DMSO cancel the effects of the lithium salts and again nearly pure cis isomers are obtained (Review: Schlosser, 1970). Conjugated ylids such as that derived from 1-bromo-2-butene give both cis and trans isomers whatever the conditions (Henrick, 1977).

$$
\underset{Br^{\ominus}}{Ph_3\overset{\oplus}{P}}\diagup\diagdown\diagup\diagdown \quad \xrightarrow[\text{2. O}\diagdown\diagup\diagdown=\diagdown\text{OEt}]{\text{1. }t\text{-BuOK, THF, 25}°} \quad \diagup\diagdown\diagup\diagdown=\diagup\diagdown\diagup=\diagdown\diagup\diagdown\text{OEt}
$$

69% yield

94% 8-cis isomer   (81)

6% 8-trans isomer

The opposite stereochemical result, the formation of only the trans isomers, is possible by equilibration of the intermediate zwitterion with base and lithium salt before the elimination of triphenylphosphine oxide (Eq. 82) (Schlosser et al., 1967, 1970).

The anion generated from the zwitterionic addition intermediate will also add to another aldehyde, leading ultimately to allylic alcohols (Eq. 83) (Corey et al., 1970a). Except with formaldehyde, the second oxygen introduced is the one eliminated with the phosphorus, and also the two aldehyde fragments become trans to each other in the allylic alcohol.

$$Ph_3\overset{\oplus}{P}Et \xrightarrow[\text{THF, ether}]{\text{PhLi}} Ph_3\overset{\oplus}{P}\diagup\diagdown\overset{\ominus}{} \xrightarrow[-78°, 5\ min]{} \left[Ph_3\overset{\oplus}{P}\overset{Br^{\ominus}\cdots Li^{\oplus}}{\diagdown}\overset{O^{\ominus}}{} \right]$$

$$Br^{\ominus}$$

threo and erythro

$$\xrightarrow[-78° \text{ to } -30°]{\text{PhLi}} \left[Ph_3\overset{\oplus}{P}\overset{Br^{\ominus}\cdots Li^{\oplus}}{\diagdown}\overset{O^{\ominus}}{} \right] \xrightarrow[t\text{-BuOH}]{t\text{-BuOK}} [\text{threo}] \rightarrow$$

$$\diagup\diagdown\diagup\diagdown$$

(82)

69% yield
99% trans
1% cis

$$Ph_3\overset{\oplus}{P}-\overset{\ominus}{C}HCH_3 \xrightarrow[\text{THF, }-78°]{\text{heptanal}} \left[Ph_3\overset{\oplus}{P}\overset{O^{\ominus}}{} \right] \xrightarrow[\text{2. }CH_3CHO, -78°]{\text{1. }n\text{-BuLi, }-78°}$$

$$\left[Ph_3\overset{\oplus}{P}\overset{O^{\ominus}}{}\overset{O^{\ominus}}{} \right] \rightarrow \overset{OH}{\diagup\diagdown}$$

67%

(83)

4.9.2 Alkylation. Phosphonium ylids may be alkylated and the resulting secondary alkyl phosphonium salt converted to a new ylid for subsequent reaction with an aldehyde (Eq. 84) (Liedtke et al., 1972; Bertele et al., 1967).

$$Ph_3\overset{\oplus}{P}Et\ \overset{\ominus}{Br} \xrightarrow[\text{ether, 25°}]{n\text{-BuLi}} Ph_3\overset{\oplus}{P}-\overset{\ominus}{C}HCH_3 \xrightarrow[\text{ether, reflux}]{} Ph_3\overset{\oplus}{P}\overset{D\ D}{\diagup\diagdown}\overset{\ominus}{Br}$$

$$\xrightarrow[\text{2. }O\diagdown\diagup\overset{O}{\diagdown}OMe]{\text{1. }n\text{-BuLi}} \overset{O}{\diagdown\diagup\diagdown\diagup\diagdown}OMe$$

$$D\ D$$

Z and E

51%

(84)

Alkylation of a phosphorus ylid followed by reductive cleavage of the phosphorus may be used to couple allylic groups to give 1,5-dienes. Lithium in ethylamine at $-76°$ cleaves allylic groups from phosphorus, but phenyl groups are also attacked; therefore in this application tributylphosphine was preferable (Eq. 85) (Axelrod et al., 1970).

$$\text{R} \diagup\diagdown\diagup \text{Br} + n\text{-Bu}_3\text{P} \xrightarrow[25°]{\text{benzene}} \xrightarrow[\text{THF}, -76°]{\text{PhLi}} \text{R} \diagup\diagdown\diagup\diagdown \text{P-}n\text{-Bu}_3$$

$$\text{R} \diagup\diagdown\text{B}^{..} \longrightarrow \text{R} \diagup\diagdown\diagup\diagdown\diagup \overset{\overset{\oplus}{\text{P-}n\text{-Bu}_3} \quad \text{Br}^{\ominus}}{} \xrightarrow[\text{EtNH}_2, -76°]{\text{Li}} \text{R} \diagup\diagdown\diagup\diagdown\diagup \text{R}$$

$$85\% \qquad\qquad\qquad\qquad \underset{\underset{65\%}{\text{squalene}}}{}$$

$$\text{R} \diagup\diagdown\diagup \text{Br} = trans, trans\text{-farnesyl bromide} \qquad\qquad (85)$$

4.9.3 Acylation. The phosphonium ylids react readily with acid chlorides to give ketones which are then converted to the new resonance stabilized ylid, consuming a second equivalent of the initial ylid as a base (Eq. 86) (Meyers et al., 1972a; Maercker, 1965). Esters without enolizable hydrogens will also acylate the ylids but the reaction is slow so that a keto ester gives the normal Wittig reaction at the ketone.

$$\overset{\oplus}{\text{Ph}_3\text{P}} \diagup\diagdown \underset{\text{Br}^{\ominus}}{} \xrightarrow[\text{benzene}\;0°]{n\text{-BuLi}} \overset{\oplus}{\text{Ph}_3\text{P}} \overset{\ominus}{} \diagup\diagdown \xrightarrow[]{\overset{\overset{O}{\|}}{\text{ClCOEt}}} \overset{O}{\diagdown}\diagup^{\text{OEt}} \underset{\overset{\oplus}{\text{Ph}_3\text{P}}\;\ominus}{}$$

$$\overset{S}{\underset{S}{\langle}} \text{-CHO} \longrightarrow \overset{S}{\underset{S}{\langle}}\diagup\diagdown\overset{O\diagdown\diagup^{\text{OEt}}}{} + \overset{S}{\underset{S}{\langle}}\diagdown\overset{O\diagdown\diagup^{\text{OEt}}}{} \qquad (86)$$

$$\begin{array}{ccc}
90 & : & 10 \text{ initially} \\
18 & : & 82 \text{ distilled}
\end{array}$$

$$85\% \text{ yield total}$$

The most efficient use of the ylid is made using acyl imidazoles as acylating agents where 1:1 stoichiometry is sufficient (Eq. 87) (Miyano et al., 1975).

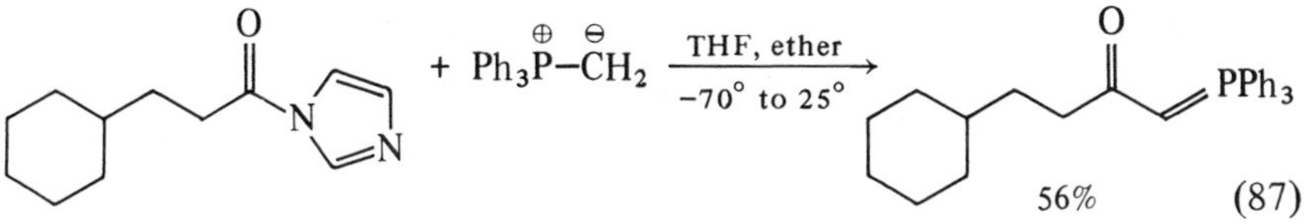

$$56\% \qquad (87)$$

These acylated products and other ylids stabilized by conjugation of the carbanionic site with an oxygen or nitrogen atom are less reactive than the simple alkylidenetriphenylphosphoranes. They will combine with aldehydes but are very sluggish with ketones. The phosphonate anions are generally superior for olefin synthesis when this conjugation is present (Section 5.10).

4.9.4 Conjugate Addition. Alkylidene phosphoranes generally react with α, β-unsaturated ketones and aldehydes by attack on the carbonyl carbon and not by conjugate addition; for instance $\Delta^{5,16}$-pregnadien-3β-ol-20-one acetate gives the usual methylene derivative (Eq. 88) (Sondheimer et al., 1957). However α,

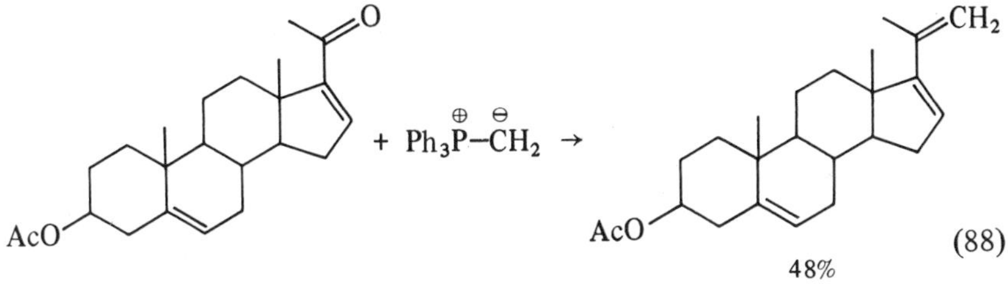

$$48\% \qquad (88)$$

β-unsaturated esters are attacked at the β position to give the enolate which then displaces triphenylphosphine and gives the cyclopropane compound (Eq. 89) (Grieco et al., 1972; Bestmann et al., 1962).

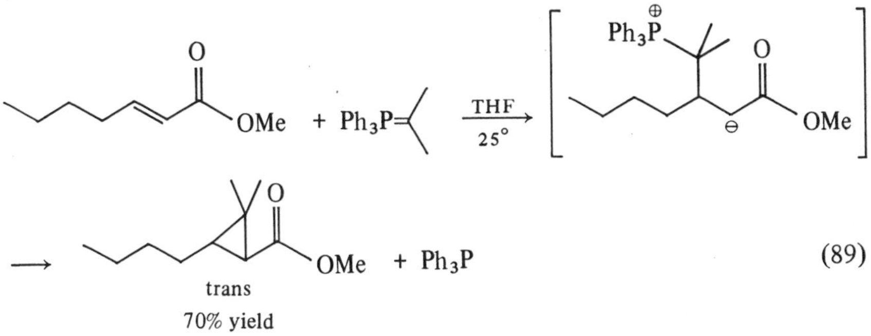

$$(89)$$

trans
70% yield

The allylidenetriphenylphosphorane reagents are exceptional in that they give conjugate addition to α,β-unsaturated esters and ketones. The products from the ketones are cyclic dienes from conjugate addition followed by intramolecular attack at the carbonyl. Dauben et al. (1973b) have applied this to the synthesis of highly strained bicyclic alkenes (Eq. 90). The fact that the allylidenephosphorane reacts initially at its own β position suggests coordination between the phosphorus and oxygen parallel to that shown for nitrile α-anions as shown in Chapter 2, Eq. 24.

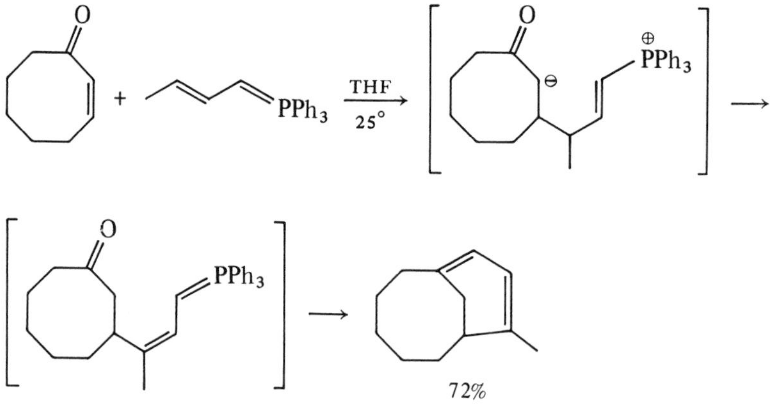

$$(90)$$

72%

Allylidenetriphenylphosphorane will attack the δ position of the longer conjugation in ethyl sorbate, again by connection with its own β position. In this case the ring closure includes formation of a cyclopropane (Eq. 91) (Dauben et al., 1973a).

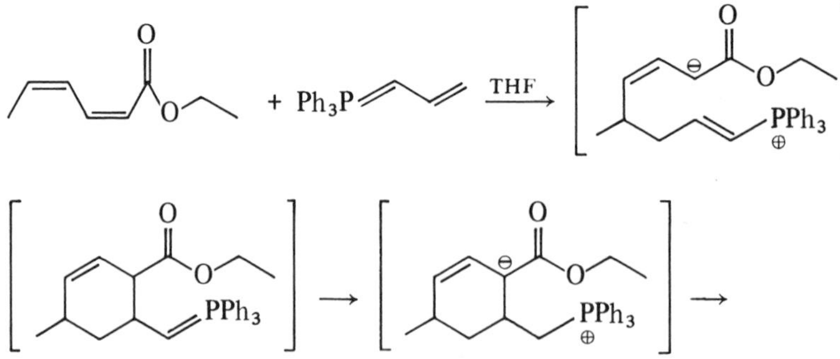

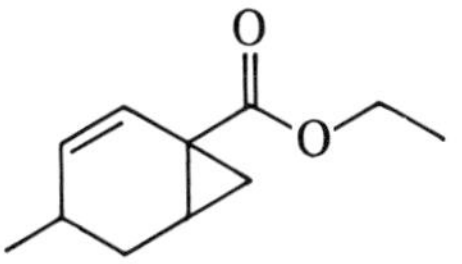

$$+ \; Ph_3P \tag{91}$$

4.10 IMINO CARBANIONS

The addition of primary, secondary, or tertiary alkyllithium reagents to the readily available 1,1,3,3-tetramethylbutyl isocyanide produces an iminocarbanion which may then be treated with various electrophiles and hydrolyzed to give ketones or aldehydes. Equation 92 shows the preparation of a β-hydroxyketone by alkylation with an epoxide. Alkylation with primary bromides, addition to benzaldehyde and carbon dioxide, and acylation with ethyl chloroformate have been demonstrated (Niznik et al., 1974). In this sequence the imino carbanion serves as an acyl anion equivalent.

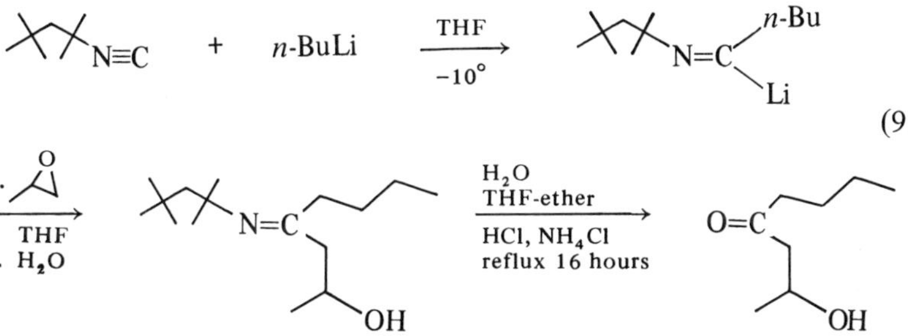

$$\tag{92}$$

90% overall

CHAPTER 5

Carbanions Stabilized by π Conjugation with One Heteroatom

T HE ACIDITY of a C-H is much greater when the resulting anion is delocalized to an oxygen or a nitrogen by conjugation. Thus hydrogen atoms α to a carbonyl, imino, or nitrile group may be removed by weaker bases than were usually applied in Chapters 3 and 4. Alkoxide and hydroxide ions will generate appreciable concentrations of carbanion, and even amines will produce enough in equilibrium to carry out aldol reactions. The anions α to carbonyl groups are generally called enolate anions, implying that the resonance form with the charge on oxygen more closely resembles the hybrid and is formally similar to the enol tautomer (Eq. 1). This is especially the case when lithium is the counterion since it shows some covalent bonding with the oxygen.

$$-CH_2-\overset{\overset{\textstyle O}{\|}}{C}-$$

carbonyl tautomer

$$\Updownarrow \quad \xrightarrow{\text{base}} \quad \left[\overset{\ominus}{-}\overset{\overset{\textstyle O}{\|}}{CH}-\overset{}{C}- \quad -CH=\overset{\overset{\textstyle O^{\ominus}}{|}}{C}- \right] \tag{1}$$

$$-CH=\overset{\overset{\textstyle OH}{|}}{C}-$$
enolate anion

enol tautomer

In spite of the fact that the negative charge resides largely on the oxygen or nitrogen, the usual reactions give carbon-carbon bond formation at the α position. Bründström (1953) attributes this to the action of the associated metal cation. The cation, especially lithium, coordinates with the electronegative atom of the carbanion and also coordinates with the electronegative end of the electrophilic reagent, be it a halide or a carbonyl compound. This polarizes the attacking reagent such that the carbon atoms are attractive to each other in a six-membered ring arrangement (Eq. 2). Reactants that are sufficiently polarized or sterically hindered would not form the cyclic transition state but would first give the metal halide salt plus an ion pair (Eq. 3) which would close at the highly charged oxygen. Acid halides and trimethylsilyl chloride give largely attachment at oxygen, that is, enol derivatives.

The six-membered ring arrangement may also be responsible for the selective α-alkylation of extended enolate anions from α,β- or β,γ-unsaturated substrates. γ-Alkylation would have produced a more conjugated product, but it does not occur.

127

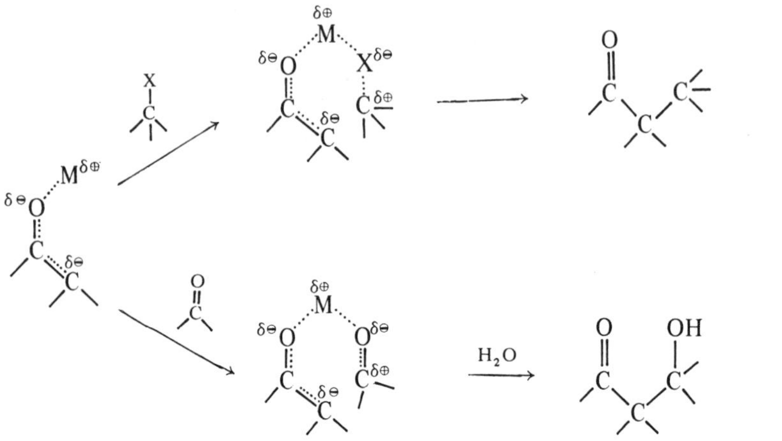

(2)

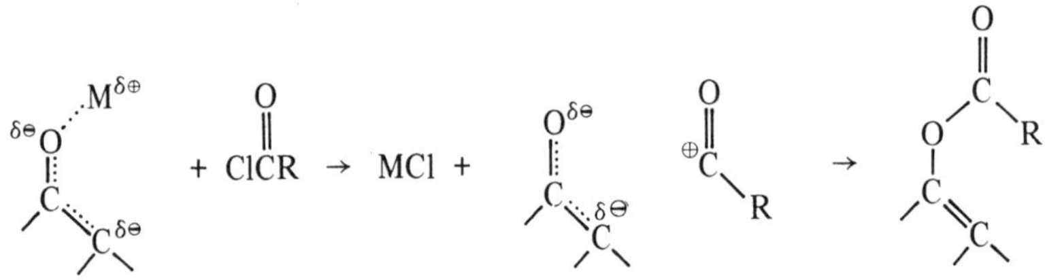

(3)

The un-ionized precursors of many of the carbanions in this chapter are active electrophiles and will react with these carbanions during their preparation. One approach for avoiding this complication is to make the carbanion less reactive by incorporating temporary charge-delocalizing groups as in the malonic ester route to substituted acetic acids (Chapter 6). Alternatively the precursor can be converted first to an imino derivative (Sections 5.7, 5.8, and 5.9) which is less electrophilic toward the corresponding carbanion. Finally, very low temperatures will suppress the electrophilic reactions and yet allow LDA and other strong, nonnucleophilic bases to convert essentially all of the precursor to the carbanion, which can be used directly without the extra steps of incorporation and removal of limiting groups.

5.1 KETONE ENOLATE ANIONS

Ketones with more than one α-hydrogen can give a mixture of products from mono- and polysubstitution at various sites. Many methods have been developed to obtain specificity in these processes as seen below (Reviews: d'Angelo, 1976; House, 1972).

5.1.1 Alkylation. The direct alkylation of ketones with more than one α-hydrogen is complicated by the formation of dialkylated or polyalkylated products. The initially monoalkylated ketone is converted to a new enolate anion by some of the unreacted original enolate and thus receives a second alkyl group mostly on the same side as the first one. This may be minimized by using a large excess of the ketone enolate anion to statistically favor monoalkylation as illustrated with 3-pentanone in Eq. 4 (Elliot et al., 1976). Of course the problem does not

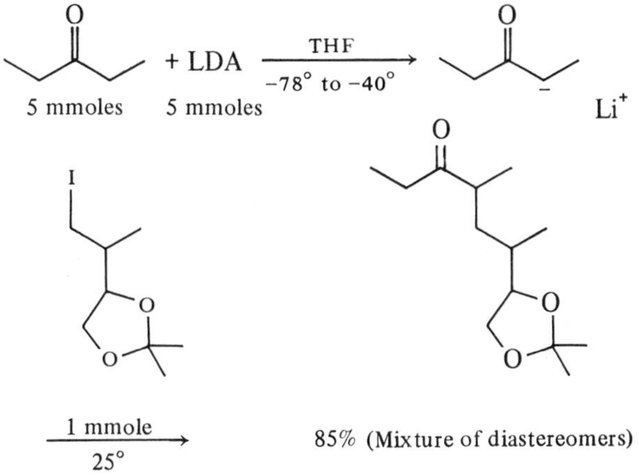

exist when peralkylation is desired; then simply a slight excess of alkyl halide is used as in Eq. 5 (Mousseron et al., 1957; Millard et al., 1978).

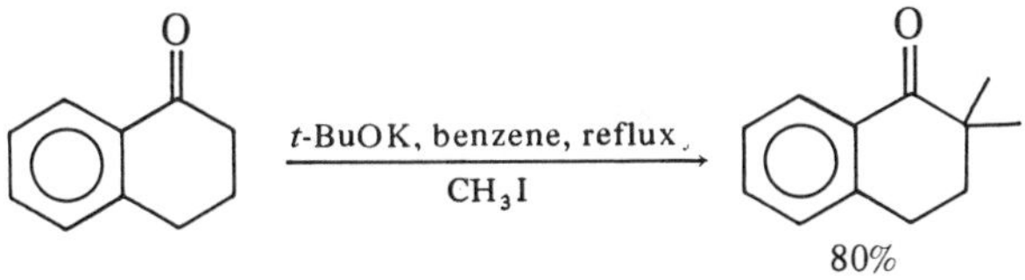

With unsymmetrical ketones having α-hydrogens on both sides, the alkylation is further complicated by a need for selectivity toward a particular side. In some cases a substituent raises the acidity on one of the two sites, favoring alkylation there. In most cases the positional selectivity and polyalkylation problems have been solved using derivatives such as imine or hydrazone enolate ions (Sections 5.7 and 5.8), enamines (Dyke, 1973), temporary activating groups, temporary blocking groups, or using specific enolate generation under nonequilibrating conditions.

A common choice in activating groups is a β-ester function which is readily made by acylation of the ketone enolate (Section 5.1.2) with dimethyl carbonate or ethyl diethoxyphosphinyl formate. This group selectively enters the least substituted side of the original ketone since the reaction is reversible, and it leads to the most thermodynamically stable β-ketoester α-anion, that is, the least substituted one. For instance 6-methyl-5-heptene-2-one was treated with sodium hydride and dimethyl carbonate to give the β-keto ester from acylation of the methyl group in 77% yield (LaLonde et al., 1977). A similar selectivity is found for methylene over methinyl sites (for example see Grieco et al., 1977). The ester group greatly acidifies the remaining hydrogen(s) on the α position, direc- ting alkylation exclusively to that site and also giving only monoalkylation. Finally the ester group is hydrolyzed and decarboxylated to give overall specific monoalkylated ketone (Chapter 6, Eqs. 15 and 16).

A 1,4-diketone was selectively monoalkylated α to one keto group using a carbethoxy group for activation (Eq. 6) (Plantema et al., 1975). Here again the selectivity is dependent on selective monoacylation which in turn arises from formation of a stable β-keto ester α-anion which will not undergo further acyla- tion. The unactivated site at the other keto group in the product is not suffi- ciently acidic to be alkylated.

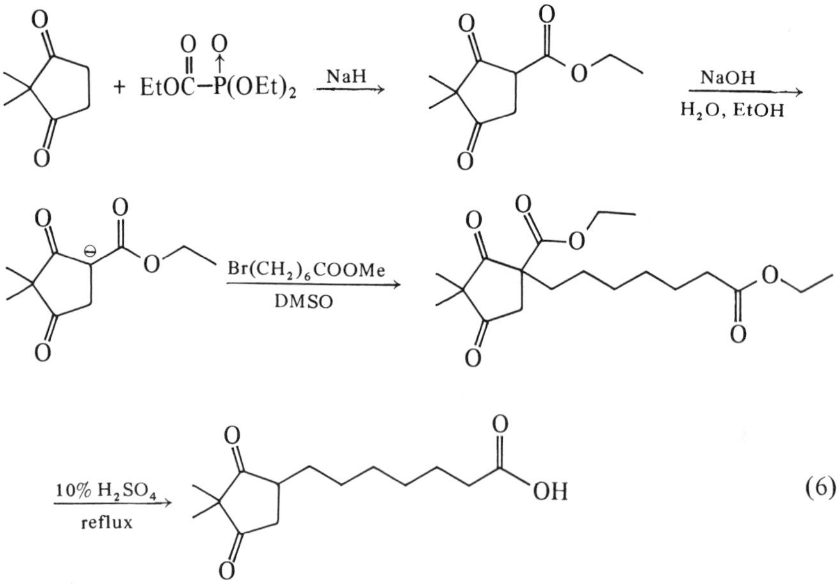

(6)

The activating group is at the same time a blocking group to prevent geminal dialkylation. It has been used for blocking alone in a β-tetralone alkylation (Eq. 7) (Palmer et al., 1977). Direct alkylation without the ester group gave geminal dimethylation in this case.

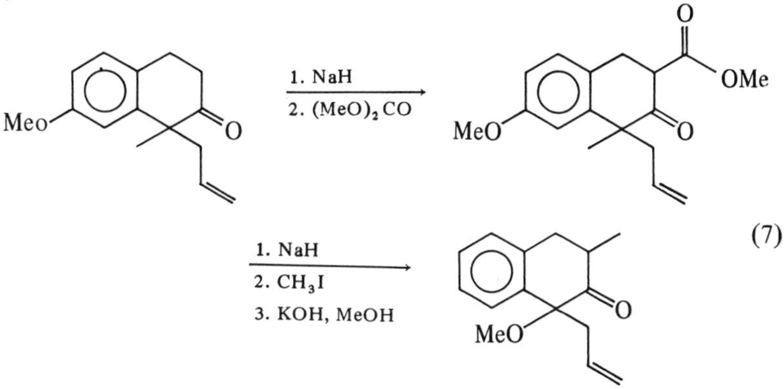

(7)

Another activator for ketone enolates is the phenylthio group. The activated ketone gives the enolate anion on the α-sulfenylated side and then may be alkylated. Moreover the sulfur may be reductively removed to give a second enolate anion on the same side which may be alkylated again with a different group (Eq. 8) (Coates et al., 1974) before equilibration can occur among possible enolate anions.

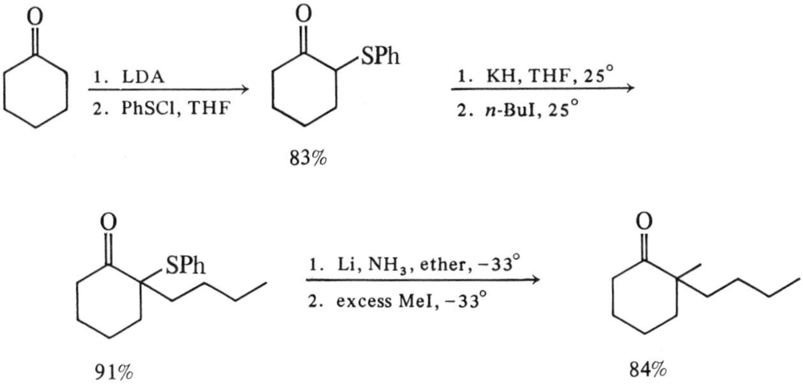

(8)

Unsymmetrical ketones with a phenylthio group on a particular side may be prepared by addition of a dithioacetal to an aldehyde followed by elimination of thiophenol. This and the enolate alkylation are illustrated in a synthesis of a butenolide (Eq. 9) (Brownbridge et al., 1977).

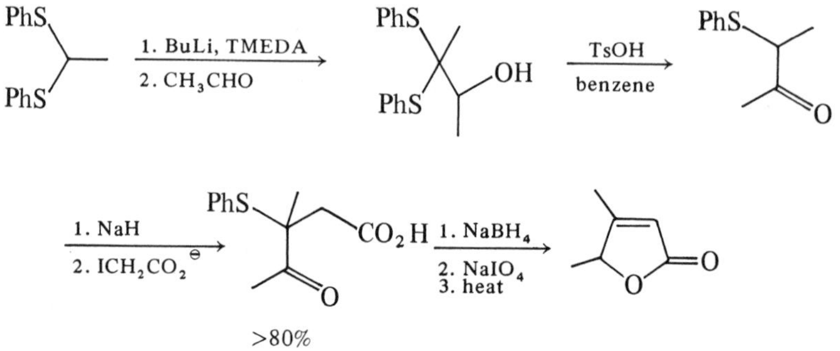

$$(9)$$

>80%

The opposite orientation from that obtained with activating groups may be obtained using blocking groups. A methylene group may be blocked, directing alkylation to a methinyl position (Review: Walton, 1973). For example the *n*-butylthiomethylene group was used on a dehydrooctalone to direct methylation to the methinyl side (Eq. 10) (Ireland et al., 1962). In this case the enol thioether was incorporated under acid conditions. If necessary, it can also be done in pyridine solution if the enol tosylate is made first. This same group may be used to construct alkyl groups on the blocked side via double conjugate additions (Chapter 3, Eq. 59).

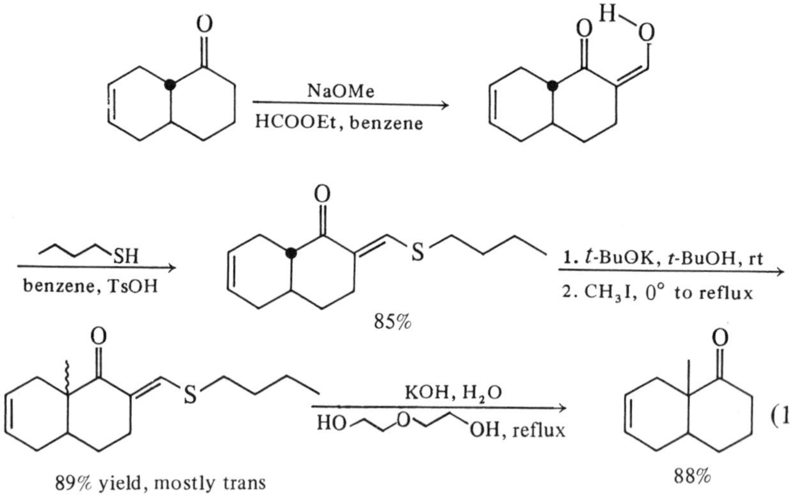

$$(10)$$

89% yield, mostly trans   88%

A benzylidine group can be used for blocking since the aldol reaction used for its incorporation is reversible. The cyclization of the tosylate in Eq. 11 gave a 1:1 mixture of both possible products (Thomas et al., 1973). If, however, a benzylidene group is attached first, it enters the least hindered side and then aklylation is exclusively on the other side (Eq. 12). Note that the temperature for the removal of this group is higher than that needed for the enol thioether.

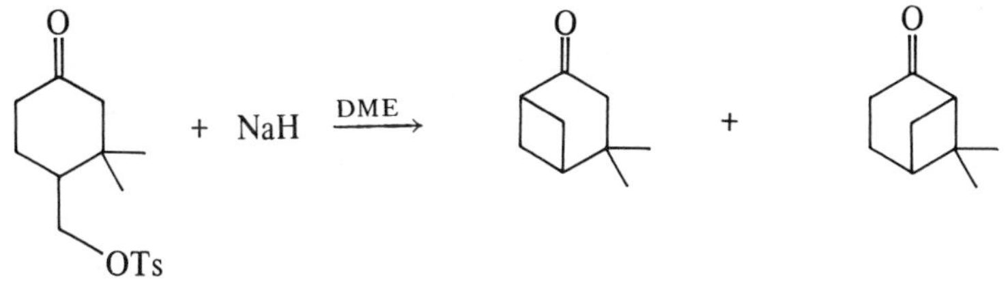

$$+ \text{NaH} \xrightarrow{\text{DME}} \qquad + \qquad \qquad (11)$$

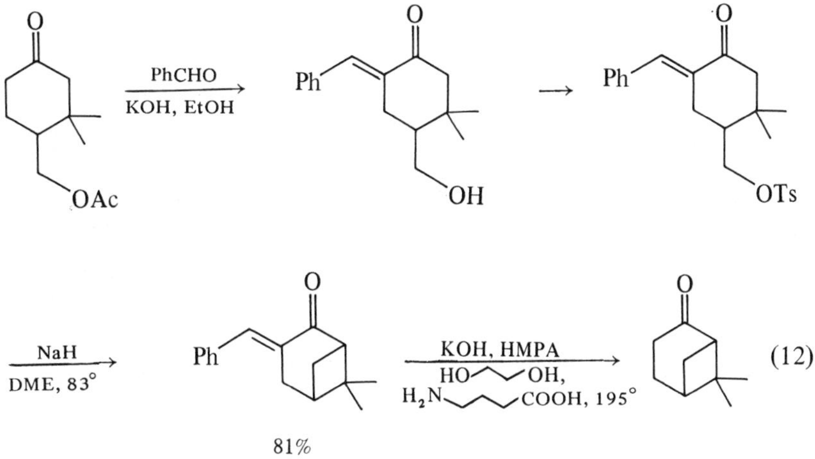

$$\xrightarrow[\text{KOH, EtOH}]{\text{PhCHO}} \qquad \longrightarrow$$

$$\xrightarrow[\text{DME, 83°}]{\text{NaH}} \qquad \xrightarrow[\substack{\text{HO}\frown\text{OH,} \\ \text{H}_2\text{N}\frown\frown\text{COOH, 195°}}]{\text{KOH, HMPA}} \qquad (12)$$

81%

Specific monoalkylation of an unsymmetrical ketone may be carried out by forming and using the enolate anion under conditions where it will not equilibriate by proton transfer to give the enolate of the alkylated product or the enolate on the other side of the carbonyl group. When the alkylating agent is added, free ketone is generated, hence the alkylation must be fast compared to proton exchange with this ketone. Specific alkylation is successful with highly reactive agents such as methyl iodide or benzyl bromide in high concentration, but with less reactive halides appreciable exchange occurs leading to other isomers or polyalkylation.

The α protons that are least sterically hindered are most rapidly removed by a bulky base, therefore adding an unsymmetrical ketone to an excess of LDA gives the enolate anion on the least substituted side. This is the less thermodynamically stable anion and is the result of kinetic control. LDA is selected because the by-product diisopropylamine is not nucleophilic enough to react with the alkylating agents, and it is too weak an acid to catalyze exchange. 2-

Methylcyclohexanone has been specifically benzylated in the 6-position in this manner (Eq. 13) (House et al., 1971; Gall et al., 1972). As described, the product is accompanied by about 5% of the 2-benzyl-2-methyl isomer and some dibenzylated products. If an excess of ketone over LDA is used, only a 9% yield of the 2,6-isomer is obtained, the rest being the products from exchange of protons.

$$42\%\text{-}45\% \quad (13)$$

Attempts to selectively alkylate the kinetic enolate of a methyl ketone have not been successful (Stork et al., 1974a).

Another method for the preparation of specific enolate anions is the treatment of specific enol acetates or enol trimethylsilyl ethers with methyllithium. This depends on the selective formation of or separation of these enol derivatives. The acid-catalyzed O-acetylation of 2-methylcyclohexanone with acetic anhydride gives a 92% yield of the isomer with the more highly substituted carbon-carbon double bond. Rapid benzylation of the enolate from this acetate gave a 54 to 58% yield of 2-benzyl-2-methylcyclohexanone, the opposite isomer from that prepared above by the kinetic deprotonation of the ketone.

The lithium tert-butoxide by-product produced from the enol acetates can lead to dialkylation, consequently the enol silyl ethers are superior in this respect. On the other hand the reaction of ketone enolate anions with trimethylsilyl chloride gives mixtures of isomeric ethers that are difficult to separate. Some cases such as 2-methylcyclohexanone give pure ethers from enolate anions generated under kinetically controlled conditions (Eq. 14) (House et al., 1969).

$$(14)$$

74% yield
99% this isomer

Treating these enol silyl ethers with methyllithium gives back the specific enolate anions which may be alkylated, but some proton transfer still competes with the alkylation reaction. If instead of lithium a quaternary ammonium counterion is present, alkylation is fast enough to completely avoid proton transfer. The carbanion can be generated in the presence of the alkylating agent by taking advantage of the high affinity of fluoride for silicon (Eq. 15) (Kuwajima et al., 1975). Even butyl iodide (10-fold excess) gave a 40% yield of the specific monoalkylation product.

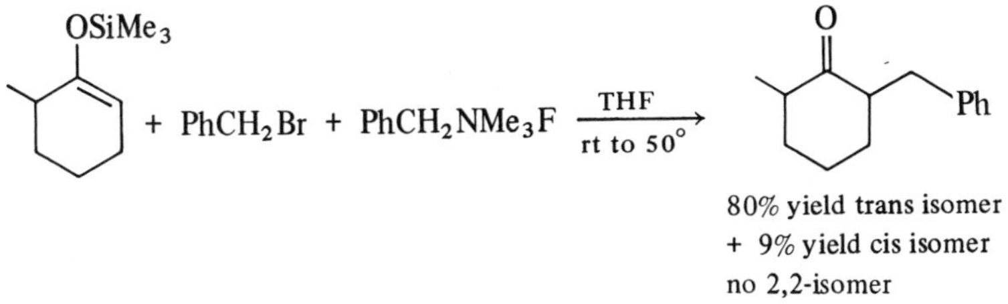

$$\text{OSiMe}_3 + \text{PhCH}_2\text{Br} + \text{PhCH}_2\text{NMe}_3\text{F} \xrightarrow[\text{rt to 50}^\circ]{\text{THF}} \qquad (15)$$

80% yield trans isomer
+ 9% yield cis isomer
no 2,2-isomer

Specific enolate anions can also be generated by reduction of or conjugate addition to α,β-unsaturated ketones. Thus 3-methyl-2-cyclohexenone was reduced and alkylated to give 2,3-dimethylcyclohexanone containing none of the 2,5-isomer (Eq. 16) (Smith et al., 1967). This contrasts with alkylation of 3-methylcyclohexanone in base, which gives only the 2,5-isomer.

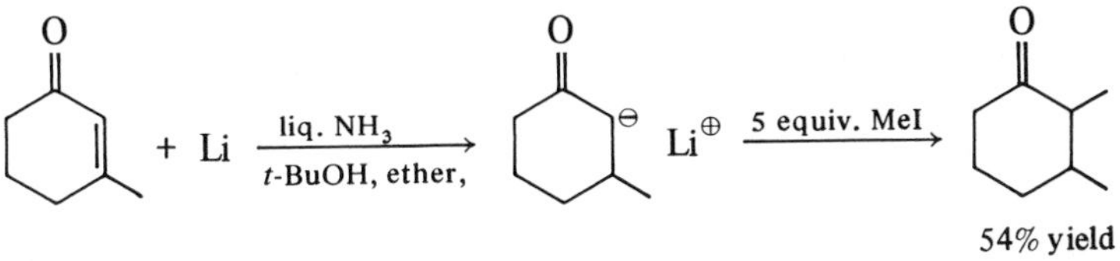

$$+ \text{Li} \xrightarrow[t\text{-BuOH, ether,}]{\text{liq. NH}_3} \quad \text{Li}^\oplus \xrightarrow{\text{5 equiv. MeI}} \qquad (16)$$

54% yield

Conjugate addition of a copper reagent to an α,β-unsaturated ketone gives a specific enolate anion which can be alkylated with active halides (Posner et al., 1975). The conversion of cyclopentenone to methyl jasmonate was accomplished beginning with this process (Eq. 17) (Greene et al., 1976b). Here again no

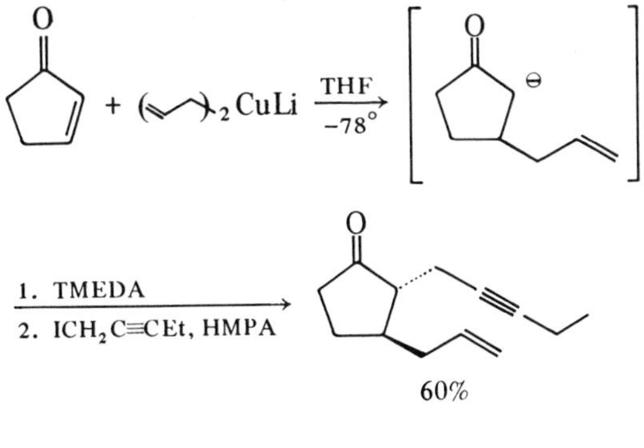

$$+ (\diagdown\diagup)_2\text{CuLi} \xrightarrow[-78^\circ]{\text{THF}} \left[\quad \right]$$

$$\xrightarrow[\text{2. ICH}_2\text{C}\equiv\text{CEt, HMPA}]{\text{1. TMEDA}} \qquad (17)$$

60%

equilibration to the other enolate which would lead to 2,5-substitution, was found. The only by-products were small amounts of polyalkylated materials. Di-*n*-butyl cuprate and cyclohexenone behave normally, but the mixed reagent *tert*-butoxy(*n*-butyl) cuprate gave extensive proton exchange.

A more delocalized enolate anion may be generated from α,β-unsaturated ketones by treatment with a base. If the anion is generated under equilibrium conditions, the more stable of two choices is formed, that is, the linearly conjugated anion from γ proton removal (d'Angelo, 1976). Alkylation of this type of anion generally occurs at the α position, and normally the α,α-dialkylated product is isolated (Eq. 18).

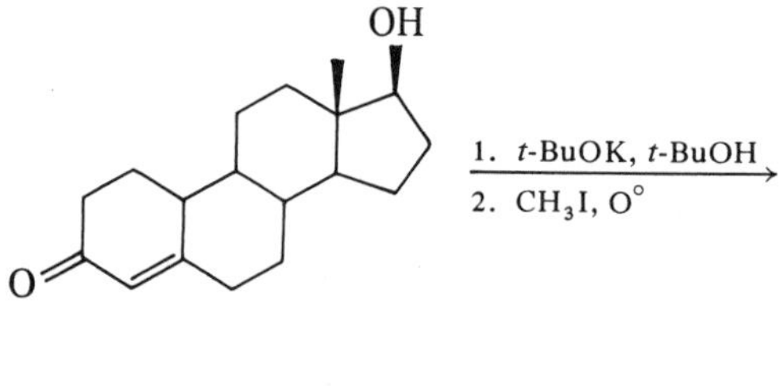

$$\text{1. } t\text{-BuOK, } t\text{-BuOH} \atop \text{2. CH}_3\text{I, O}°$$

(18)

57%

Kinetically the α'-hydrogens are first to be removed; so if alkylation is carried out quickly at low temperature, α'-substitution is the result (Eq. 19) (Nedelec et al., 1974). The β configuration of the 2-methyl group is a typical example of the general preference for axial attack on rigid cyclohexyl enolate anions.

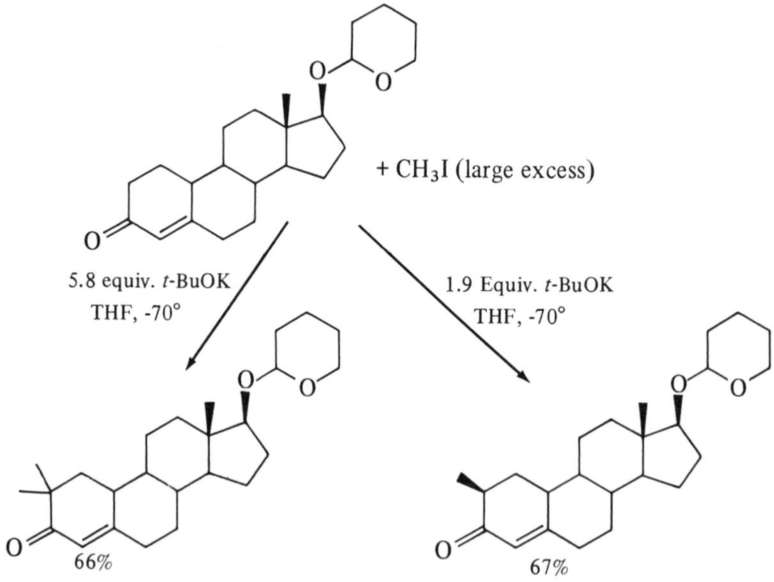

+ CH₃I (large excess)

(19)

5.8 equiv. *t*-BuOK
THF, -70°

1.9 Equiv. *t*-BuOK
THF, -70°

66%

67%

5.1.2 Acylation. When a solution of a ketone in a simple ester is treated with a base such as sodium ethoxide, the major product is a 1,3-diketone from acylation of the ketone enolate anion. This is often called the Claisen reaction (Review: Hauser et al., 1954). The yields are highest when there is least α-substitution on the ketone and ester. The various equilibria are drained by the final formation of the weakest base, the anion of the 1,3-diketone. This requires a molar equivalent of base and also final acidification to obtain the diketone. A few out of a great many known examples are given in Eqs. 7, 10, and 20 (Sprague et al., 1934).

$$CH_3COEt \ + \qquad + \ NaOEt \ \xrightarrow{rt}$$

$$3 \quad : \quad 1 \quad : \quad 1$$

$$\xrightarrow{HOAc} \qquad\qquad (20)$$

$$56\%$$

At ordinary temperatures, acid chlorides or anhydrides plus enolate anions give mostly *O*-acylation (House et al., 1973a) but at −78° the reaction of acid chlorides with lithium enolates gives good yields of 1,3-diketones (Beck et al., 1977; Watanabe et al., 1977). The process is most efficient when the carbanion is added slowly to the acid chloride because the consumption of half of the carbanion for the formation of the anion of the acidic diketone product is avoided. It is best to prepare the enolate anions free from amines using the methods shown in Eqs. 21 and 22.

$$\begin{array}{l} 1. \ \text{—}\langle\text{O}\rangle\text{—Li}, -78° \\ \hline 2. \ \text{PhCOCl}, -78° \end{array}$$

$$71\%$$

$$\qquad\qquad\qquad\qquad (21)$$

$$\begin{array}{l} 1. \ \text{MeLi, THF} \\ \hline 2. \ \text{Cl}\overset{O}{\underset{}{\|}}\text{NO}_2, \text{THF}, -78° \end{array}$$

$$73\%$$

$$\qquad\qquad\qquad\qquad (22)$$

Copper enolates can give good yields of C-acylation even at ordinary temperatures. The conjugate addition of lithium di-n-butylcuprate to cyclohexenone
gives a specific enolate which is readily converted to a 1,3-diketone (Eq. 23)
(Tanaka et al., 1975).

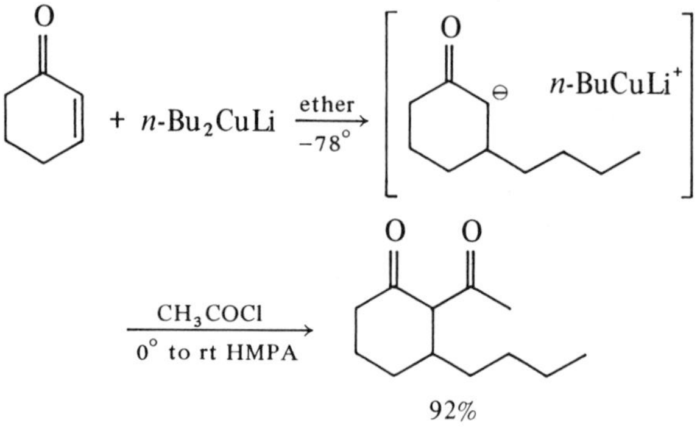

(23)

92%

If dimethylcopper lithium is used instead, only O-acylation is found because
methylcopper separates leaving only the lithium enolate (House et al., 1976).
Lower yields were found starting with cyclopentenone owing to O-acylation of
the β-diketone product.

5.1.3 Addition (Aldol Condensation). Ketone enolate anions will add to
aldehydes or ketones to give α-hydroxy ketones or the dehydration products
therefrom (Review: Nielsen et al., 1968).

Simple ketones will self-condense when treated with a catalytic amount of
base to give β-hydroxy ketones (Eq. 24). These will often eliminate water under

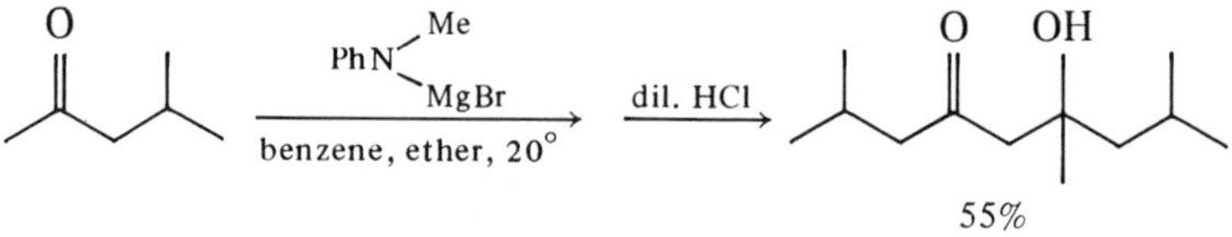

(24)

55%

the conditions of the condensation reaction to give α,β-unsaturated ketones.
Many of these same products are formed with acid catalysis also. In some cases
the product undergoes further condensations and dehydrations to give a
variety of other products. The combination of two different ketones can give
a mixture of products derived from each possible enolate ion reacting with

either ketone. A high yield of a single product may be obtained if only one enolate ion is possible as exemplified in Eq. 25 (Yates et al., 1958). In this case two aldol condensations and one dehydration led to the cyclopentenone product.

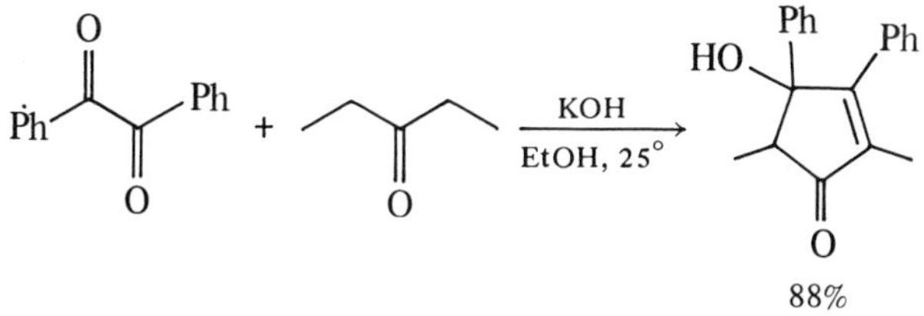

(25)

88%

The addition of ketone enolate anions to aldehydes is a widely used reaction. Where the aldehyde has no α-hydrogens the process is straightforward (Eq. 26) (Drake et al., 1932) but where the aldehyde can form an enolate anion, self-

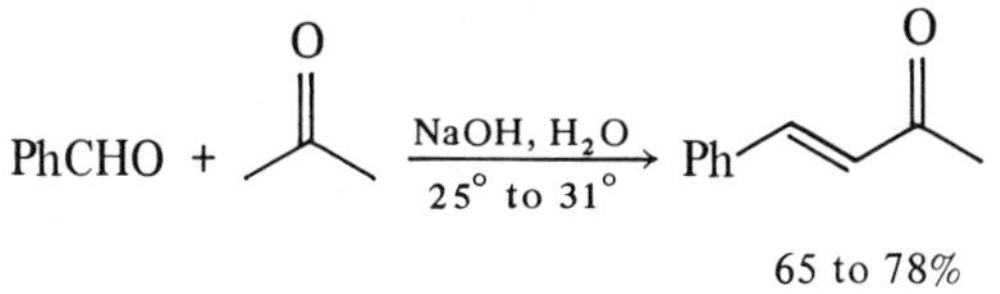

(26)

65 to 78%

condensation of the aldehyde competes. The addition of ketone enolate to aldehyde carbonyl may be favored by the slow addition of the aldehyde to an excess of the ketone with a small amount of base. In this manner acetaldehyde and 2-butanone gave 3-methyl-2-pentanol-4-one in 85% yield (Kyrides, 1933). Alternatively the ketone can be completely converted to the enolate anion with a strong base at low temperature and then added to an aldehyde still at low temperature where proton transfer is slower than the addition. This was applied to the synthesis of a cholestene derivative as shown in Eq. 27 (Byon et al., 1977; Stork et al., 1974a) where a specific kinetic enolate was used.

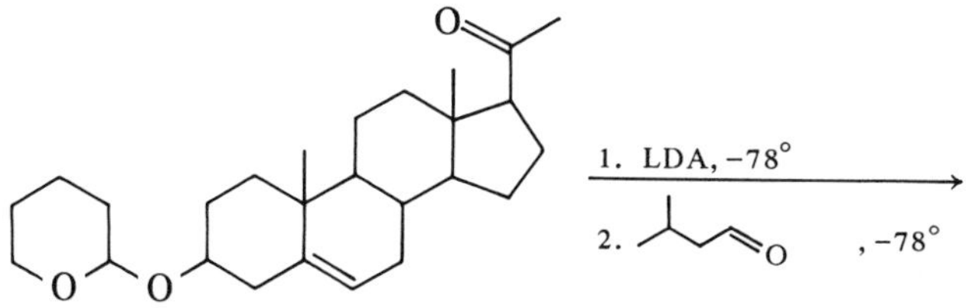

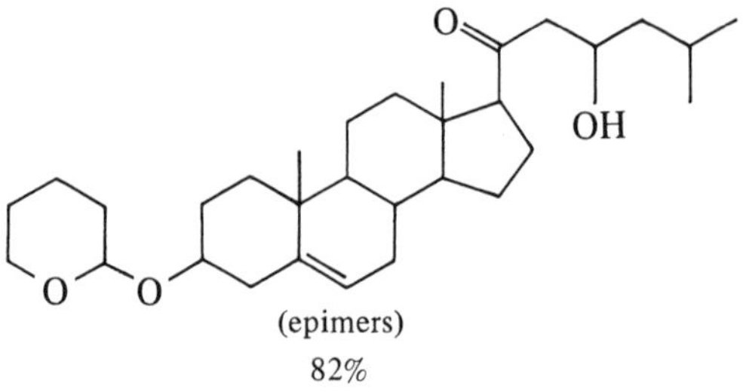

(27)

(epimers)
82%

Methyl ketone enolates can be generated selectively even in the presence of the aldehyde when the sterically hindered base lithium 1,1-bis(trimethylsilyl)-3-methylbutoxide is used (Eq. 28). Here the methyl group is more accessible than the methylene groups of aldehyde or ketone so that the kinetically controlled process gives the desired addition from only a slight excess of ketone premixed with the aldehyde (Kuwajima et al., 1976a).

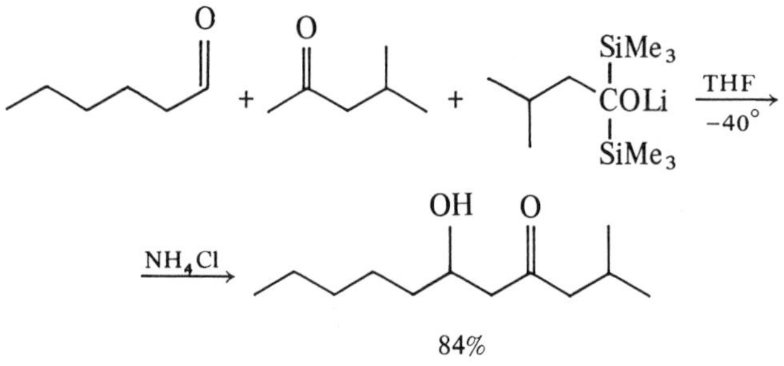

(28)

84%

The specific enolates from conjugate addition of lithium dimethylcuprate to α,β-unsaturated ketones will add to acetaldehyde in the presence of zinc chloride to give specific aldol products (Eq. 29) (Heng et al., 1975). No equilibration of enolates occurred as shown by the absence of product from what would be the more stable enolate conjugated with the phenyl group.

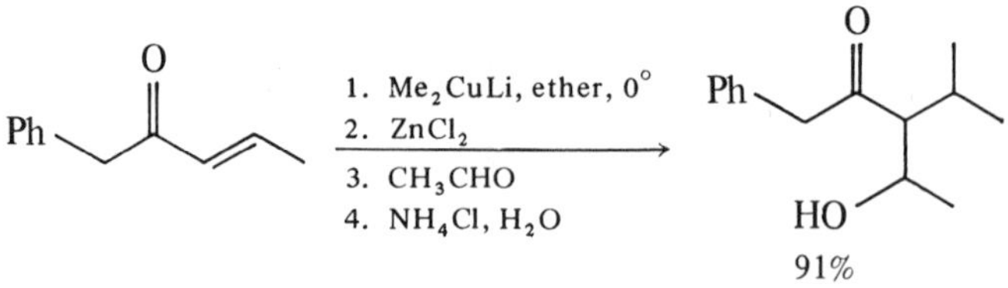

(29)

91%

Intramolecular additions that lead to five- and six-membered rings are very common. The yields are often high since the intramolecular reactions are much favored over the multiplicity of conceivable intermolecular reactions. If the carbonyl groups are arranged 1,4, the product is the cyclopentenone as in Eq. 30 (Büchi et al., 1966). If they are arranged 1,5, a cyclohexenone is the

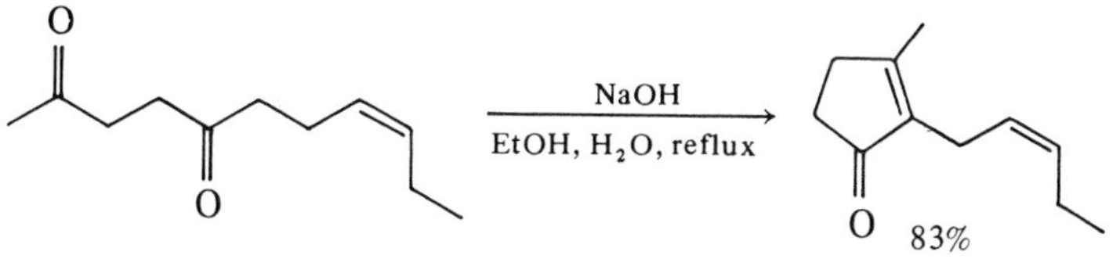

$$\text{(30)}$$

product (Danishefsky et al., 1976; Larchevêque, et al., 1977) as in the Robinson annulation reactions (Section 5.1.4). If they are 1,6, a cyclopentenyl ketone is formed (Eq. 31) (Poos et al., 1955).

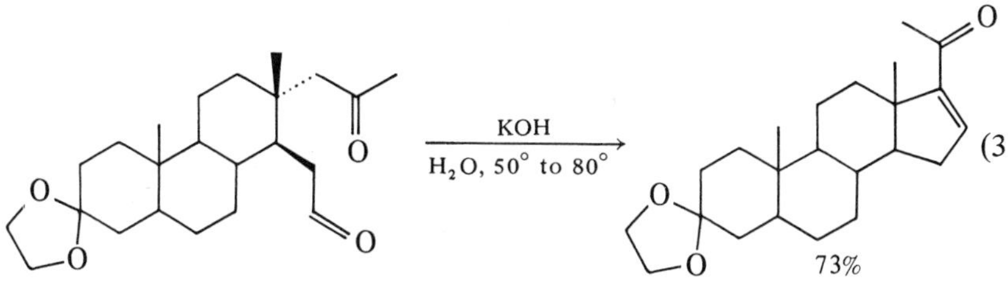

$$\text{(31)}$$

5.1.4 Conjugate Addition. Ketone enolate anions readily undergo conjugate addition to α,β-unsaturated ketones and nitriles (Review: Bergmann et al., 1959). The most common case is the addition to methyl vinyl ketone. If hydroxide is used as the enolate generating base, the 1,5-diketone product will undergo aldol cyclization and dehydration (Eq. 32) (Sher et al., 1977).

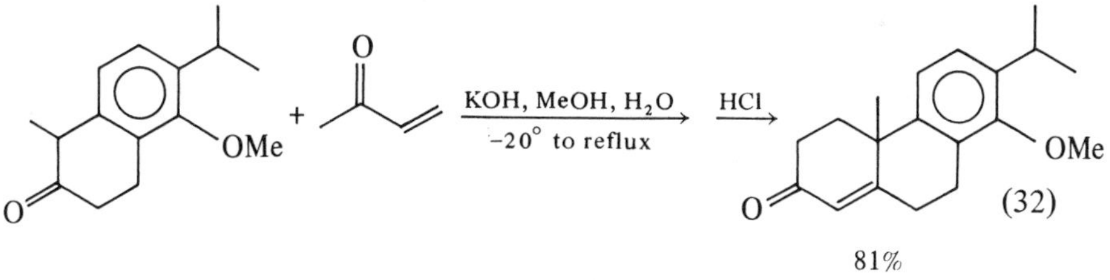

$$\text{(32)}$$

This frequently used process for the preparation of fused cyclohexenones is called the Robinson annulation (Reviews: Gawley, 1976; Jung, 1976). The methyl vinyl ketone is readily polymerized by base so it is added last and slowly.

It may also be generated in situ by elimination of a quaternary ammonium salt. Alternatively, it can be added as a vapor diluted with nitrogen. When DBN is used as the base, the intermediate 1,5-diketone is isolated (Eq. 33) (DeBoer,

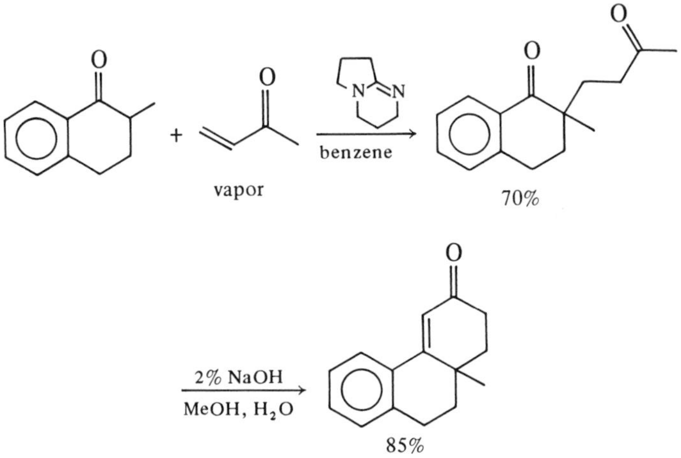

$$(33)$$

1974). The yield of conjugate addition may also be improved by further activating the ketone α position with a temporary formyl group and then using a very weak base (Section 6.3.3).

In some cases it may be desirable to use completely preformed enolate anion rather than the low equilibrium concentrations used above. For instance, a specific enolate generated under aprotic conditions could be annulated. The polymerization problem is more severe here so the stable α-silylated vinyl ketone is used instead (Eq. 34) (Stork et al., 1974e; Boeckman, Jr., 1974). The silicon stabilizes the anion resulting initially from the conjugate addition, thus preventing it from competing with the remainder of the original enolate anion for the α,β-unsaturated substrate. It is cleaved under the basic conditions of the subsequent aldol cyclization.

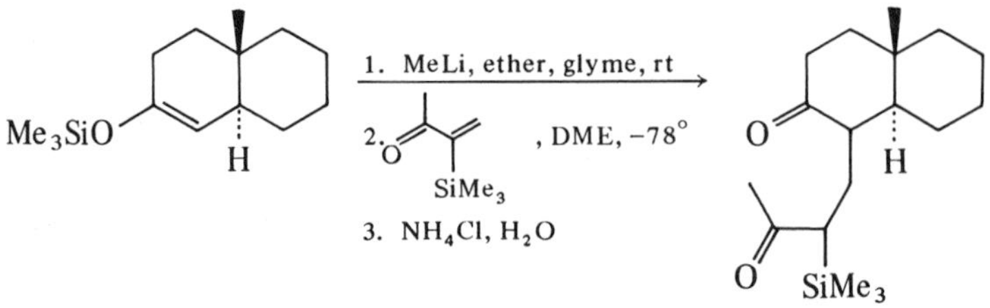

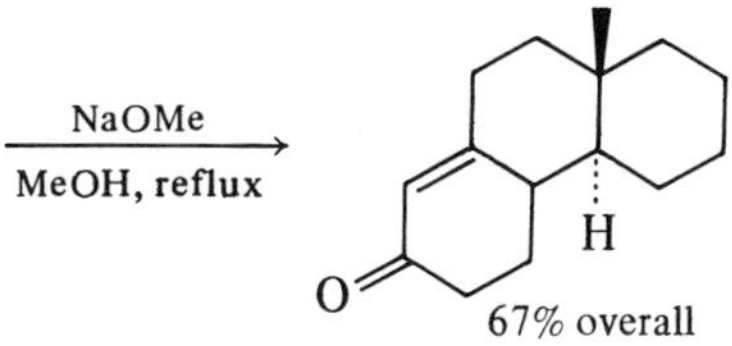

$$\xrightarrow[\text{MeOH, reflux}]{\text{NaOMe}}$$

(34)

67% overall

5.2 ALDEHYDE ENOLATE ANIONS

Although aldehydes have only one enolate site, their higher tendency to undergo aldol condensations limits their use in other carbanion reactions. Here, as with many ketones, better results are obtained with the derivative enamines, imine and hydrazone enolate anions (Sections 5.7 and 5.8), and dihydrooxazines (Section 5.9).

5.2.1 Alkylation. Addition of a base to simple aldehydes gives rapid polymerization before alkylation could be attempted. The enolate of acetaldehyde can be prepared by treatment of THF with n-BuLi, but subsequent treatment with alkylating agents then gives only polymeric material (Jung et al., 1977). Aldehydes having only one enolizable hydrogen may be alkylated with reactive halides under phase transfer catalysis (Eq. 35) (Dietl, et al., 1973). The halide

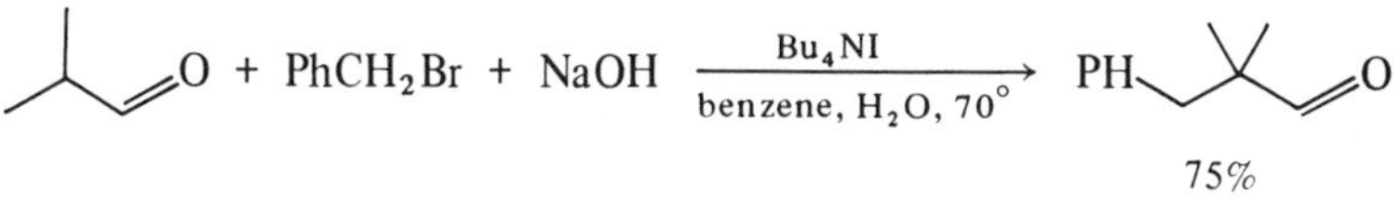

$$\text{\Large$\curlyvee$}\!\!\!=\!\!O + PhCH_2Br + NaOH \xrightarrow[\text{benzene, } H_2O, 70°]{Bu_4NI} PH\!\!\diagdown\!\!\diagup\!\!\diagdown\!\!=\!\!O \qquad (35)$$

75%

and aldehyde are added together dropwise to the base and catalyst to minimize the aldol condensation. Higher molecular weight aldehydes are less prone to condensation, thus the example in Eq. 36 could be completely converted to the anion by potassium triphenylmethide. Alkylation with methyl iodide then gave an abietatrienal in good yield (Huffman et al., 1977b).

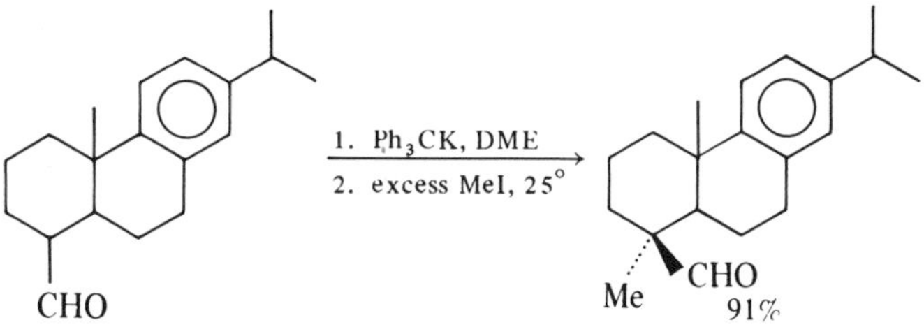

$$\xrightarrow[\text{2. excess MeI, 25°}]{\text{1. Ph}_3\text{CK, DME}}$$

(36)

CHO   Me   CHO 91%

α,β-Unsaturated aldehydes may be converted to the enolate anions at low temperature and then alkylated in the α position (Eq. 37) (Isobe et al., 1977).

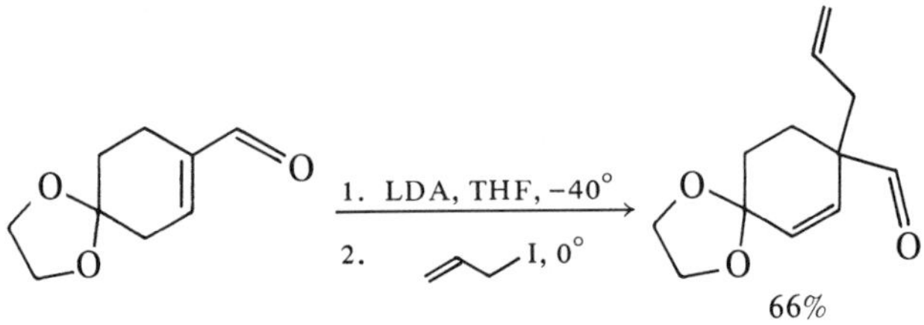

$$(37)$$

66%

5.2.2 Addition (Aldol Condensation). n-Aldehydes of up to six carbons give normal aldol condensations, but the higher temperatures needed for the larger aldehydes generally lead to elimination of water (Review: Nielsen et al., 1968). Combining two simple aldehydes would lead to four products, although one may predominate. If one of the two aldehydes has no enolizable hydrogen, the other one may be added slowly to it plus base to give nearly all one aldol product (Eq. 38) (Kraft, 1948). If dehydration joins the conjugation of a phenyl ring to the carbonyl, it occurs quickly.

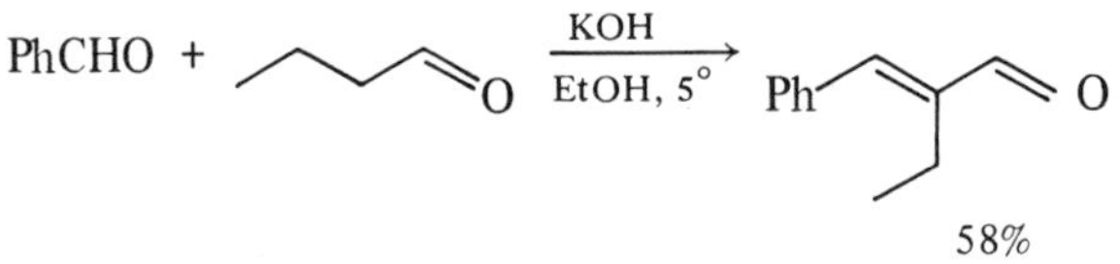

$$(38)$$

58%

5.3 ESTER ENOLATE ANIONS

Simple ester enolate anions can be prepared in THF solution using lithium-amide bases and stored at $-78°$, but if they are warmed to room temperature they give β-keto ester enolates. This occurs even though no free ester was present; therefore it is proposed that elimination of alkoxide gives a ketene which is susceptible to attack by enolate (Sullivan et al., 1977). However a mere trace of relatively acidic protons may be sufficient for the conventional Claisen mechanism to convert all of the material to the β-keto ester enolate. This instability is quite in contrast with the behavior of ketone enolate anions. As a consequence, the reactions of simple ester enolates will generally be carried out at $-78°$, except when they can be prepared in the presence of the next reactant, as in the classical Claisen or Reformatsky reactions.

The *tert*-butyl group in *tert*-butyl acetate hinders attack on the carbonyl group sufficiently to disfavor the Claisen condensation and allow complete formation of the anion using isopropylmagnesium chloride at $-20°$ to $+30°$ (Dubois et al., 1963) or lithium N-isopropylcyclohexylamide (Rathke et al., 1973). The lithio ester is stable in hydrocarbon solutions (but not THF) for a few hours at $25°$ or in the solid state for months.

The sodium enolates of esters that have only one α-hydrogen, such as ethyl isobutyrate, are reluctant to undergo self-condensation (because of steric hindrance and the inability to form the weak base β-keto ester α-enolate) and therefore can be prepared and used at room temperature. An equivalent amount of triphenylmethylsodium in ether is used to generate the anion which is then used without delay (Hudson, Jr., et al., 1941). The corresponding bromomagnesium enolate can be prepared from ethyl α-bromoisobutyrate in ether at $10°$ to $20°$ (P.Y. Johnson, et al., 1976).

The bromozinc enolates form stable solutions in dry methylal. They can be prepared in good yield from ethyl α-bromoacetate and ethyl α-bromoisobutyrate but in poor yield from ethyl α-bromopropionate (Gaudemar et al., 1966).

5.3.1 Alkylation. Ester enolates prepared at $-78°$ with lithium N-isopropyl-cyclohexylamide may be alkylated by adding to a solution of an alkyl halide in DMSO at $20°$. In this way ethyl hexanoate was converted to ethyl 2-methylhexanoate in 83% yield. Ethyl acetate gives only condensation, but *tert*-butyl acetate gives good yields of alkylated products (Rathke et al., 1971c). With a more reactive benzylic halide, alkylation could be carried out in THF at low temperature, even adding the halide to the enolate (Eq. 39) (Ewing et al., 1975).

(39)

93.5%

HMPA promotes the reaction sufficiently to allow alkylation at −78° with equivalent amounts of simple primary and secondary bromides and iodides with no competition from condensation reactions (Eq. 40) (Cregge et al., 1973).

$$96\% \qquad (40)$$

These ester products were classically made by the double alkylation of malonic ester (Section 6.1.1), but the direct ester alkylation is shorter and in much higher yield. In addition, a second alkylation produces an ester with a quaternary α-carbon (Eq. 41), which is not possible with the malonic ester synthesis.

$$(41)$$

$$92\%$$

Lactone enolates may be alkylated and dialkylated in the same manner (Herrmann et al., 1973).

Some alkylations of esters containing only one α-hydrogen were carried out earlier at room temperature using potassium triphenylmethide as base (Levine et al., 1944).

Similar alkylations were performed on the O,α-dianions of 2- and 3-hydroxy esters (Herrmann et al., 1973b; Ciochetto et al., 1977). Likewise a δ-halo ester was converted to a cyclobutyl ester with LDA in THF-HMPA at 0° (Babler, 1975).

α,β-Unsaturated esters may serve as sources of α-enolate anions. For instance treatment of methyl 2-nonenoate with lithium tri-sec-butylborohydride at −70° gave the α-enolate, which was alkylated with allyl bromide in 50% overall yield (Ganem et al., 1975). Conjugate addition of a carbanion could also be followed by alkylation as shown in Eq. 42 (Ziegler et al., 1978).

If an ester has a heteroatom on the α position that can further stabilize the anion, alkylation is possible at ordinary temperatures (Eq. 43) (Hirai et al., 1977). An α-thiothiazoline group gives this stabilization and is then reductively

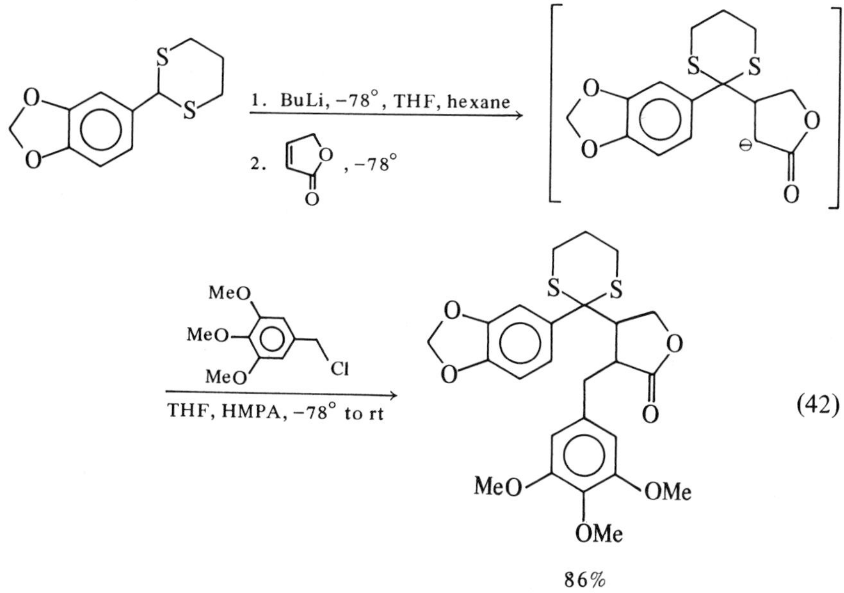

$$(42)$$

86%

61%

$$(43)$$

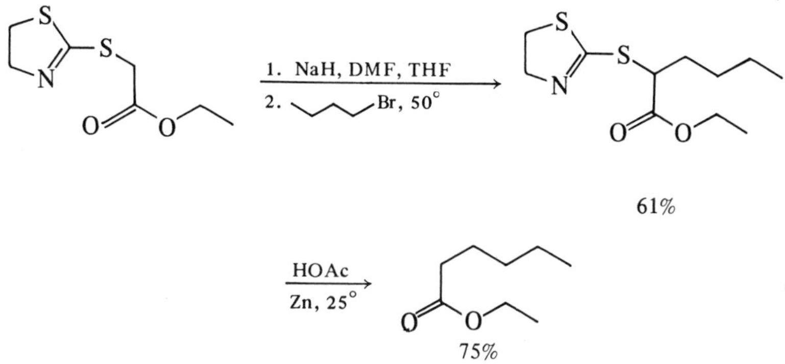

75%

$$(44)$$

78%

90 to 100%

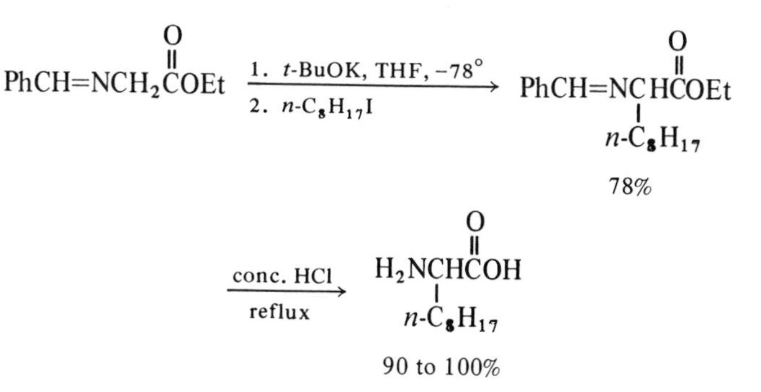

147

removed with zinc. The benzylideneimine of glycine ethyl ester can be deprotonated by potassium *tert*-butoxide and alkylated with primary iodides. The products
may then be hydrolyzed to the amino acids in high yield (Eq. 44) (Stork et al.,
1976).

When α,β-unsaturated esters are treated with a 1:1 complex of LDA and
HMPA, a γ proton is removed giving the conjugated enolate anion. Without the
HMPA, the result is Michael addition of diisopropylamine. This anion is readily
alkylated exclusively at the α position giving the deconjugated ester. It may be
alkylated a second time, again at the α position (Eq. 45) (Herrmann et al.,
1973c; Smith, III, 1975). Starting from a β,γ-unsaturated ester, the same kind
of enolate was formed and α-alkylated as part of a synthesis of (±)-perhydro-

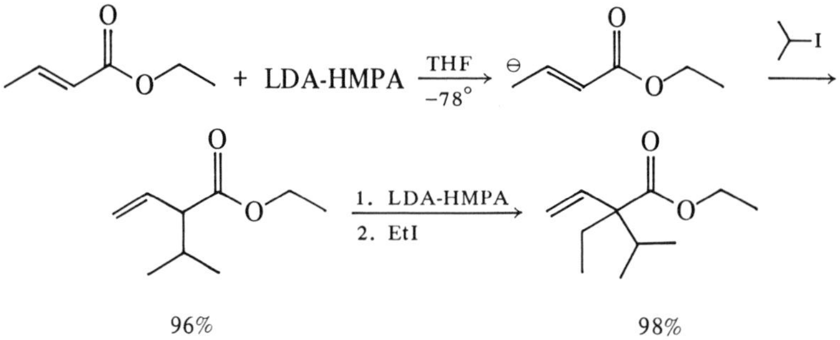

(45)

96%   98%

histrionicotoxin (Eq. 46) (Corey et al., 1975b).

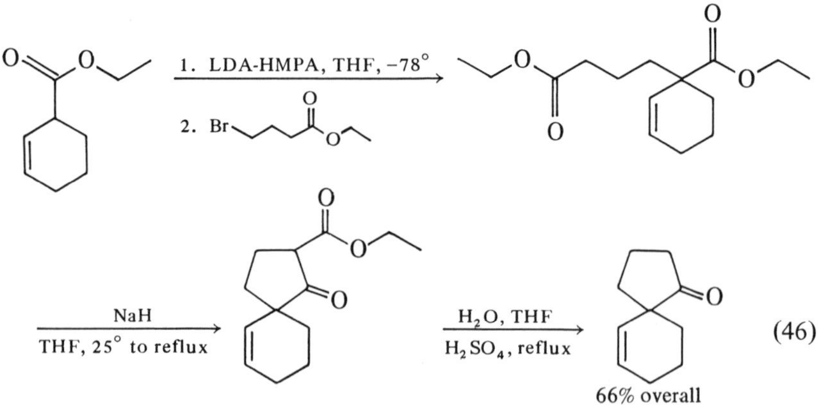

(46)

66% overall

5.3.2 Acylation. Ester enolate anions may be acylated with esters or acid
chlorides. In simple cases where the acylating agent can be the same ester as

the substrate, we have the classic Claisen condensation. This can be carried out using a low concentration of enolate generated in equilibrium with sodium alkoxide (Reviews: Hauser et al., 1942; Henecka, 1952b). The steps of product formation are reversible so the reaction will proceed favorably if the final product is the weakest base present (Eq. 47). The enolate anion of the β-

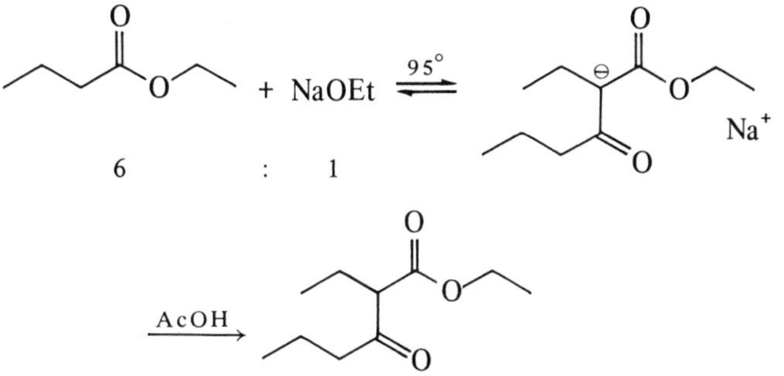

42% yield based on NaOEt

(47)

dicarbonyl compound is a weaker base than sodium ethoxide or than the enolate of ethyl butyrate. The yield can be raised if ethanol is distilled from the reaction as it is formed. When there is only one α-hydrogen in the starting ester, an α-anion cannot be formed from the β-keto ester, therefore ethyl isobutyrate does not condense using sodium ethoxide. However the stronger base tri-phenylmethylsodium or potassium hydride will give condensation. With these bases the γ-anion of the β-keto ester forms and is protonated in the workup to give for example ethyl α-isobutyryl isobutyrate (Hauser et al., 1937; Brown, 1975).

If two different esters are combined, a mixture of four different products is obtained, but if one of the esters has no α-hydrogen (e.g. ethyl benzoate, ethyl formate, or ethyl oxalate), cross Claisen products may be obtained in fair yield (Eq. 48) (Korte et al., 1960).

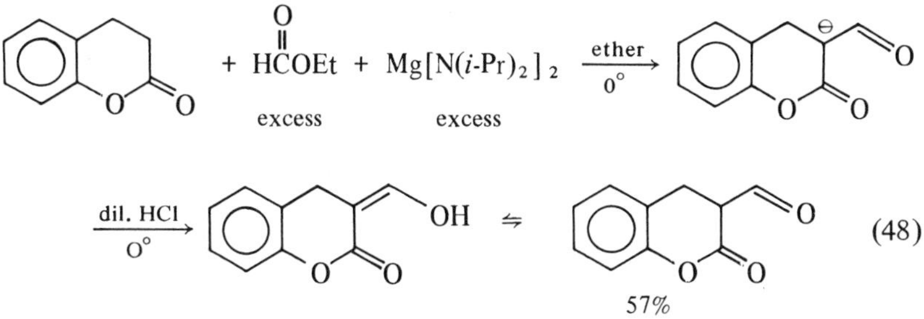

(48)

57%

Intramolecular examples that lead to five- or six-membered rings proceed more favorably than the intermolecular cases. This cyclization of diesters is called the Dieckmann condensation (Review: Schaefer et al., 1967). Where the ester is unsymmetrical, one may find two different products, but where only one product can form an α-enolate, it is the sole product (Eqs. 46, 49, and 50). Larger rings have been prepared using high-dilution conditions.

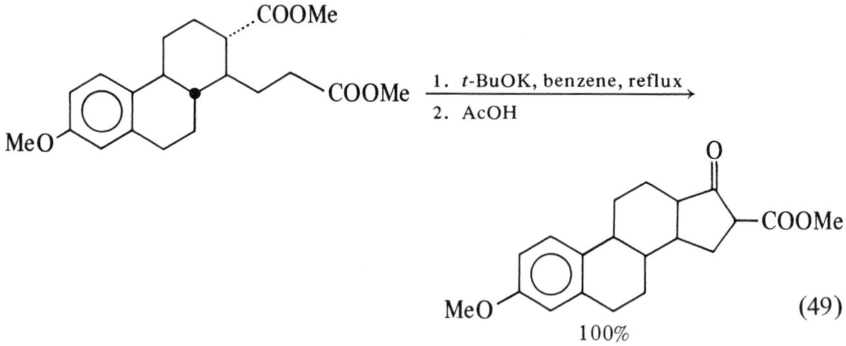

(49)

100%

(Meyer et al., 1962)

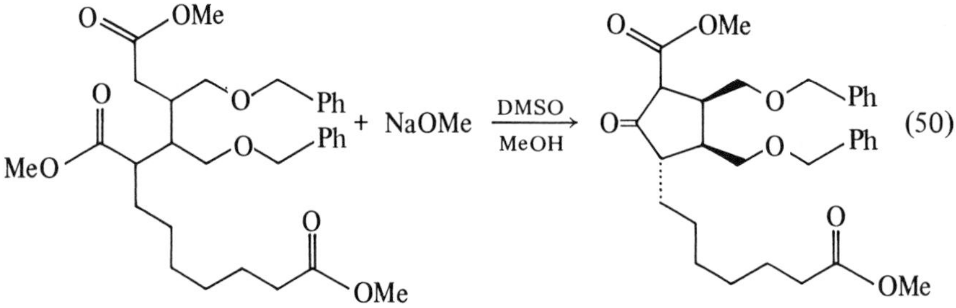

(50)

(Ide et al., 1976)   81%

The β-keto ester products are often hydrolyzed and decarboxylated to give ketones as exemplified in Eq. 46.

Acid chlorides may be used to acylate ester enolates that are fully preformed by strong bases. The by-product sodium chloride forms irreversibly, so only a single β-keto ester is expected, in contrast with mixed ester condensations. This may be carried out at room temperature with those esters that are reluctant to self-condense. Butyryl chloride gave ethyl 2,2-dimethyl-3-ketohexanoate (Eq. 51), and benzoyl chloride gave ethyl α-benzoylisobutyrate in 65% yield (Hudson, Jr., et al., 1941). The lithium enolate of *tert*-butyl isobutyrate behaves similarly (Logue, 1974).

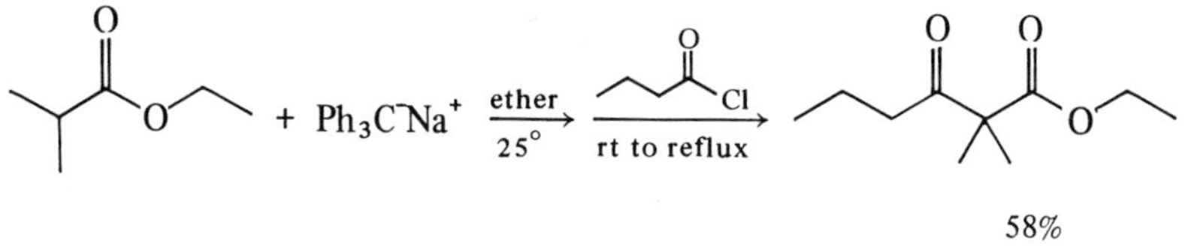

58%   (51)

Those esters that readily self-condense at room temperature may be converted to the enolate and acylated at $-78°$ as exemplified in Eq. 52 (Rathke et al., 1971a; Olofson et al., 1973).

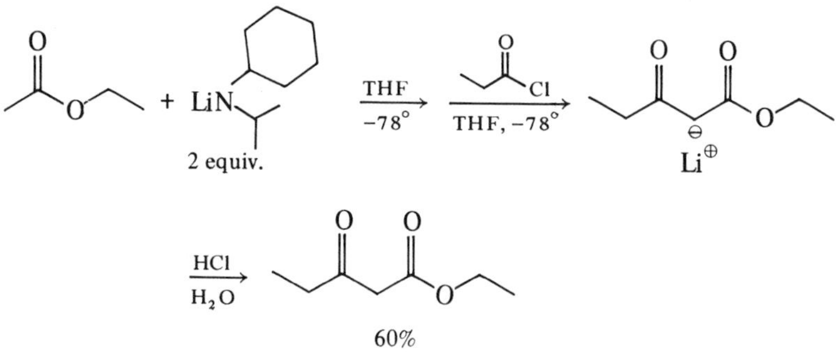

60%   (52)

5.3.3 Addition. The addition of ester enolate anions is carried out classically by treating an α-halo ester with zinc in the presence of a ketone or aldehyde. This is called the Reformatsky reaction (Review: Rathke, 1975). There are many recent modifications which improve the yields by minimizing the exposure of the starting materials to the basic products, and minimizing the exposure of the product to heat. Passing the reactants dropwise through a heated column of granular zinc gives the alkoxide which is neutralized with cold acid to afford a β-hydroxy ester in high yield (Eq. 53) (Ruppert et al., 1974). A zinc-copper couple works well in refluxing ether or THF (Santaniello et al., 1977). Fresh

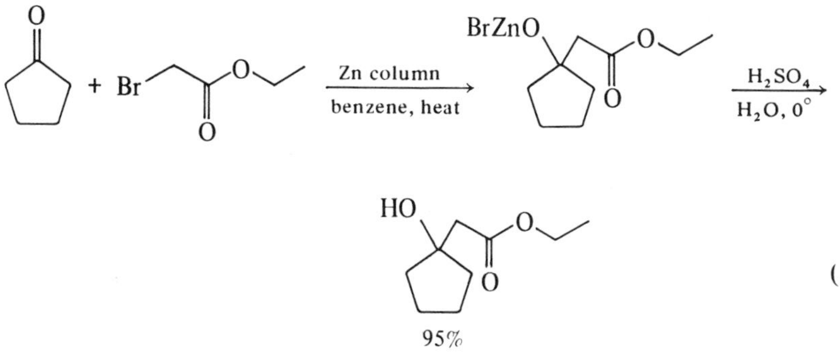

95%   (53)

finely divided zinc can be prepared by potassium reduction of zinc chloride, and this gives Reformatsky products even at $0°$ to $25°$ (Reike et al., 1975). Alternatively the mildly acidic trimethyl borate may be used in situ with zinc in THF to neutralize the alkoxide as it is formed (Rathke et al., 1970).

The organozinc reagent may be prepared in methylal solution and in a separate step treated with the aldehyde or ketone (Curé et al., 1969). The more stable zinc reagent from *tert*-butyl α-bromoacetate can be prepared in THF and then added to the carbonyl compound (Liu et al., 1978).

A vinylogous Reformatsky reaction occurs with γ-bromocrotonate. With most ketones the γ-carbon of the ester joins to the carbonyl carbon giving δ-hydroxy α,β-unsaturated esters (Eq. 54) (Eiter et al., 1967). With aldehydes and unhindered methyl ketones the α-carbon of the ester attacks, and a β,γ-unsaturated hydroxyester is formed.

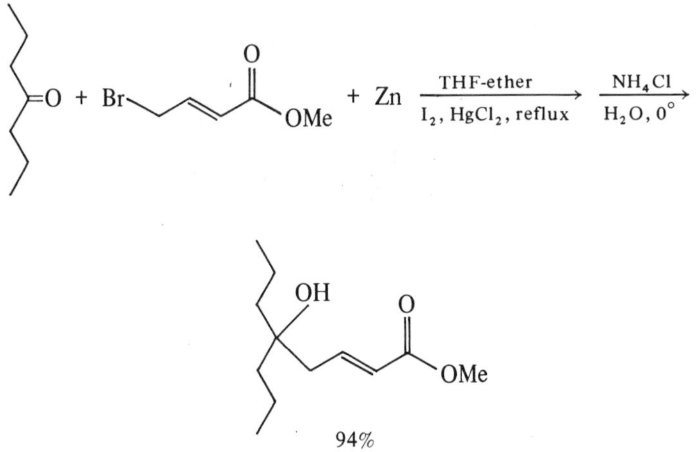

$$(54)$$

94%

The Reformatsky addition to nitriles gives β-iminoesters which are hydrolyzed to β-keto esters. Esters of α-bromopropionic and α-bromobutyric acid gave yields of 30 to 60% (Cason et al., 1953). The recent modifications to the Reformatsky reaction given above should make this a good yield alternative to the acylations in Section 5.3.2.

Ester enolates, prepared from esters using the amide anion bases, will readily add at $-78°$ (Rathke, 1970; Rathke et al., 1973). For example methyl oleate has been used in the synthesis of prostaglandins (Eq. 55) (Stork et al., 1977a).

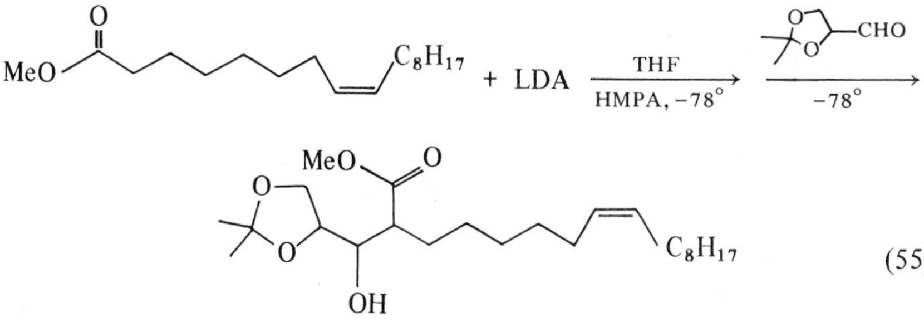

(55)

76%

An addition to a ketone is shown in Eq. 56 (Bruderer et al., 1977). In these cases

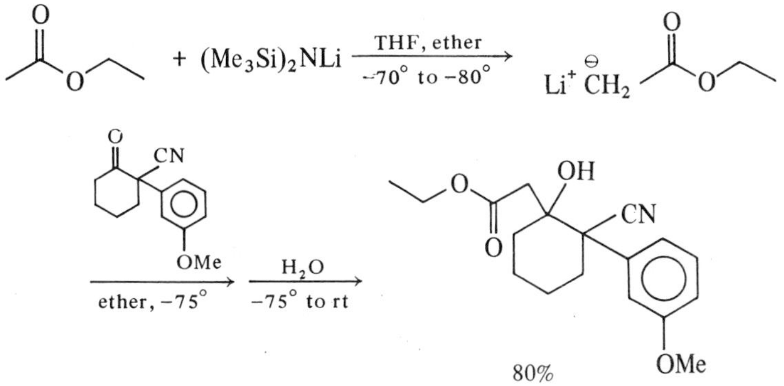

(56)

80%

the anion is prepared first and then the aldehyde or ketone is added, otherwise the base would produce the anion of the ketone or aldehyde instead. In the special case of ethyl acetate addition to aldehydes, the hindered base lithium 1,1-bis(trimethylsilyl)-3-methylbutoxide will kinetically remove the ester α-hydrogen, even in the presence of the aldehyde, and give good yields of β-hydroxy ester (Eq. 57) (Kuwajima et al., 1976b).

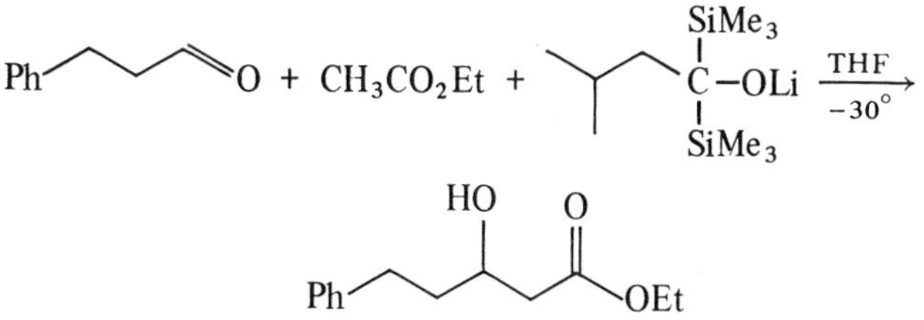

(57)

77%

The α-enolate of a γ-butyrolactone could be prepared at 20° in benzene using lithium hexamethyldisilylamide as base. Addition to an aldehyde then gave a high yield of the hydroxy lactone. This was superior to the use of LDA in THF at −70° (Brown et al., 1978).

The dehydration of ester addition products gives mostly the α,β-unsaturated ester, but it is often accompanied by substantial amounts of the β,γ-unsaturated isomer. If an α-trimethylsilyl group is eliminated instead of a proton, the conjugated isomer is the exclusive product (Eq. 58) (Taguchi et al., 1974a).

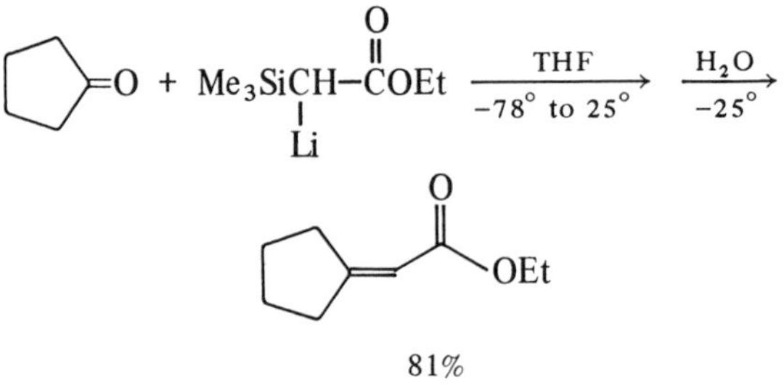

$$\text{(58)}$$

81%

This specific placement of the double bond is analogous to the Horner-Emmons reaction (Section 5.10) and is superior to the latter with readily enolized ketones. With aldehydes, mixtures of Z and E isomers were obtained. Similar results were obtained using an α-dithiocarbonate group (Tanaka et al., 1976).

If an α-halo atom is present in the enolate anion, the alkoxide produced in the addition displaces the halogen giving an epoxide (Eq. 59) (Burness, 1963). This

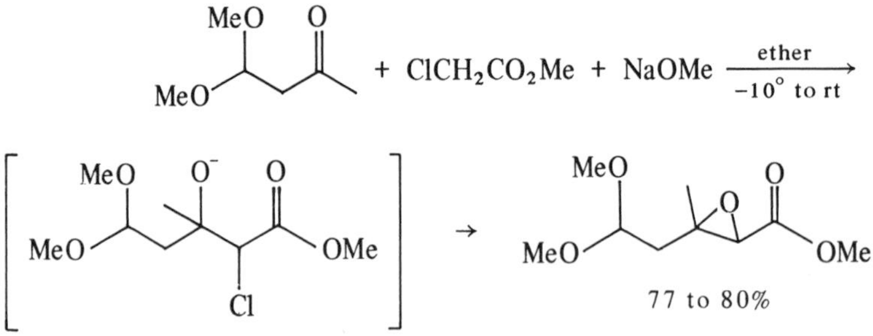

$$\text{(59)}$$

77 to 80%

is called the Darzens glycidic ester condensation (Review: Newman et al., 1949). Cainelli et al. (1972) prepared the α-halo enolate for this purpose at −78° by metal-halogen exchange starting with ethyl dibromoacetate and butyllithium.

A vinylic enolate was prepared by the reaction of lithium hexynylmethyl-cuprate and ethyl α-bromoacrylate. This copper enolate has unusual selectivities at $-78°$. Of the various alkylating agents, it reacts only with allyl and propargyl halides. It adds quantitatively to simple cyclic ketones (Eq. 60), and with α,β-unsaturated ketones and aldehydes it gives only 1,2-addition. This behavior contrasts with that of simple cuprates (Section 3.3.3) (Marino et al., 1974, 1975a).

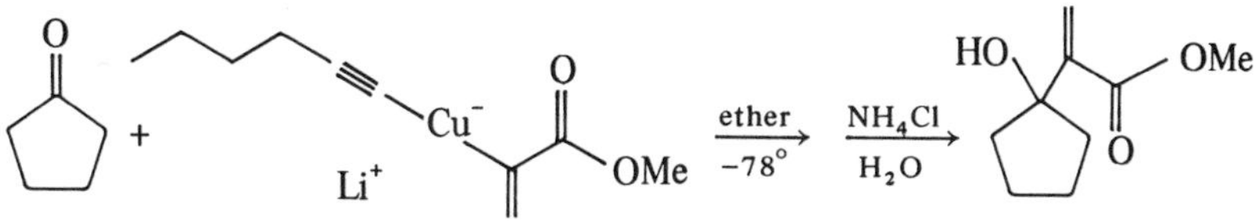

(60)

100%

5.3.4 Conjugate Addition. The Reformatsky reaction of esters of the type $RCHBrCO_2Et$ with α,β-unsaturated ketones generally gives 1,2-addition. The more hindered ethyl α-bromoisobutyrate however gives some 1,4-addition. Similarly, the lithium enolates of unhindered esters give 1,2-addition (Eq. 61)

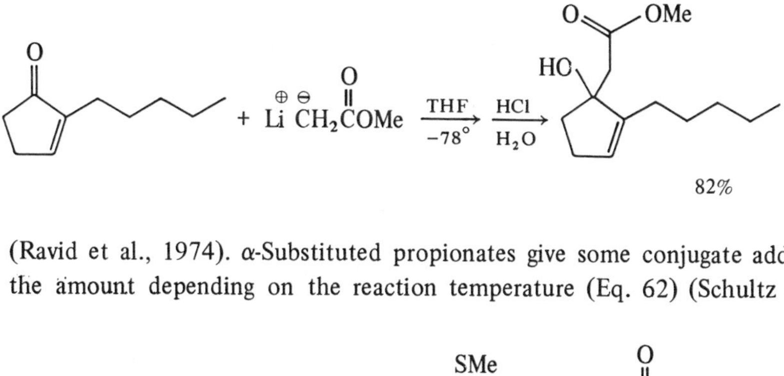

(61)

82%

(Ravid et al., 1974). α-Substituted propionates give some conjugate addition, the amount depending on the reaction temperature (Eq. 62) (Schultz et al.,

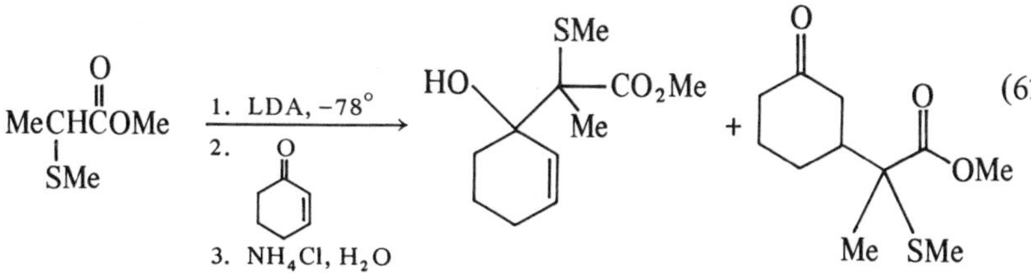

(62)

step 2 at $-78°$	63%	7%
step 2 at $-78°$ then $25°$	0%	85%

1976). The kinetic product is 1,2-addition, but if the initial adduct is allowed to equilibrate at 25° before quenching, 1,4-addition is the result. The reversible nature of it is shown by the conversion of a 1,2-adduct to the 1,4-adduct with base (Eq. 63). Thus the thermodynamically more stable product is from conjugate addition (Section 2.6).

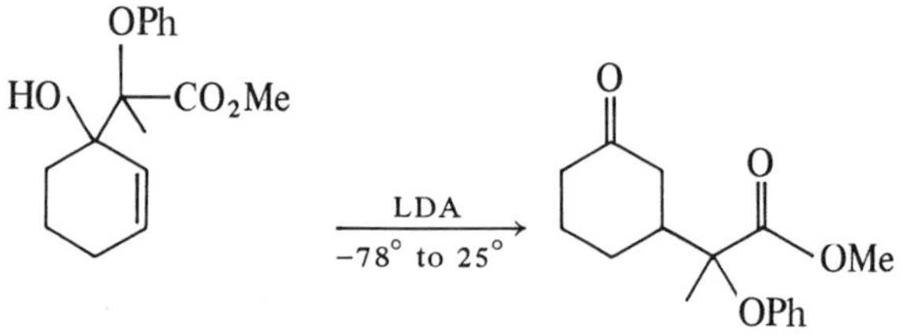

$$(63)$$

The reversibility is apparently absent with the α-anion of methyl dimethoxy-acetate. It gives 84% 1,2-addition and 14% 1,4-addition with cyclohexenone, and the distribution is independent of the reaction temperature. This anion will give more 1,4-addition where steric hindrance to 1,2-addition is high (Neef et al., 1977).

An α,β-unsaturated lactone is a suitable substrate for exclusive conjugate addition (Eq. 64) (Damon et al., 1975).

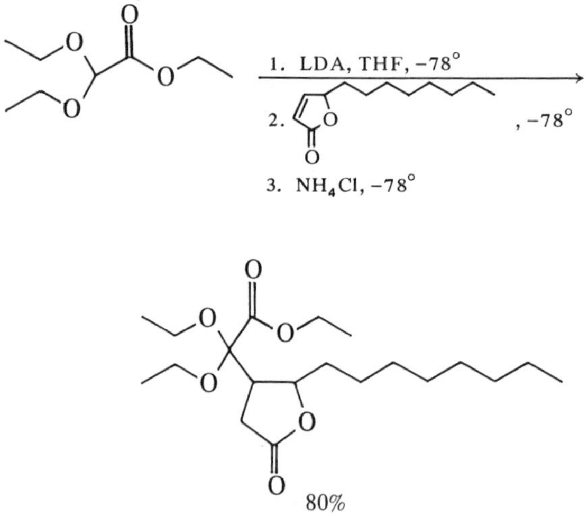

$$(64)$$

The more resonance-stabilized extended anion from angelica lactone gives only conjugate addition to α,β-unsaturated ketones, esters, and nitriles (Kraus et al., 1977). The stabilized anions from imine derivatives of α-amino esters do likewise. In fact, it can even be done in ethanol using a catalytic amount of sodium ethoxide as base (Stork et al., 1976; Fitt et al., 1977).

5.4 CARBOXYLIC ACID α-ANIONS

The dilithio O,α-dianions of carboxylic acids may be prepared in THF-HMPA solution using LDA as the base. The straight-chain carboxylic acids are 94% converted in 30 minutes at 25°, while the α-branched acids and lithium acetate require 2 hours at 50° for α-anion formation (Pfeffer et al., 1972). The α-hydrogens of arylacetic acids are sufficiently acidic to be removed with an isopropyl Grignard reagent or aryllithium reagent without nucleophilic attack on the carbonyl carbon (Ivanov et al., 1975). These dianions are far more stable than ester α-anions since the negatively charged carboxylate is not at all attractive toward enolate anion attack.

5.4.1 Alkylation. The solutions of carboxylic acid O,α-dianions in THF-HMPA may be treated with primary alkyl halides or epoxides to give α-alkylated products in good yield (Eqs. 65 through 68). This is more convenient than the alkylation of esters in that low temperatures are not required.

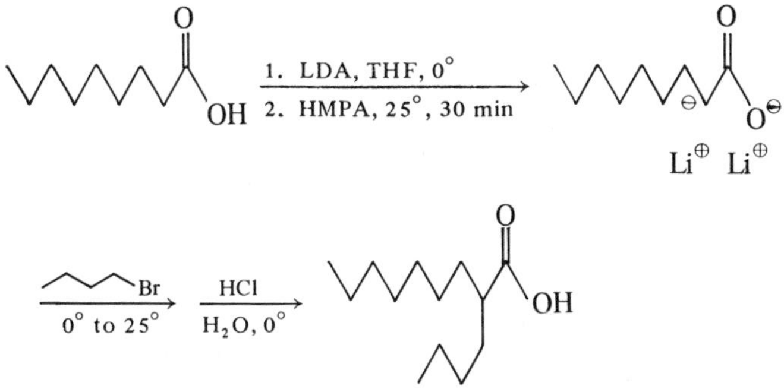

(65)

97%

(Pfeffer et al., 1972)

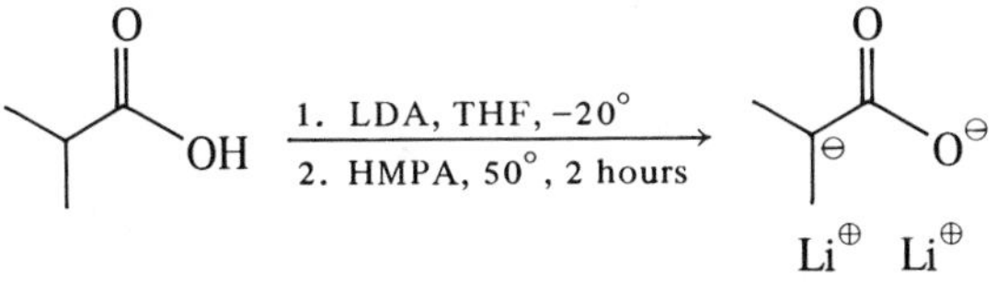

$$\xrightarrow[0° \text{ to } 40°]{\text{Br(CH}_2)_5\text{Br}} \xrightarrow[\text{H}_2\text{O}]{\text{HCl}} \quad \text{HO} \underset{O}{\overset{O}{\parallel}} \text{OH} \qquad (66)$$

62%

(Newkome, 1975)

$$\text{HOAc} \xrightarrow[\text{2. HMPA, 50°, 2 hours}]{\text{1. LDA, THF, }-20°} \quad \overset{\ominus}{\text{CH}_2}\overset{O}{\overset{\parallel}{\text{C}}}\text{O}^{\ominus} \\ \quad\quad\quad\quad\quad\quad\quad\quad\quad\quad\quad \text{Li}^{\oplus} \quad \text{Li}^{\oplus}$$

$$\xrightarrow[\text{2. HCl, H}_2\text{O}]{\text{1.} \quad \diagdown\diagup\diagdown \text{Br}} \quad \overset{O}{\diagdown\diagup\diagdown\diagup}\text{OH} \quad + \quad \overset{O}{\diagdown\diagup\diagdown\diagup}\text{OH} \qquad (67)$$

82%   8%

(Pfeffer et al., 1972)

$$\overset{O}{\underset{\ominus}{\diagup\diagdown}}\text{O}^{\ominus} + \quad \underset{\text{Li}^{\oplus} \quad \text{Na}^{\oplus}}{} \quad \text{(epoxide)} \quad \xrightarrow[0° \text{ to } 40°]{\text{THF}} \xrightarrow[15°]{\text{H}_2\text{O}} \quad \overset{\text{OH}}{\text{(cyclohexane)}}\text{OH} \qquad (68)$$

91%

(Creger, 1972; see also Bonjouklian et al., 1977).

5.4.2 Acylation. Acetic acid and disubstituted acetic acids may be acylated in high yields but not the monosubstituted cases. The dianion of acetic acid reacts with a variety of esters to give β-keto acids which are commonly thermolyzed to methyl ketones (Eq. 69) (Angelo, 1973). With higher homologues of acetic

$$\text{HOAc} + \underset{}{\text{(naphthalenide)}}^{\ominus} \text{Li}^{\oplus} \xrightarrow[-10° \text{ to } 60°]{\text{THF}} \overset{\ominus}{\text{CH}_2}\overset{O}{\overset{\parallel}{\text{C}}}\text{O}^{\ominus} \xrightarrow[\text{hexane, }-20° \text{ to } 55°]{\diagup\diagdown\diagup\diagdown\diagup\diagdown\diagup\text{OEt}}$$
$$\quad\quad\quad\quad\quad\quad\quad\quad\quad \text{Li}^{\oplus} \quad \text{Li}^{\oplus}$$

$$\xrightarrow{\text{HCl}} \quad \diagup\diagdown\diagup\diagdown\diagup\diagdown\overset{O \quad O}{\overset{\parallel\quad\parallel}{}}\text{OH} \xrightarrow{\text{heat}} \quad \diagup\diagdown\diagup\diagdown\diagup\diagdown\overset{O}{\overset{\parallel}{}} + \text{CO}_2 \qquad (69)$$
$$\quad\quad\quad 85\%$$

acid the yields are poor. The more reactive acid chlorides work well with disubstituted acetic acid dianions. The keto acids are isolated as the diisopropyl-ammonium salts, which then may be thermolyzed to the ketones (Eq. 70)

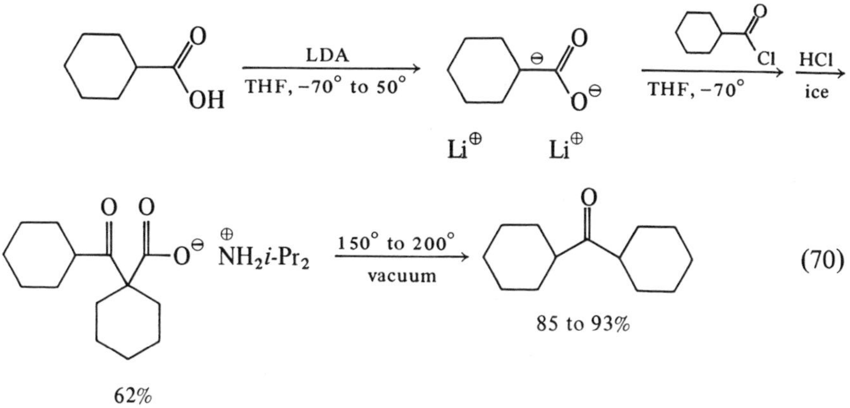

62%

85 to 93%

(70)

(Krapcho et al., 1977). With less substituted cases such as propionic acid and acetic acid itself, the yields are low due in part to formation of amides of diisopropylamine.

Considering the overall process, the R group of RCOOH is equivalent to R⁻ in an acylation process and thus is an alternative to the acylation of simple carbanions in Section 3.1.2.

5.4.3 Addition. Addition of carboxylic acid O,α-dianions to ketones or alde-hydes gives β-hydroxy acids directly. This may be superior to the Reformatsky route via esters since the hydrolysis of β-hydroxy esters is often accompanied by retro aldol and dehydration reactions. Various examples are given in Eqs. 71 through 74.

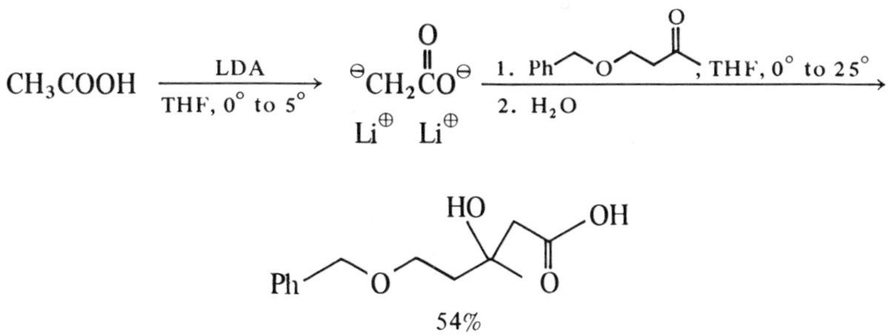

54%

(71)

(Lawson et al., 1975)

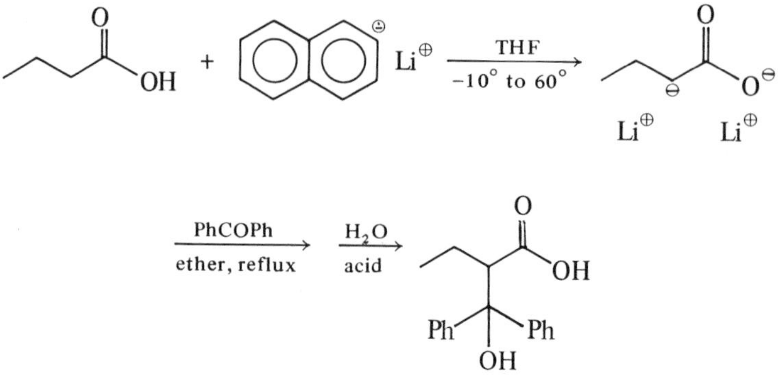

(72)

60%

(Angelo, 1970)

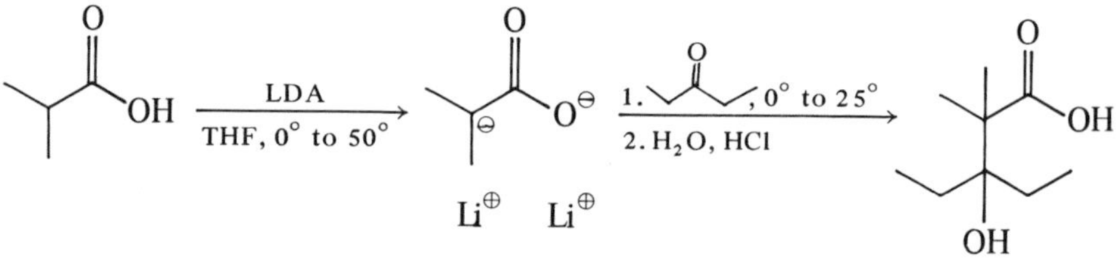

77%   (73)

(Moersch et al., 1971)

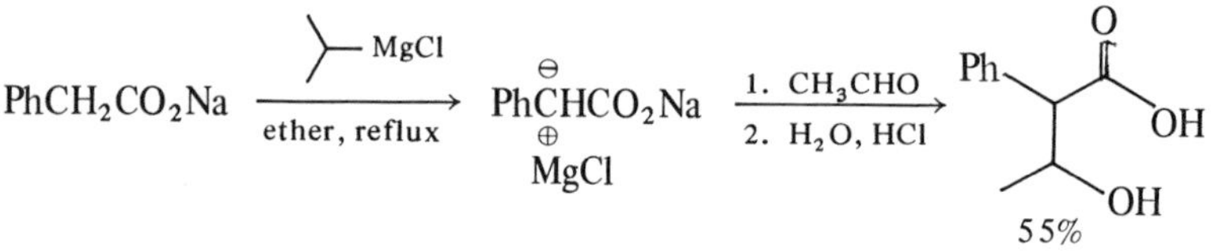

(74)

(Blagoev et al., 1970)

An analog of the Reformatsky reaction can be carried out using a zinc salt of an α-bromo acid (Eq. 75) (Bellassoued et al., 1973). This can be done in one step or in separate steps by preparing the enolate and then adding the ketone.

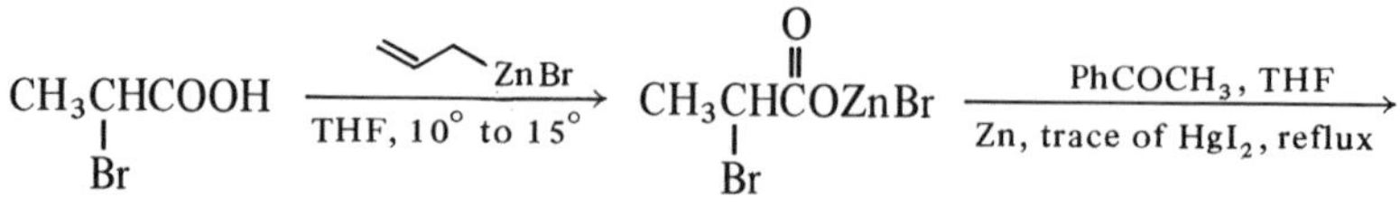

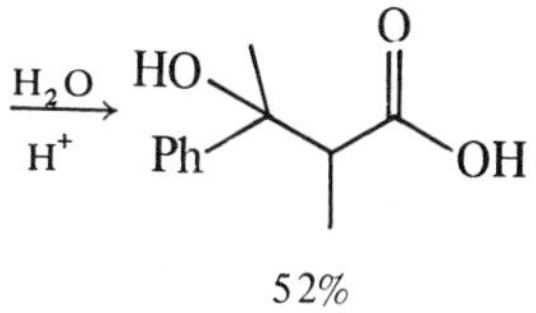

$$(75)$$

52%

These same zinc salts will add to nitriles, giving initially an imine which is hydrolyzed to a β-keto acid which then decarboxylates to a ketone (Eq. 76) (Bellassoued, et al., 1974). The overall process is again an alternative to the acylation of unstabilized carbanions (Section 3.1.2).

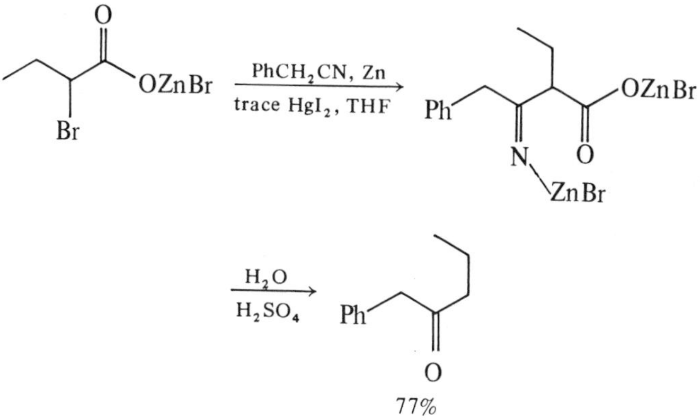

$$(76)$$

77%

5.5 AMIDE α-ANIONS

Amides lacking an N-H, especially lactams, can be converted to the α-anion by treatment with LDA in THF. α-Lithio-*N,N*-dimethylacetamide can be prepared in solution in THF at 0°. After 60 hours at 25° the anion remains, with only a small amount of the condensation product *N,N*-dimethylacetoacetamide being formed. This shows far greater stability than ester enolate anions (Woodbury et al., 1977).

5.5.1 Alkylation. *N,N*-Dimethylamides, *N*-methylpyrrolidones, and *N*-methylpiperidones have been alkylated with iodides, bromides, or epoxides as illustrated in Eqs. 77 through 80.

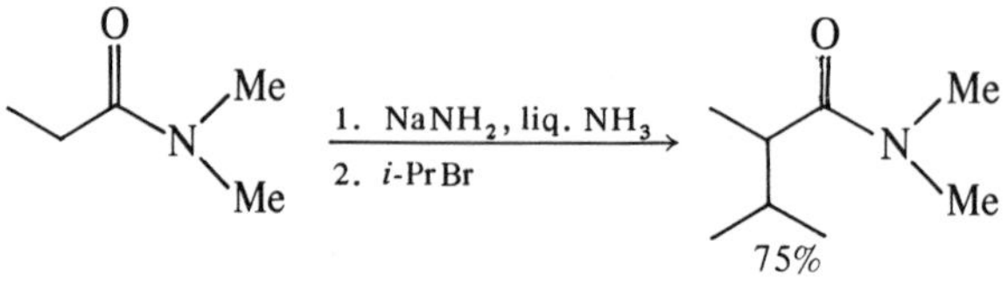

$$(77)$$

(Gassman et al., 1966a; see also Needles et al., 1966)

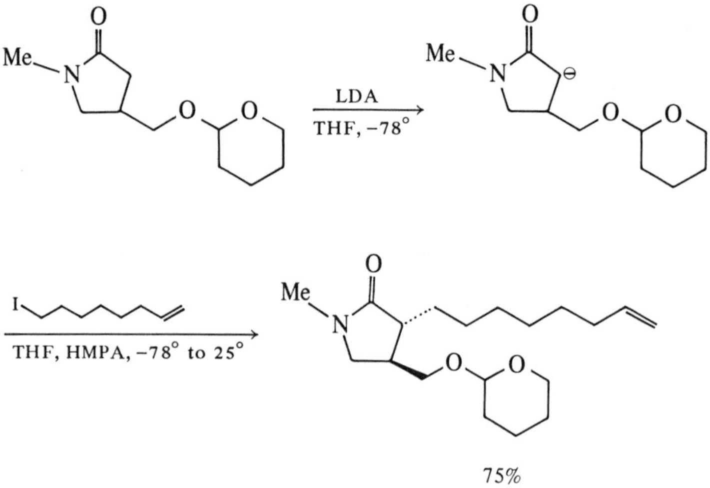

$$(78)$$

(Zoretic et al., 1977a; see also Kühlein et al., 1976)

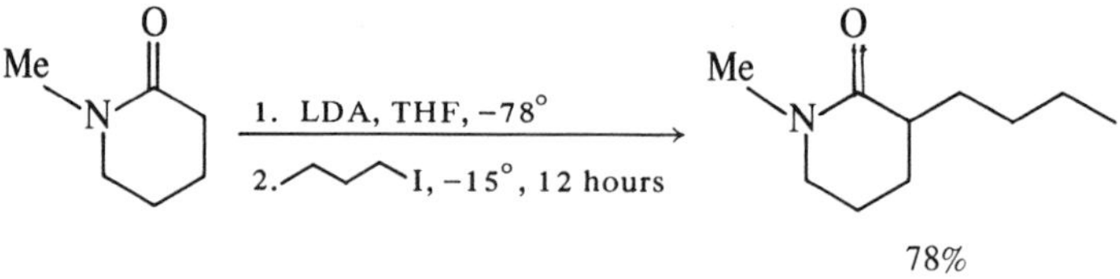

$$(79)$$

(Trost et al., 1974)

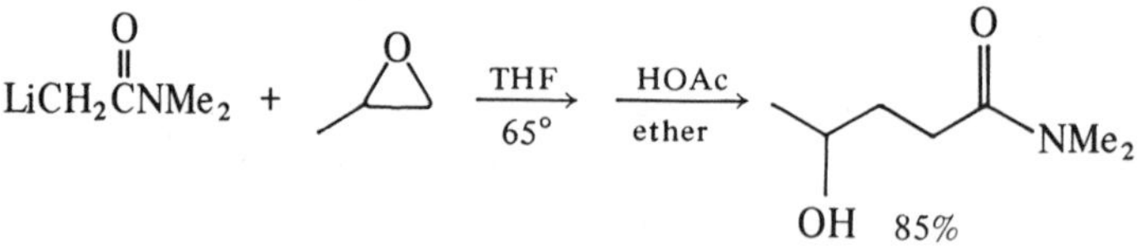

$$(80)$$

(Woodbury et al., 1977)

The enolate ion of prolylproline anhydride is further stabilized by a β-nitrogen and by the ring structure so that it can be generated and alkylated at high temperatures (Eq. 81) (Poisel et al., 1972).

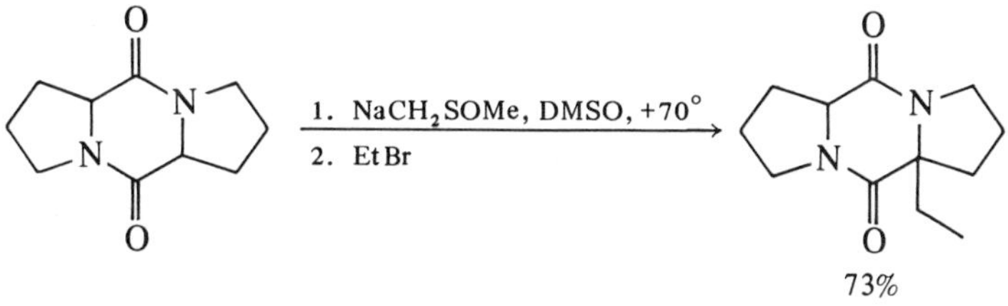

$$\xrightarrow[\text{2. Et Br}]{\text{1. NaCH}_2\text{SOMe, DMSO, +70}^\circ}$$

(81)

73%

Lactams with an N-H can still be alkylated if the N,α-dianion is prepared, in analogy with the alkylation of carboxylic acids. For example, see Eq. 82 (Deprès et al., 1978).

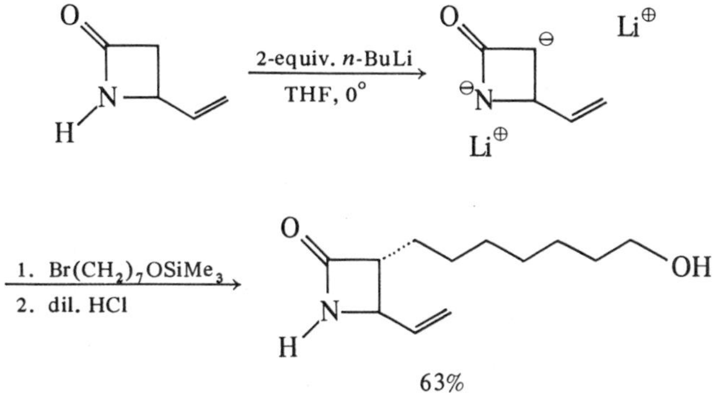

$$\xrightarrow[\text{THF, 0}^\circ]{\text{2-equiv. } n\text{-BuLi}}$$

$$\xrightarrow[\text{2. dil. HCl}]{\text{1. Br(CH}_2)_7\text{OSiMe}_3}$$

(82)

63%

5.5.2 Acylation. Several β-lactams have been α-benzoylated via the LDA-generated anion using methyl benzoate (Durst et al., 1972).

5.5.3 Addition. α-Lithio-N,N-dimethylacetamide will add to ketones at 0° to give tertiary alcohols. At this temperature aldehydes will self-condense, but at −78° the addition of the amide α-enolate proceeds in high yield (Eq. 83) (Woodbury et al., 1977). An addition to methyl vinyl ketone at low temperature

$$\text{LiCH}_2\overset{\overset{\text{O}}{\|}}{\text{C}}\text{NMe}_2 + \text{CH}_3\text{CHO} \xrightarrow[-78°]{\text{THF}} \xrightarrow{\text{HOAc}} \quad \overset{\text{HO}\quad\text{O}}{\diagup\diagdown\diagup\diagdown}_{\text{NMe}_2} \qquad (83)$$

98%

gave no 1,4-addition, only 1,2-addition (Eq. 84) (Trost et al., 1974).

$$(84)$$

1. LDA, THF, −78°

2. (acetone), −78°

57%

5.6 NITRILE α-ANIONS

The α-anions of nitriles may be generated with alkali amides in liquid ammonia, lithium diethylamide with HMPA, or with n-BuLi in THF. Low concentrations of the anion are produced by potassium hydroxide.

5.6.1 Alkylation.
As with ketones, the alkylation of nitriles is complicated by multiple alkylation owing to formation of the α-anion of the initially alkylated product. However selective monoalkylation may be achieved using lithium dialkylamide bases and low temperatures (Watt, 1974). For example propionitrile was monoalkylated and then alkylated again with a different halide in a route to citronellol (Eq. 85) (Debal et al., 1976). Note that the cyano group may be reductively removed with sodium or potassium in HMPA. With this reaction, the nitrile α-anion RCH⁻CN becomes synthetically equivalent to the

$$\diagup\diagdown_{\text{CN}} + \diagup\diagdown\diagdown-\text{Br} \xrightarrow[\text{ether, }-75° \text{ to } 10°]{\text{LiNEt}_2, \text{HMPA}} \diagdown\diagup\diagdown\diagdown_{\text{CN}}$$

85%

$$\xrightarrow[\text{ether, Cl}\diagdown\diagup_{\text{O}}\diagdown_{\text{O}}]{\text{LiNEt}_2, \text{HMPA}} \quad \diagdown\diagup\diagdown\diagdown\underset{\text{CN}}{\diagup}\diagdown\diagup_{\text{O}}\diagdown_{\text{O}}$$

80%

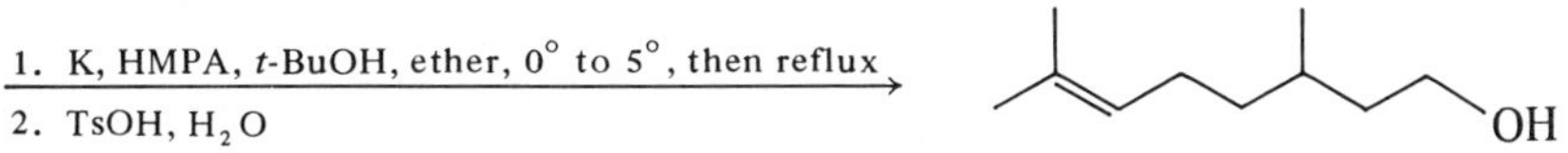

$$\begin{array}{l}\text{1. K, HMPA, } t\text{-BuOH, ether, } 0^\circ \text{ to } 5^\circ \text{, then reflux}\\ \text{2. TsOH, H}_2\text{O}\end{array} \longrightarrow$$

(±)-citronellol
91%

(85)

anion RCH_2^- with the advantages of easy preparation and a facile alkylation reaction.

Acetonitrile may be selectively monoallylated if it is converted to the copper enolate. Allyl bromides and tosylates react (Eq. 86), but the reagent is not sufficiently nucleophilic to attack benzylic or unactivated bromides (Corey et al., 1972a; Ibuka et al., 1976).

$$CH_3CN + n\text{-BuLi} \xrightarrow[-78^\circ]{THF} \xrightarrow[-25^\circ]{CuI} CuCH_2CN$$

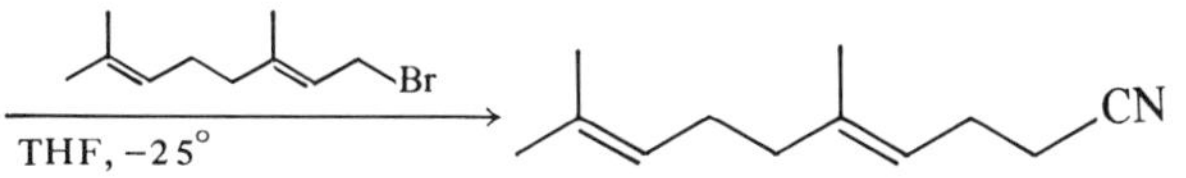

(86)

95%

Phenylacetonitrile and other α-phenylnitriles have been alkylated in good yield using aqueous hydroxide base under phase transfer conditions (Chapter 1, Eq. 12) (Makosza et al., 1976a).

Intramolecular alkylation of nitriles has been carried out using halides and epoxides and is particularly valuable for the synthesis of cyclobutanes. The requisite δ-halonitriles are prepared by alkylating nitriles with 1-chloro-3-bromopropane and then cyclized with LiNEt$_2$-HMPA (Larchevêque et al., 1973). Intramolecular attack of a nitrile α-anion on a δ,ε-epoxide group involves a choice of attack on the δ- or the ε-carbon. Usually the attack will occur on the least substituted site, but if substitution is equal, attack will occur on the near side giving a cyclobutane rather than a cyclopentanol (Eq. 87) (Stork et al., 1974b,c).

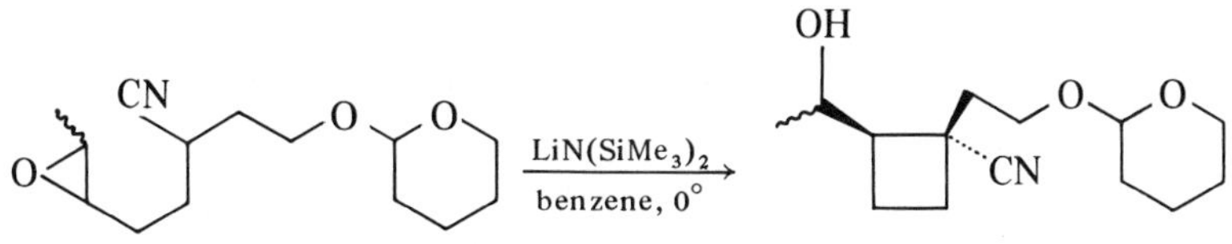

$$\xrightarrow[\text{benzene, } 0^\circ]{\text{LiN(SiMe}_3)_2}$$

(87)

> 52%

The extended anion derived from a β,γ- or an α,β-unsaturated nitrile is dialkylated in the α position leading to β,γ-unsaturated products in high yield (Marshall et al., 1977b). Selective monoalkylation is possible if again one uses LDA at $-78°$ (Eq. 88). Allyl and benzyl bromides as well as unactivated primary and

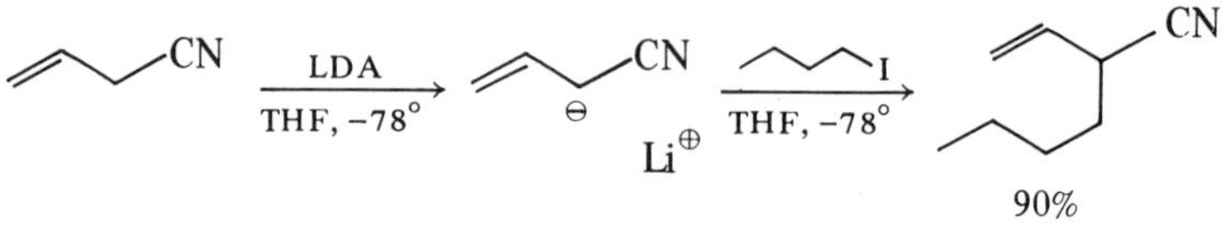

$$\tag{88}$$

secondary iodides gives yields of 72 to 98% (Kieczykowski et al., 1975). Similar results were found with a $\beta,\gamma,\delta,\epsilon$-unsaturated nitrile (Borch et al., 1977).

As discussed in Section 4.6, certain derivatives of aldehydes allow formation of a carbanion at the carbonyl carbon. Cyanohydrin acetals are examples of such derivatives. The anion may be generated at low temperature and alkylated with activated and unactivated primary and secondary bromides (Eq. 89) (Stork et al., 1971). Hydrolysis then affords ketones. If the starting aldehyde is

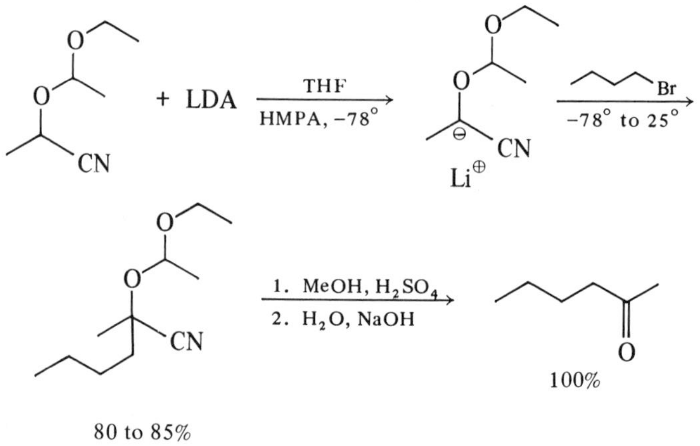

$$\tag{89}$$

an α,β-unsaturated one, the alkylation still occurs at the carbon that bears the oxygen atom. Thus acrolein was converted to n-hexyl vinyl ketone in 75% yield. Aromatic aldehydes can be converted to ketones similarly.

Aryl cyanohydrin silyl ethers can likewise be converted to the anion and alkylated. They can be made directly from the aldehyde using trimethylsilyl cyanide (Eq. 90) (Hünig et al., 1975a). The ketone may be freed by acid and

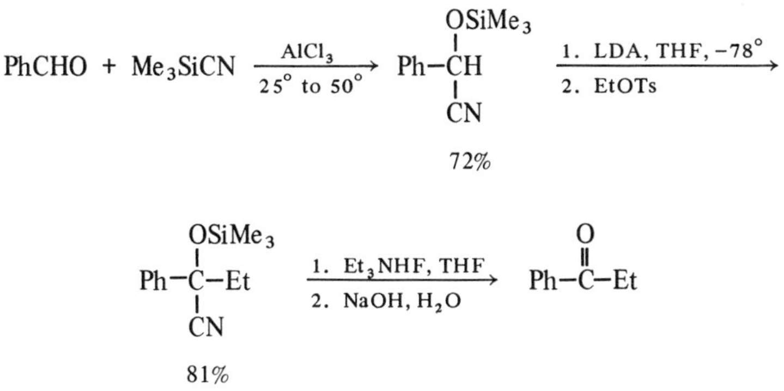

PhCHO + Me$_3$SiCN $\xrightarrow[25° \text{ to } 50°]{\text{AlCl}_3}$ Ph$-$CH$\begin{smallmatrix}\text{OSiMe}_3\\|\\\\|\\\text{CN}\end{smallmatrix}$ $\xrightarrow[\text{2. EtOTs}]{\text{1. LDA, THF, }-78°}$

72%

Ph$-$C$-$Et with OSiMe$_3$ above and CN below $\xrightarrow[\text{2. NaOH, H}_2\text{O}]{\text{1. Et}_3\text{NHF, THF}}$ Ph$-$C$-$Et with O above $\qquad$ (90)

81%

base or, taking advantage of the affinity of fluoride for silicon, by treatment with triethylamine hydrofluoride. The alkylation gives high yields with a wide variety of agents; even *tert*-butyl iodide gave an 85% yield.

5.6.2 Acylation. Nitriles may be acylated with esters using a low equilibrium concentration of carbanion as generated with sodium ethoxide or using preformed α-anion. The relatively acidic phenylacetonitrile may be warmed with ethyl acetate and sodium ethoxide to give a 65 to 70% yield of α-phenyl-β-ketobutyronitrile. The less acidic alkyl nitriles such as valeronitrile may be α-benzoylated using ethyl benzoate and sodium ethoxide (59% yield), but heating to 160° and removal of the ethyl alcohol product is required to force the reaction (Henecka, 1952c).

A strong base may be used to completely convert a nitrile to the α-anion first, and then an acid chloride or an ester will acylate it (Eqs. 91 and 92) (Albarella, 1977; Taylor et al., 1978). In these cases an extra equivalent of base or carbanion is required since the product is more acidic than the starting nitrile.

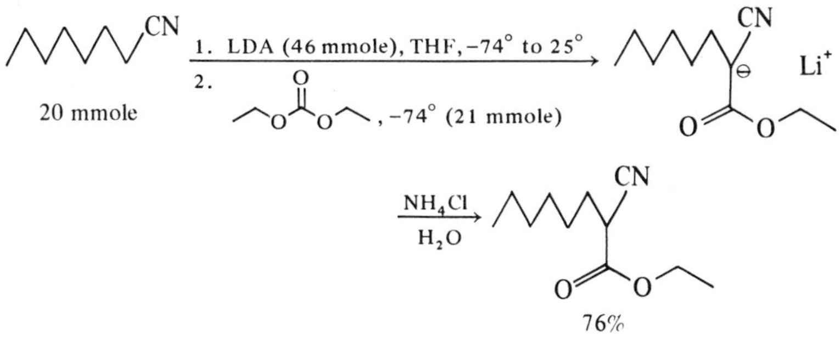

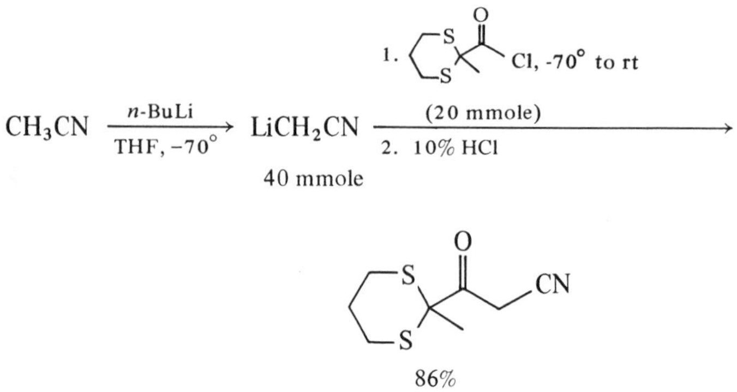

$$CH_3CN \xrightarrow[\text{THF, }-70°]{n\text{-BuLi}} LiCH_2CN \xrightarrow[\text{2. 10\% HCl}]{\text{(20 mmole)}}$$

1. [structure: 1,3-dithiane with C=O, Cl, -70° to rt]

40 mmole

[structure: 1,3-dithiane bearing C(=O)CH₂CN group]   (92)

86%

5.6.3 Addition. *n*-Butyllithium in THF will convert acetonitrile to the α-anion, which will then add to various ketones or benzaldehyde to afford β-hydroxynitriles (Eq. 93) (Kaiser et al., 1968) (Review: Ivanov et al., 1975). Since the

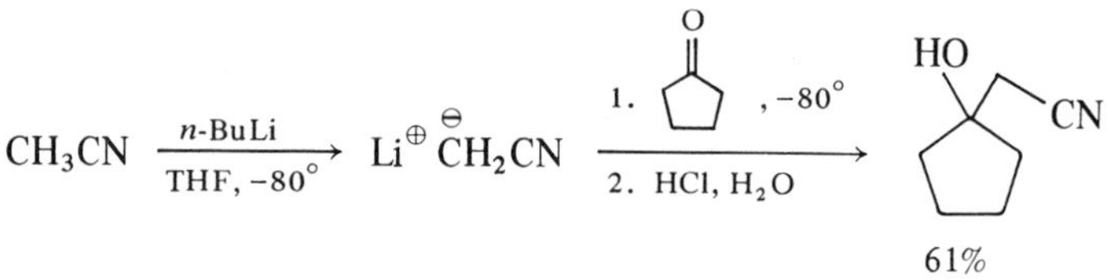

$$CH_3CN \xrightarrow[\text{THF, }-80°]{n\text{-BuLi}} Li^{\oplus} \overset{\ominus}{C}H_2CN \xrightarrow[\text{2. HCl, H}_2O]{1. \text{ cyclopentanone, }-80°}$$

[structure: cyclopentane ring with HO and CH₂CN substituents]   (93)

61%

nitrile α-anions are among the most basic of those with heteroatom stabilization, they may cause ketone enolate formation in competition with the addition reaction. Thus the addition to benzophenone, which has no α-hydrogens, proceeds in higher yield (89%).

Potassium hydroxide will generate a low equilibrium concentration of the anion from acetonitrile in excess acetonitrile, and this gives α,β-unsaturated nitriles from addition to ketones or benzaldehydes (Eq. 94) (Gokel et al., 1976c; DiBiase et al., 1977). This fails with the more readily enolized cyclopentanone or simple aldehydes.

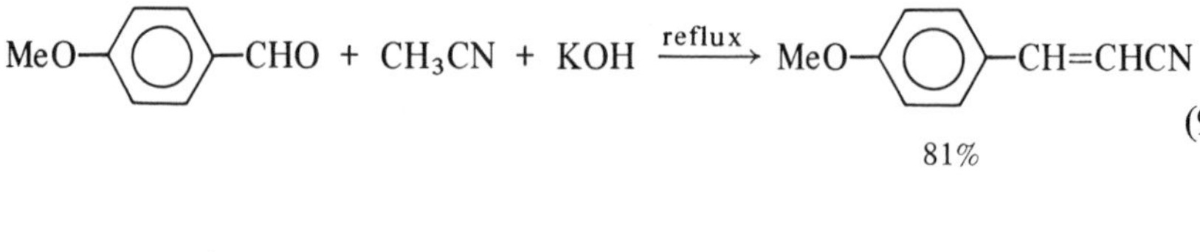

$$MeO-\langle\bigcirc\rangle-CHO + CH_3CN + KOH \xrightarrow{\text{reflux}} MeO-\langle\bigcirc\rangle-CH=CHCN$$

(94)

81%

The carbanions derived from aryl cyanohydrin silyl ethers will add to ketones and aldehydes to give, after deprotection, α-hydroxy ketones (Eq. 95) (Hünig et al., 1975b).

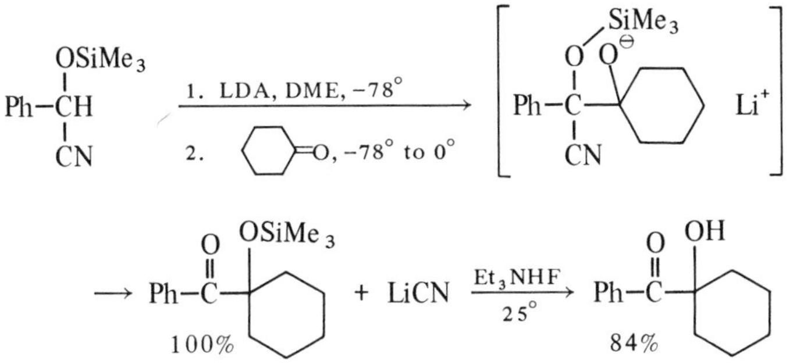

$$\text{(95)}$$

Nitriles will self-condense to dimers, trimers, and so forth, when treated with strong bases at ordinary temperatures. The reaction may be controlled to give fair yields of the dimers (enaminonitriles) which can then be hydrolyzed to β-ketonitriles (Reynolds et al., 1951). The cyclization of dinitriles by this process is called the Thorpe-Ziegler condensation (Review: Schaefer et al., 1967). Five- and six-membered rings are produced in high yield (Eq. 96) (Brown, 1975), and high dilution conditions allow preparation of larger rings.

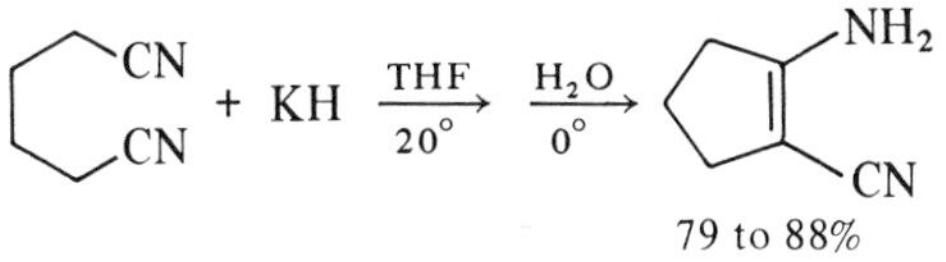

$$\text{(96)}$$

5.6.4 Conjugate Addition. Nitrile α-anions will add conjugatively to β-aryl α,β-unsaturated ketones (Eq. 97) and to acrylonitrile (Review: Bergmann et al., 1959):

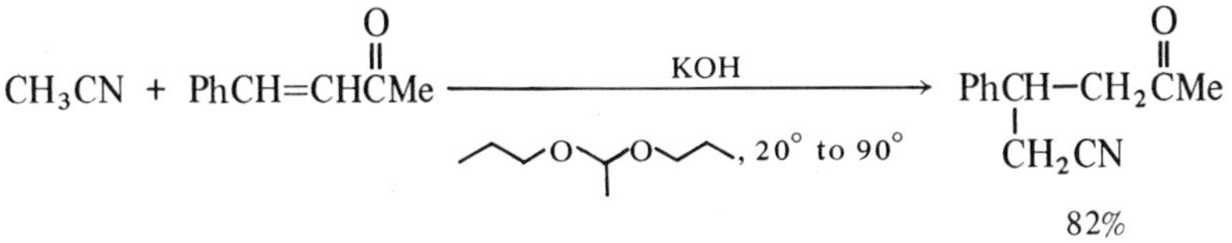

$$\text{(97)}$$

(Weizmann et al., 1950)

Simple aldehyde-derived cyanohydrin acetals give mixtures of 1,2- and 1,4-additions with cycloalkenones (Chapter 2, Eq. 24); but with the additional resonance stabilization in those derived from crotonaldehyde, clean 1,4-addition is observed (Eq. 98) (Stork et al., 1974d). Any competing 1,2-addition is

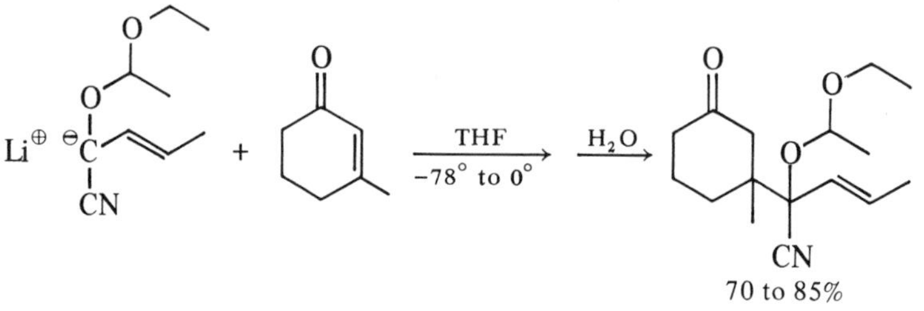

(98)

70 to 85%

reversible and changes to the 1,4-adduct as discussed in Section 2.6. A morpholino analog of a cyanohydrin acetal also serves as an acyl anion equivalent, giving conjugate addition to acrylonitrile and to phenyl vinyl ketone (Eq. 99). The protection may be removed to give the ketone by simply warming in aqueous acetic acid (Leete, 1976).

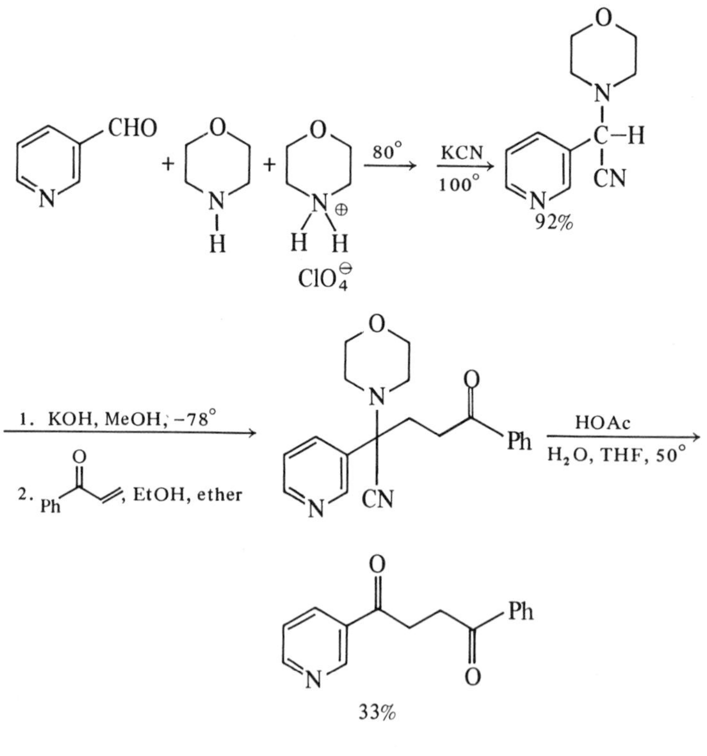

92%

1. KOH, MeOH, −78°

2. Ph—⧸=O, EtOH, ether

HOAc
H₂O, THF, 50°

(99)

33%

5.7 IMINE α-ANIONS

Ketones and aldehydes may be converted to imines which will lose a proton to a Grignard reagent or lithium dialkylamide base. The anions are superior to the corresponding anions of the parent carbonyl compounds because they give only monoalkylation, and most of them do not self-condense but will add readily to other ketones or aldehydes. Since the imine function is readily hydrolyzed, the imine α-anions are synthetically equivalent to the ketone or aldehyde α-anions but with superior characteristics. See also enamines for a similar synthetic advantage with lower reactivity (Dyke, 1973).

5.7.1 Alkylation. As pointed out in Section 5.2.1, the alkylation of aldehydes is successful only in a few special circumstances but it is effectively carried out via the aldimine as shown in Eq. 100 (Stork et al., 1974f). Some low molecular

weight aldimines do self-condense when the Grignard reagent is added; but if the alkyl halide is added before the Grignard reagent, alkylation is successful. For example the *tert*-butylimine of propionaldehyde gave overall 2-methyl-hexanal in 60% yield (Stork et al., 1963). Self-condensation of aldimines was not a problem when the anion was generated with lithium dialkylamide (Eqs. 101 and 105). This may then be alkylated with primary halides (Eq. 101) (LeBorgne et al., 1976).

The alkylation of ketimines gives selective monoalkylation (Eq. 102) (Pearce et al., 1976), and where the ketone is unsymmetrical, the new group enters on the side which had the least original substitution (Eq. 103) (Cuvigny et al., 1975). This regiospecific behavior was explained (Stork et al., 1963) in terms of the stability of the intermediate anion where the planar carbon-carbon double

(101)

(102)

(103)

bond brings α-substituents into steric interference with the group on the nitrogen at least in some conformations. Thus the less substituted anion would have the greater stability and be the one formed and alkylated. This same factor may be

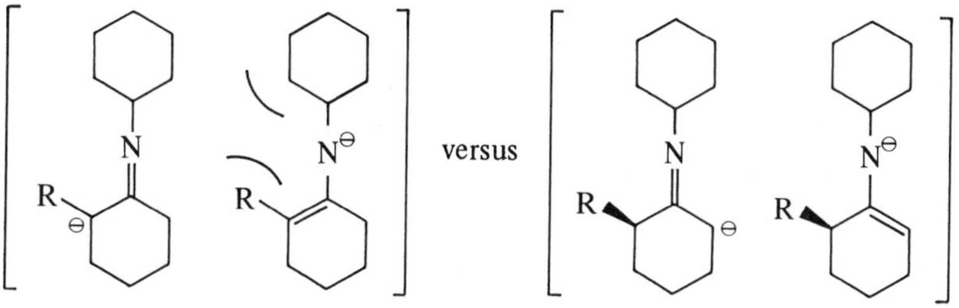

responsible for the absence of polyalkylation because the unalkylated carbanion would not generate an appreciable amount of a new, more hindered anion from the alkylated imine.

The extended anion from N-cyclohexylcrotonaldimine, like other extended anions, is specifically alkylated in the α position (Eq. 104) (Kieczykowski, et al., 1976). A second alkyl group may be introduced in the α-position in the same manner to give a β,γ-unsaturated aldehyde in high yield.

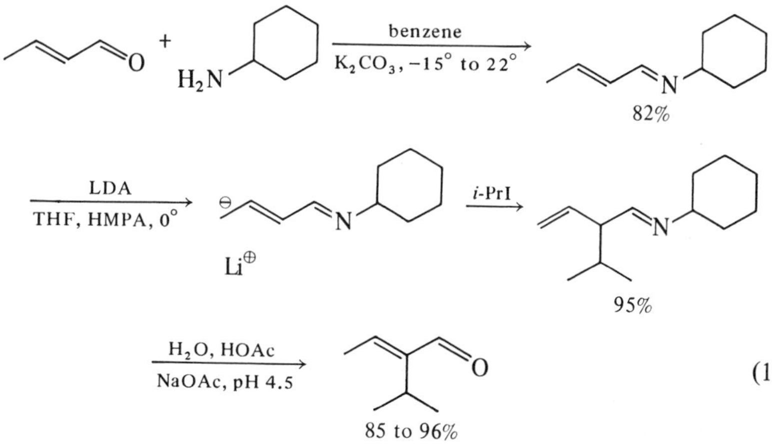

(104)

5.7.2 Acylation. The cyclohexylimine of acetone gives the C-carbethoxylated product from ethyl chloroformate in 90% yield. Ethyl benzoate and N,Ndimethylbenzamide also react at carbon, but benzoyl chloride gives mostly the N-benzoyl product (Reiff, 1971).

5.7.3 Addition. Since nitrogen is less electronegative than oxygen, the imines are less attractive toward carbanions than ketones or aldehydes are. Therefore the imine enolate anions can be prepared without self-condensation, and yet they will rapidly add to the carbonyl compounds. This amounts to directed cross aldol condensations and can even be used to add aldehyde enolate equivalents to ketones. (Review: Reiff, 1971; Wittig et al., 1970). These reactions are carried out at low temperature to suppress competing proton transfer which would give ketone or aldehyde enolate anions instead of addition. This method was used as part of a synthesis of (±)-cembrene as shown in Eq. 105 (Dauben

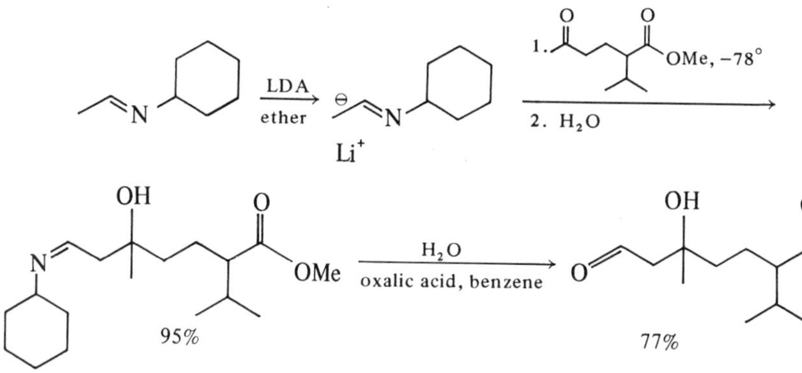

et al., 1975). Similarly the anion from ethylidene-tert-butylamine was added to isobutyraldehyde at −70° to afford, after acidification, trans-4-methyl-2-

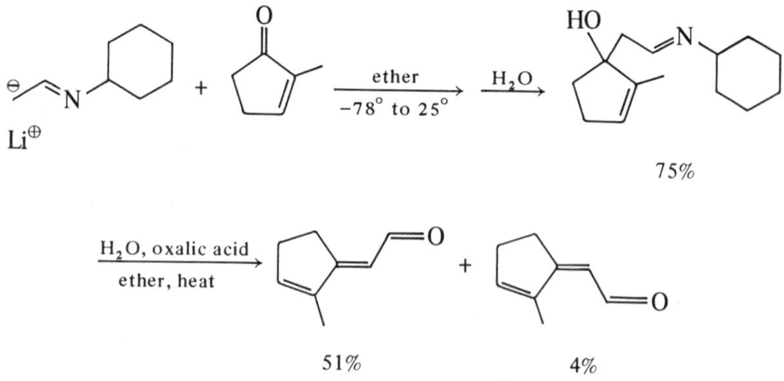

pentenal (43%) (Borch et al., 1977). An addition to a cyclopentenone is illustrated in Eq. 106 (Gilbert et al., 1976). Note that only 1,2 addition and not 1,4 addition has occurred. In this case the tertiary alcohol was eliminated by steam distillation from acid. Milder conditions suffice when a silyl derivative is eliminated instead (Eq. 107) (Corey et al., 1976b).

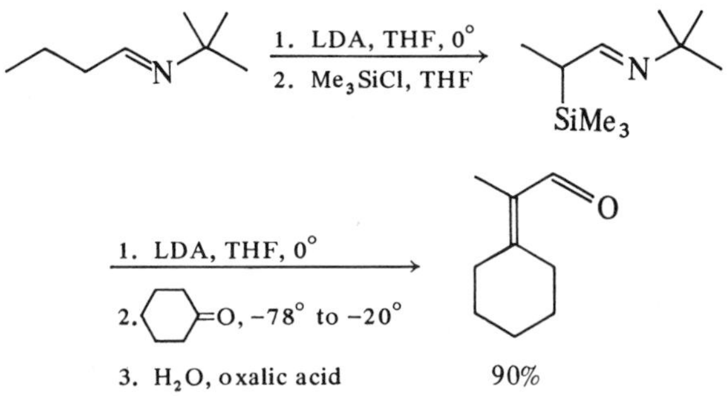

$$(107)$$

5.8 HYDRAZONE α-ANIONS

Aldehydes and ketones may be protected as *N,N*-dimethylhydrazones. The α-anions may then be generated with LDA at 0° or *n*-BuLi at −78° and used in the same way as the imines in Section 5.7. The advantages are that the hydrazones can be formed in quantitative yield even from hindered ketones, that the anions are more reactive than the imine anions, and finally that the alkylation and addition reactions proceed in higher yield. On the other hand, they are less easily hydrolyzed than imines, but they can be removed readily under neutral oxidative conditions or with cupric salts (Corey et al., 1976a,b,d).

5.8.1 Alkylation. The selective alkylation of 2-methylcyclohexanone is shown in Eq. 108 (Corey, et al., 1976a). Here, as with the imines, the anion forms on the least hindered side to give only the 2,6-dimethyl isomer. The product is 97%

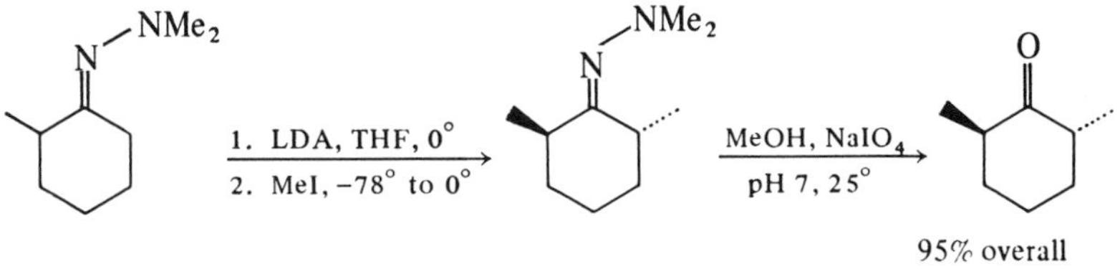

$$(108)$$

95% overall

trans and 3% cis. The removal of the hydrazone group was done in a phosphate buffer at pH 7, and the similar aldehyde derivatives were removed at pH 4.5 in THF-water. γ-Hydroxy ketones are available from alkylation with epoxides (Eq. 109) (Corey et al., 1976a).

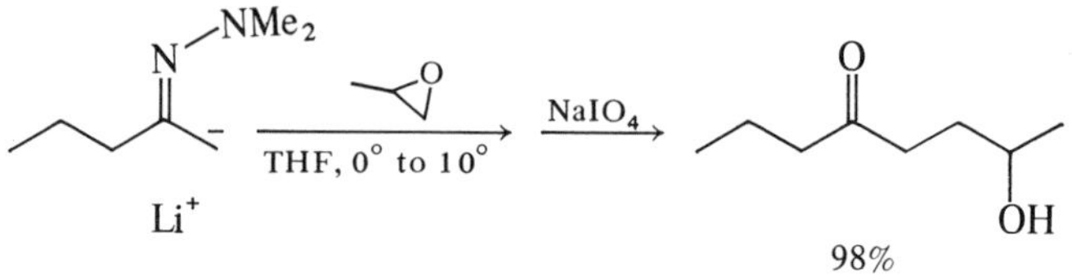

(109)

98%

If instead of *N,N*-dimethylhydrazine the chiral (*S*)-1-amino-2-methoxymethyl-pyrrolidine is used, the alkylation can produce ketones and aldehydes with chiral centers at the α position in enantiomeric excess of 30 to 87% (Enders et al., 1976, 1977).

α,β-Unsaturated ketones may be alkylated at the α position via the hydrazone derivative using NaH as the base (Eq. 110) (Stork et al., 1977b).

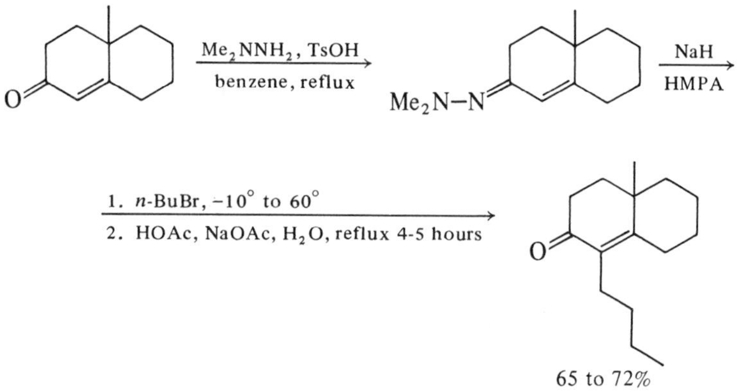

(110)

65 to 72%

5.8.2 Addition. As with the imines, the directed crossed aldol condensation gives β-hydroxy ketones and aldehydes (Eq. 111) (Corey et al., 1976a):

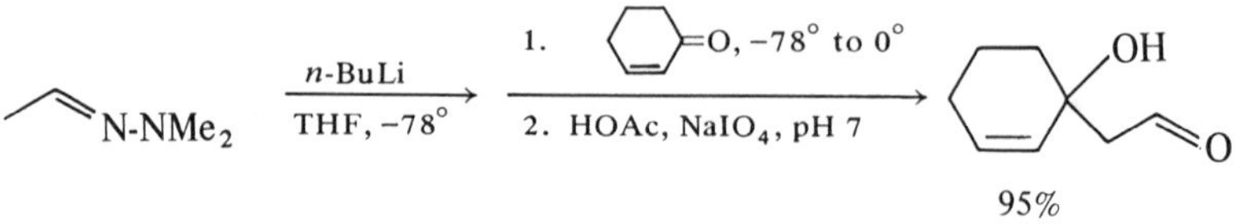

(111)

95%

5.8.3 Conjugate Addition. The lithium dimethylhydrazone enolates give only 1,2-addition (Eq. 111), but if cuprous salts are added, conjugate addition occurs (Eq. 112) (Corey et al., 1976a). Here again the hydrazone anions react at the

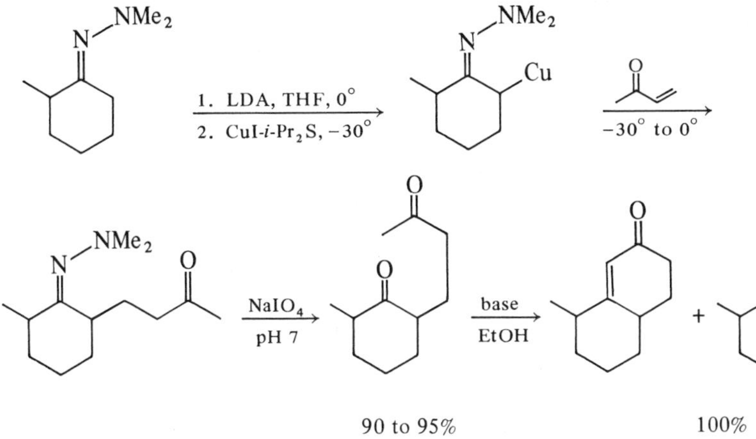

90 to 95%   100%   (112)

least substituted side. The usual Robinson annulation procedure using the free ketone (Section 5.1.4) gives attachment on the side with the methyl group.

5.9 HETEROCYCLIC ANALOGS OF IMINOESTER α-ENOLATE ANIONS

5.9.1 Dihydro-1,3-oxazines. A carbanion can be generated α to certain heterocyclic rings where the charge is resonance delocalized to a nitrogen atom as shown for dihydro-1,3-oxazines in Eq. 113 (Review: Meyers, 1974). These

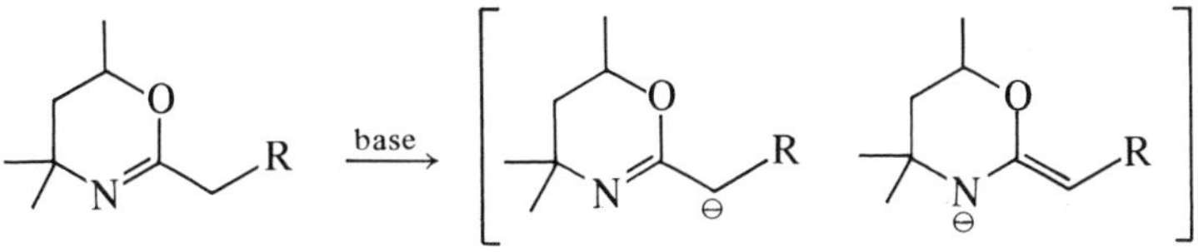

(113)

carbanions are readily alkylated or added to ketones or aldehydes. The elaborated dihydrooxazine may then be reduced to the tetrahydro compound and then hydrolyzed to an aldehyde (Eq. 114) (Meyers et al., 1973a). The overall process is an aldehyde synthesis involving two-carbon homologation of electrophiles. Dialkylation does not compete because the monoalkylated product is not acidic enough to give a carbanion with n-BuLi below $-50°$. Dialkylation will occur if the second one is intramolecular as in preparing rings from 1,2-,

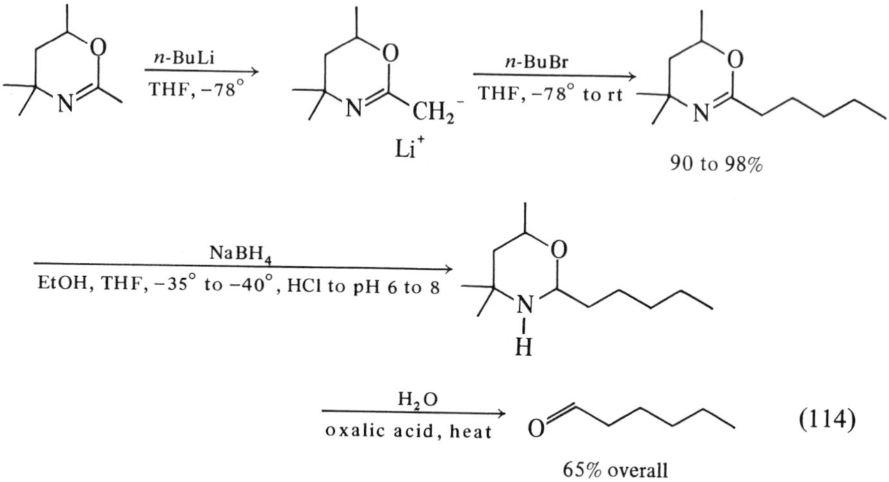

$$(114)$$

65% overall

1,3-, and 1,4-dihalides. This requires warming the reaction to $-50°$ for 2 hours. Alkylation with an epoxide gives γ-hydroxy aldehydes. Addition to an aldehyde or ketone gives ultimately α,β-unsaturated aldehydes.

The original dihydrooxazine may have carbanion-stabilizing groups on the α position (Eq. 113, R = Ph, COOEt, or Cl) which lead to α-substituted aldehydes. The dihydrooxazines are prepared by treating nitriles with a 1,3-diol in concentrated sulfuric acid (Eq. 115) (Meyers et al., 1973a).

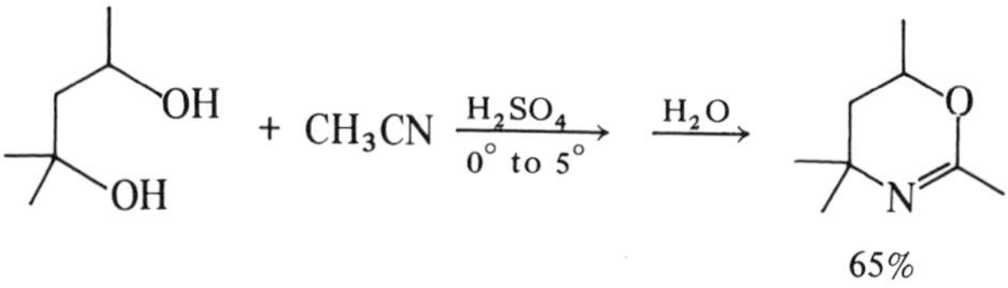

$$(115)$$

65%

The overall process may be carried out more conveniently using the N-quaternized dihydrooxazine. Removal of a proton from this gives a very reactive enamine which can be alkylated, reduced, and hydrolyzed quickly without the use of lithium reagents or Dry-Ice temperatures (Eq. 116) (Meyers et al., 1972c).

Dihydrooxazines are also precursors to ketones if, instead of reducing the ring with sodium borohydride, a Grignard reagent is added to the C=N bond in the N-methyloxazinium salts. (The dihydrooxazines themselves are inert toward Grignard addition.) The dihydrooxazinium salt with a simple methyl

(116)

(117)

group in the 2-position gives poor yields of Grignard addition owing to competing proton removal to give the enamine, but higher homologs such as the 2-ethyl or even 2-benzyl compounds function well in ketone syntheses (Meyers et al., 1972b). This process, along with a conjugate addition of the enamine discussed above, was used to prepare methyl jasmonate (Eq. 117) (Meyers et al., 1973b).

5.9.2 2-Thiazolines. 2-Methyl-2-thiazolines may be converted to the α-anion and used in alkylation and addition reactions as was done with the dihydro-oxazines above (Meyers et al., 1975b,c). There are two significant ways in which the thiazolines differ, however. They can be alkylated a second and a third time to give highly α-branched aldehydes, and the aldehydes can be released from the heterocyclic derivative under neutral conditions using mercuric chloride (Eq. 118). The addition to simple aldehydes and ketones leads to β-hydroxy aldehydes (Eq. 119), but more complex cases can give dehydrated products or retro-aldol reactions.

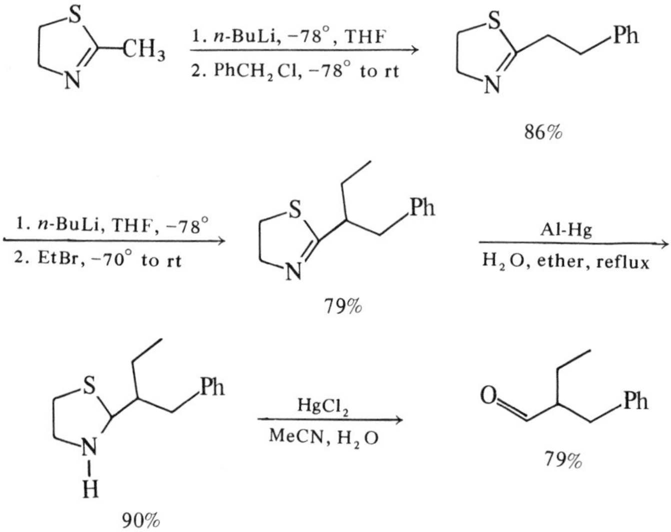

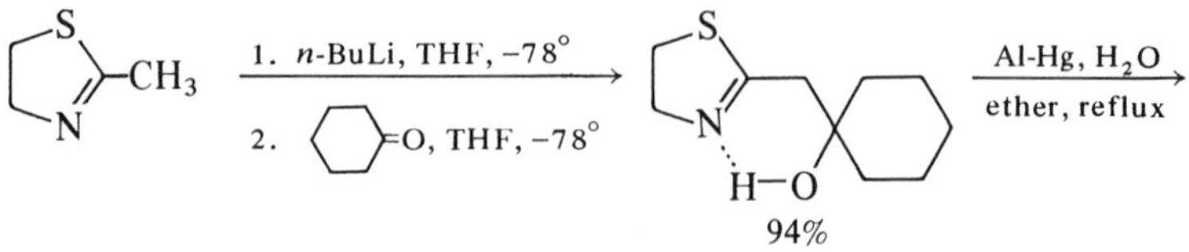

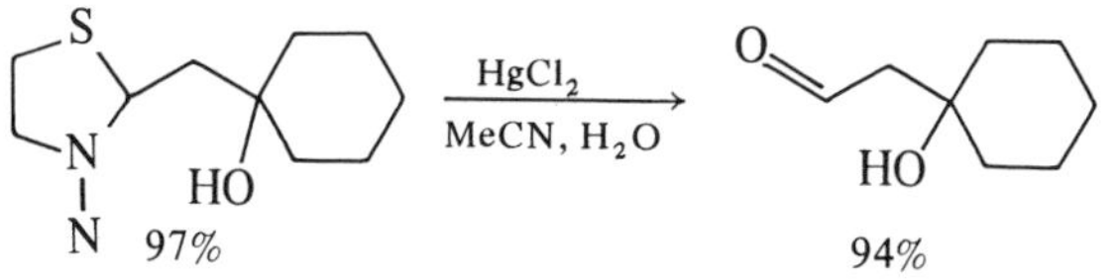

(119)

5.9.3 2-Oxazolines.

The analogous 2-oxazolines can be converted to α-anions which then react with electrophiles (Review: Meyers et al., 1976a). Unlike the oxazines and thiazolines, they cannot be reduced to derivatives of aldehydes; however they are readily hydrolyzed or alcoholyzed to carboxylic acids or esters. They are thus useful for preparation of substituted acetic acids. The readily available 2,4,4-trimethyl-2-oxazoline (Eq. 120) can be alkylated once or twice but not a third time. Alkyl halides give longer-chain carboxylic acids, while monosubstituted epoxides afford γ-butyrolactones (Eq. 121) (Meyers et al., 1974a). The oxazoline α-anions will add to carbonyl compounds,

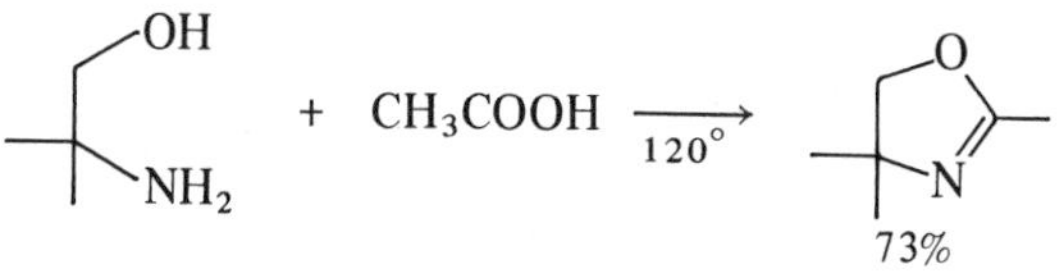

(120)

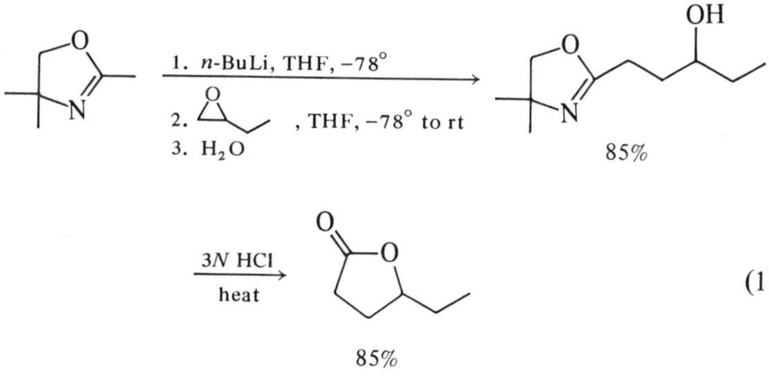

(121)

after which ethanolysis with dilute sulfuric acid gives β-hydroxy esters. Stronger acid gives dehydration to α,β- and β,γ-unsaturated acids or esters.

An optically active oxazoline can give asymmetric induction in reactions at the α position. The commercially available (+)-1-phenyl-2-amino-1,3-propanediol can be converted to an oxazoline which may be alkylated and then hydrolyzed to carboxylic acids with a new chiral center in the α position. These can be made with up to 78% enantiomeric excess (Eq. 122) (Meyers et al., 1974b, 1976b).

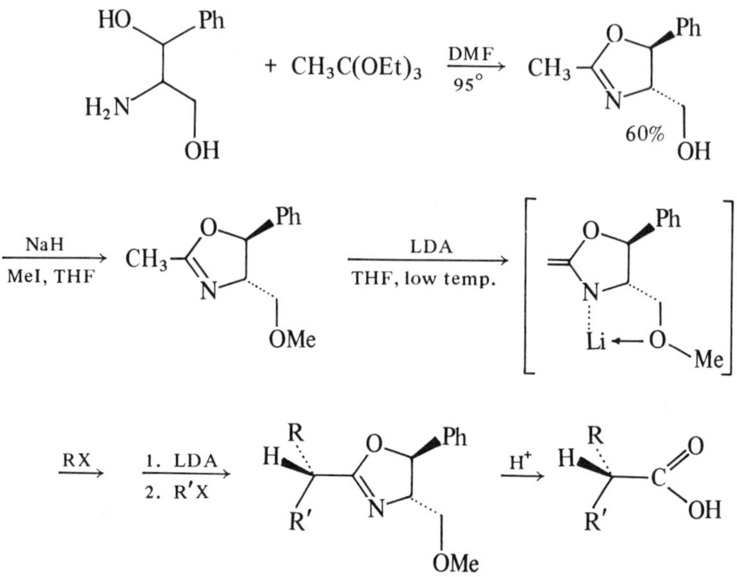

(122)

Furthermore the optically active α-anion will kinetically select from racemic alkyl iodides to give, ultimately, optically active carboxylic acids with β-chiral centers in 30 to 47% enantiomeric excess (Meyers et al., 1976b). Optically active butyrolactones were produced similarly in 64 to 73% enantiomeric excess (Meyers et al., 1975a).

5.10   PHOSPHONATE CARBANIONS

Many of the stabilized carbanions discussed thus far in this chapter have been used to prepare α,β-unsaturated compounds via addition and dehydration. If an α-phosphonate group is present in these carbanions, two advantages are gained. First, the carbanion is further stabilized enough to prevent self-condensation during carbanion preparation, and weaker bases may be used. Second, the elimination of phosphate occurs rapidly under mild conditions without rearrangements. The process, commonly called the Horner-Emmons reaction, has been demonstrated in the preparation of α,β-unsaturated esters, ketones, nitriles, imines, and oxazines (Reviews: Boutagy et al., 1974; Wadsworth, Jr., 1977).

This reaction is similar to the Wittig reaction of alkylidenephosphoranes where triphenylphosphine oxide is eliminated to generate the unsaturation (Section 4.8). When the carbanionic character of a Wittig reagent is delocalized to a heteroatom, the ylid is very sluggish in reactions with ketones and aldehydes, thus the corresponding phosphonates are preferred. The ylids do find use in reactions with aromatic aldehydes but require prolonged boiling in benzene (Cooke et al., 1977). Furthermore, the phosphate salt by-product from the Horner-Emmons reaction may be washed out with water, which is more convenient than the separation of triphenylphosphine oxide from the Wittig reaction.

The preparation of an α,β-unsaturated ester is exemplified in Eq. 123 (Bullivant et al., 1976b). The addition-elimination with aldehydes such as this

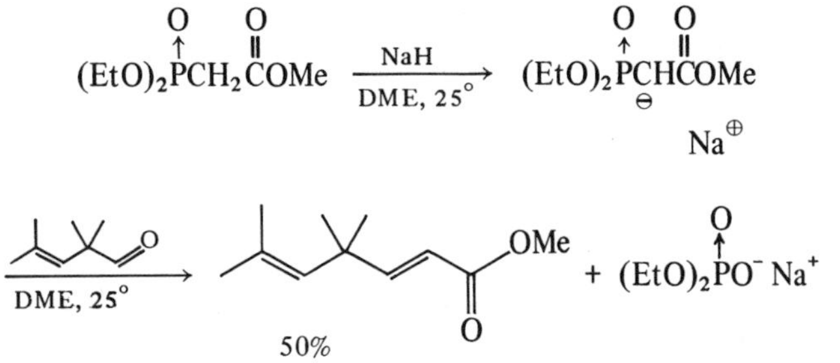

$$(EtO)_2\overset{\overset{O}{\uparrow}}{P}CH_2\overset{\overset{O}{\parallel}}{C}OMe \xrightarrow[\text{DME, 25}^\circ]{\text{NaH}} (EtO)_2\overset{\overset{O}{\uparrow}}{P}\overset{\ominus}{CH}\overset{\overset{O}{\parallel}}{C}OMe$$

$$Na^\oplus$$

$$+ (EtO)_2\overset{\overset{O}{\uparrow}}{P}O^- Na^+ \qquad (123)$$

50%

gives predominantly the trans product, in contrast with the Wittig reaction which gives more of the cis product.

An α,β-unsaturated ketone was required in the synthesis of $(\pm)$-cembrene, therefore it was prepared using a β-ketophosphonate α-anion (Eq. 124) (Dauben et al., 1975). This gave greater than 95% pure trans product. The same process was used in the synthesis of prostaglandins (Eq. 125) (Corey et al., 1969).

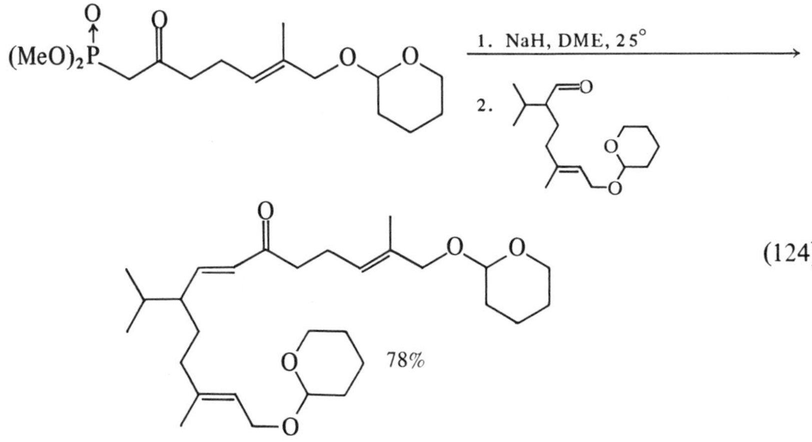

$$(124)$$

78%

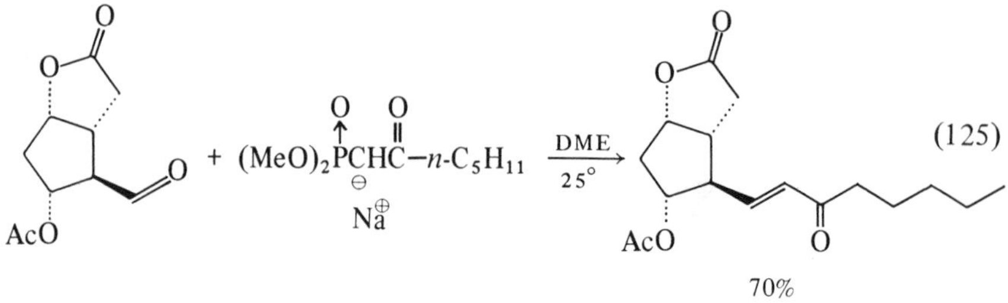

(125)

70%

The preparation of an α,β-unsaturated ester is illustrated in Eq. 126 (Morizur et al., 1975). Reactions with ketones show low stereoselectivity; for example, a 35:65 ratio of Z to E isomers was found in this case.

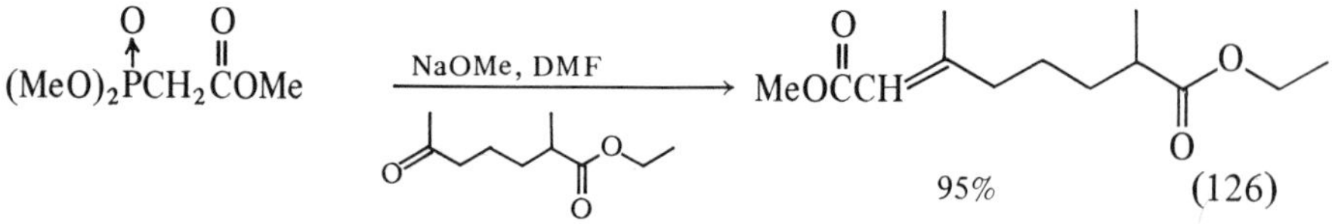

(126)

α,β-Unsaturated nitriles were prepared similarly (Eq. 127) (Raggio et al., 1976; see also Wroble et al., 1976).

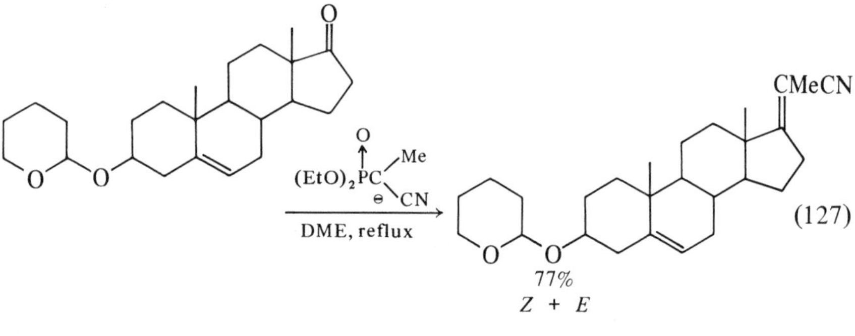

(127)

77%
Z + E

Imine enolate anions bearing a phosphonate group will likewise add to carbonyl compounds and give α,β-unsaturated imines. These are then hydrolyzed to give α,β-unsaturated aldehydes (Eq. 128) (Nagata et al., 1969; see also Nagata et al., 1973).

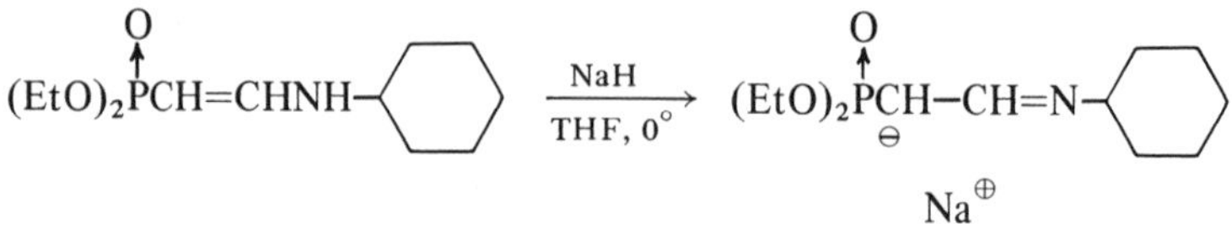

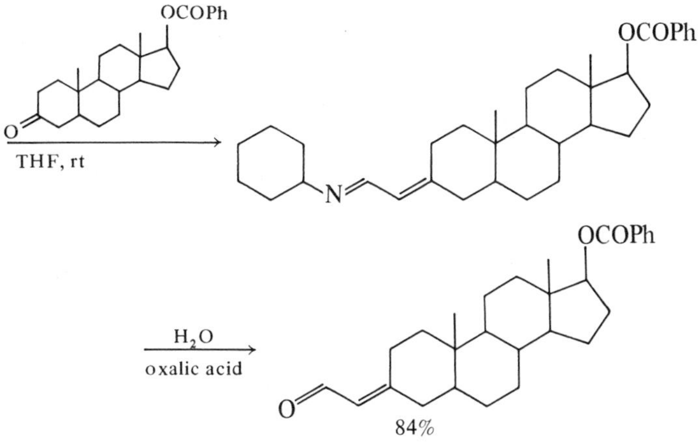

(128)

84%

A similar but more lengthy procedure gives α,β-unsaturated aldehydes from 2-(diethylphosphonomethyl)-4,4,6-trimethyl-5,6-dihydrooxazine. This reagent may also be used to prepare vinyl ketones and α,β-unsaturated acids by reactions analogous to those in Section 5.9.1 (Malone et al., 1974).

The requisite phosphono esters or nitriles are prepared by the Arbuzov reaction of trialkyl phosphites with α-halo esters or nitriles. This reaction is not suitable for making the keto derivatives, so they are instead prepared by acylating alkylphosphonate carbanions. For example, the phosphono ketone used in Eq. 124 was prepared using an ester to acylate lithium dimethyl methylphosphonate (Eq. 129) (Dauben et al., 1975).

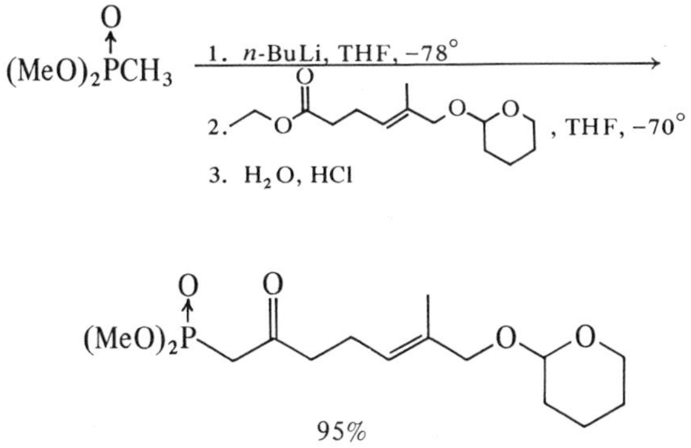

(129)

95%

When an enol lactone is used as the acylating agent, a cyclic enone is formed from in situ addition and elimination (Eq. 130) (Dauben et al., 1977).

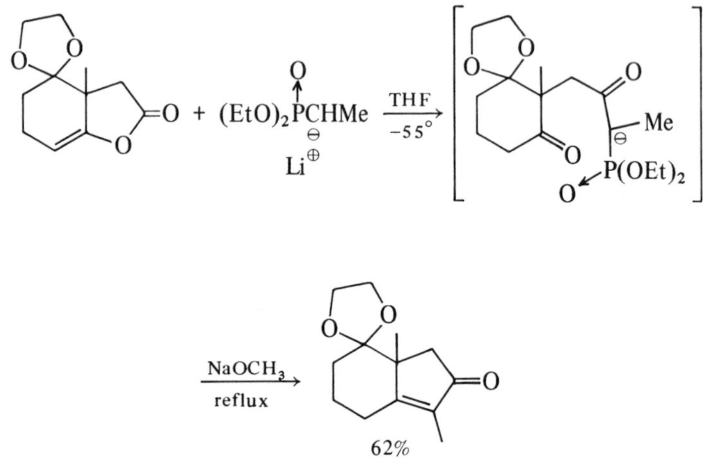

(130)

62%

Acid chlorides may be used as acylating agents if the lithium compound is first converted to the copper compound. The yields are high even with sterically hindered acid chlorides, and only one equivalent of alkylphosphonate is needed as opposed to the acylation with esters where an extra equivalent is consumed generating the enolate anion of the α-phosphono ketone (Savignac et al., 1976).

The α-phosphono enolate anions may be alkylated at the α position to give new reagents which can in turn be used in the Horner-Emmons reaction (Eq. 131) (Wadsworth, Jr., et al., 1961). Diethyl cyanomethylphosphonate can be

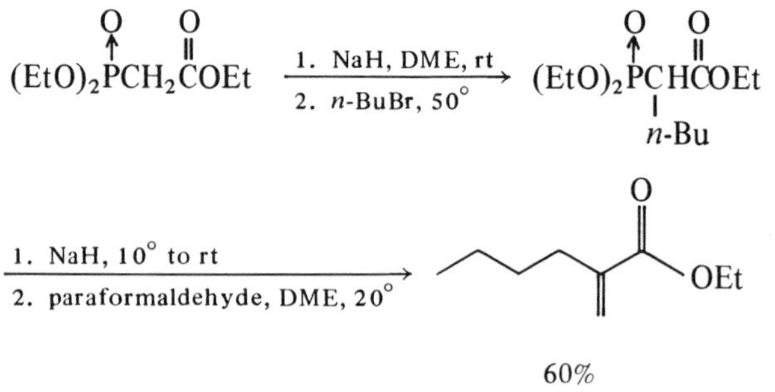

(131)

60%

alkylated in good yields with alkyl halides using aqueous sodium hydroxide under phase transfer conditions (D'Incan et al., 1975). More elaborate ketophosphonates may be prepared by alkylating the α,α'-dianion as shown in Eq. 132 (Grieco et al., 1973).

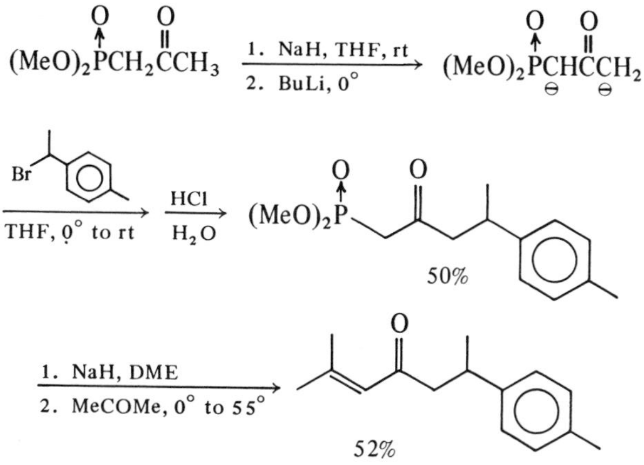

$$(MeO)_2\overset{O}{\overset{\uparrow}{P}}CH_2\overset{O}{\overset{\parallel}{C}}CH_3 \xrightarrow[\text{2. BuLi, 0°}]{\text{1. NaH, THF, rt}} (MeO)_2\overset{O}{\overset{\uparrow}{P}}\underset{\ominus}{C}H\underset{\ominus}{C}CH_2$$

$$\xrightarrow[\text{THF, 0° to rt}]{} \xrightarrow[\text{H}_2\text{O}]{\text{HCl}}$$

$$\xrightarrow[\text{2. MeCOMe, 0° to 55°}]{\text{1. NaH, DME}} \qquad\qquad (132)$$

50%

52%

CHAPTER 6

Carbanions Stabilized by π Conjugation with Two Heteroatoms

AS A GROUP, those carbanions with resonance delocalization to two hetero-atoms are the most stabilized, that is, the conjugate carbon acids are the most acidic of those covered in these chapters. These carbanions may be prepared with weak bases including alkoxides, hydroxides, and even amines. With such delocalized charges, the carbanions are less nucleophilic, so that alkylation often requires more vigorous conditions than those shown in earlier chapters, typically several hours at reflux in alcohol solvents. Where at least one of the delocalizing groups is a keto or nitro group, the oxygen has sufficient nucleophilicity to compete as the reactive site; therefore enol ethers are common by-products in alkylations. C-alkylation is generally favored by less polar solvents and with more covalent, small cations such as lithium which coordinate with the oxygens. The more reactive halides (allylic, propargylic, and benzylic) give more C-alkylation than the saturated ones.

The reactions are often carried out at temperatures and in solvents where the reversible transfer of protons is fast, and therefore dialkylation may compete with monoalkylation. This may be minimized by using an excess of the enolate salt.

Conjugate addition is very favorable with these highly stabilized anions. The reactions are fast, conditions are mild, and there is no competition from 1,2-addition to the α,β-unsaturated substrate.

In most synthetic applications, the double stabilization is a temporary device, and in subsequent reactions a carboxyalkyl or formyl group is removed. Where this removal begins with an ester hydrolysis, particularly in basic solution, a competing process is the retro-Claisen condensation (Eq. 1). Techniques for avoiding this alternative, such as neutral hydrolysis or hydrogenolysis (or for favoring it where desired) are shown in the following sections (Review: Henecka, 1952a).

(1)

Many of these ions have been known for a half century or more, and a wealth of examples of their useful reactions are available.

6.1 MALONIC ACID AND ESTERS

If a new bond is formed to the α position of malonates and then decarboxylation occurs, the overall result is an alkylated acetic acid. The malonate enolate anion is thus synthetically equivalent to the acetate α-anion, but it differs from the acetate α-anion in stability, lower basicity, ease of generation, and high selectivity toward conjugate addition.

6.1.1 Alkylation. The alkylation of diethyl malonate, along with hydrolysis and decarboxylation, is a classic method for the two-carbon homologation of primary and secondary halides (Review: Cope et al., 1957). Sodium ethoxide in ethanol is a good medium for monoalkylation since it is basic enough to produce the enolate anion of diethyl malonate but not much of the anion of the less acidic alkylated product (Eq. 2) (LeGoff et al., 1958). A second alkylation may

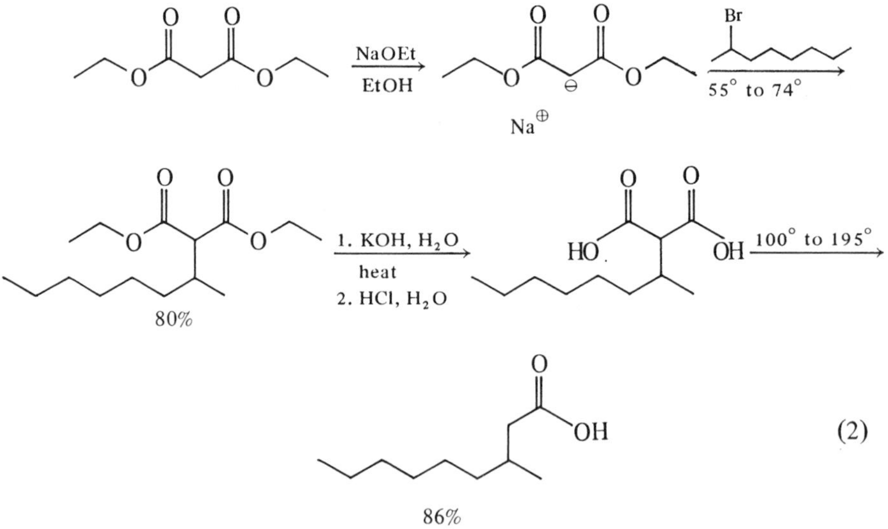

be accomplished if the relatively acidic ethanol solvent is replaced with benzene;

for example see Eq. 3 (Meyer et al., 1977b). Alternatively *tert*-BuOK in *tert*-butyl alcohol may be used for the second alkylation.

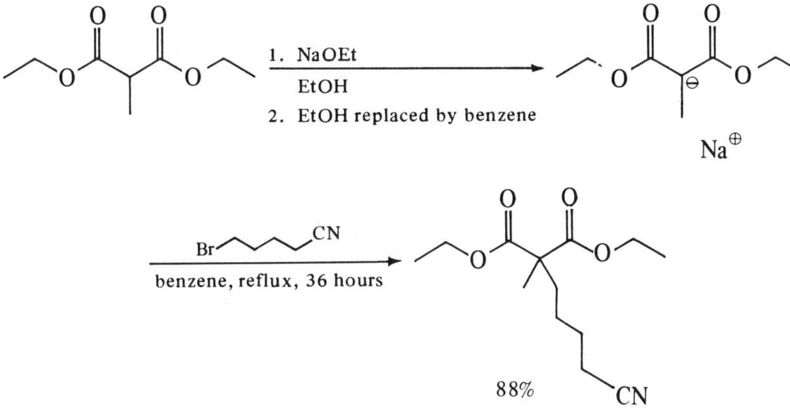

(3)

88%

Decarbalkoxylation may be carried out under neutral conditions to give the monoester rather than the free acid. Heating diethyl diethylmalonate with lithium chloride in DMSO gives ethyl α-ethylbutyrate in 95% yield, and diethyl isopropylmalonate gave ethyl β-methylbutyrate in 90% yield (Krapcho et al., 1978). The hydrolysis step is unnecessary if the dianion from monoethyl malonate is alkylated and then decarboxylated (Eq. 4) (McMurry et al., 1975). In this case all the steps may be conducted in one flask.

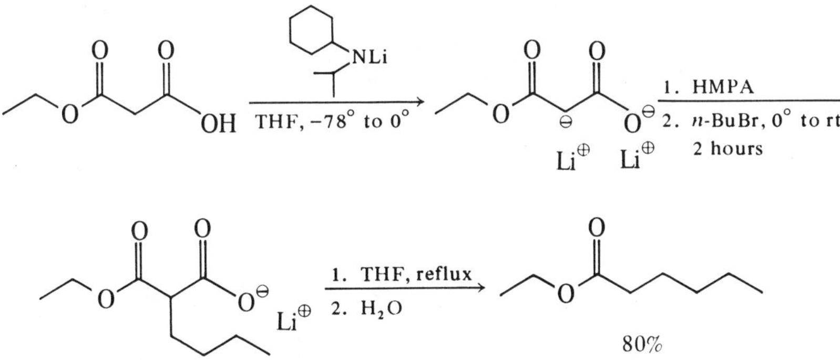

(4)

80%

α-Heterosubstituted malonates may be alkylated and decarboxylated in the usual manner. The acetamido compound leads to α-amino derivatives as shown in Eq. 5 (Dean et al., 1978).

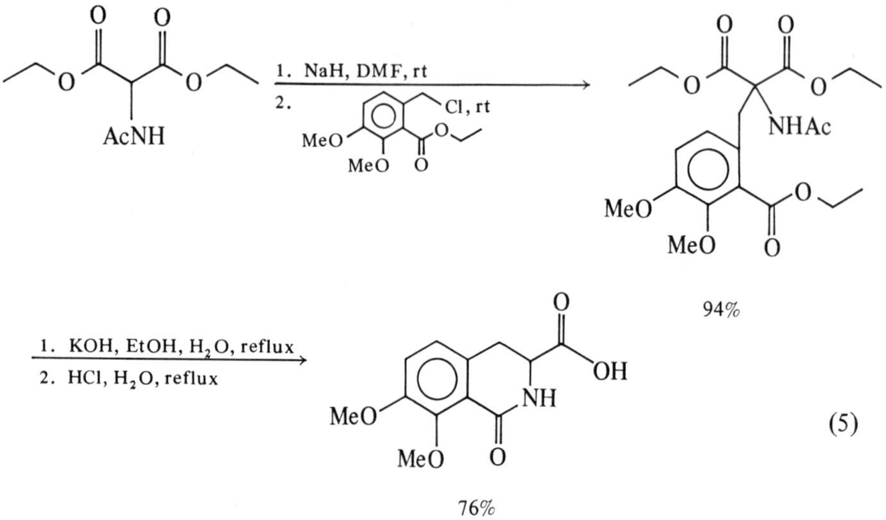

94%

76%

6.1.2 Acylation. Acylation of the enolate anion of malonic esters may be accomplished using acid chlorides. The presence of the β-keto group allows removal of both of the carboxylate groups to give methyl ketones. Base- or acid-catalyzed hydrolysis fails to do this in most cases and instead cleaves the acyl group from the malonate. However one may use acid-catalyzed trans esterification to free the malonic acid (propionic acid becomes esterified) (Bowman, 1950). This process was used by Sih et al. (1973) to prepare 9-ketodecanoic acid in 95% yield from 8-carbomethoxyoctanoyl chloride, for use in the synthesis of prostaglandins.

Starting from an α-substituted malonate, higher ketones are accessible; but the ester groups must be removed by hydrogenolysis since acidolysis fails (Bowman, 1950). For example, Eq. 6 shows the preparation of 8-heptadecanone.

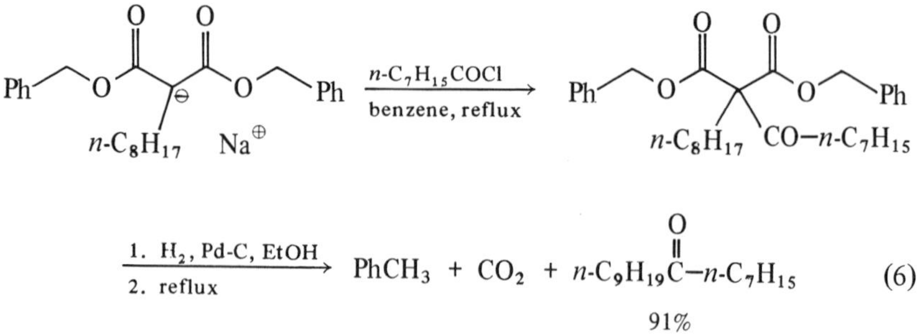

$$\xrightarrow[\text{2. reflux}]{\text{1. } H_2, \text{Pd-C, EtOH}} \quad PhCH_3 + CO_2 + n\text{-}C_9H_{19}\overset{\overset{\displaystyle O}{\displaystyle \|}}{C}\text{-}n\text{-}C_7H_{15} \qquad (6)$$

91%

Acylation of malonic esters can be used to prepare β-ketoesters if one of the ester groups can be removed selectively. This may be done using ethyl *tert*-butyl malonate where the *tert*-butyl group is removed by acid-catalyzed elimination of isobutylene (Eq. 7) (Breslow et al., 1944). Ethyl γ-(1,3-dithian-

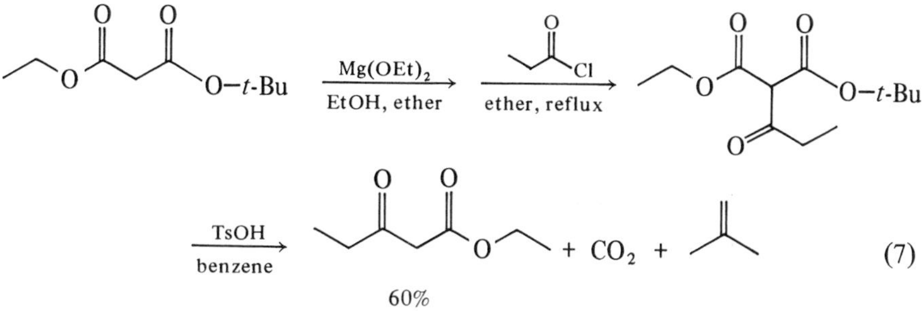

$$\text{(7)}$$

60%

2-yl)acetoacetate was prepared in 44% overall yield by acylation of the magnesium ethoxide salt of ethyl *tert*-butyl malonate with (1,3-dithian-2-yl)acetyl chloride. The acidity of the dithiane ring foiled attempts to prepare the β-ketoester by acylating lithio *tert*-butyl acetate or lithio *tert*-butyl trimethylsilylacetate (Taylor et al., 1978). Thus the lower basicity of the highly resonance-stabilized anions gives better selectivity.

Ethyl propionylacetate could be prepared, starting with diethyl malonate, by boiling the keto diester with water. This gave the β-keto ester contaminated with 17% diethyl malonate from cleavage of the propionyl group (Paine, III et al., 1976).

Once again the use of the dianion of monoethyl malonate simplified the preparation of β-keto esters since the selective decarboxylation occurs directly (Eq. 8) (Banerji et al., 1976). In this case N-acetylimidazole was used as the

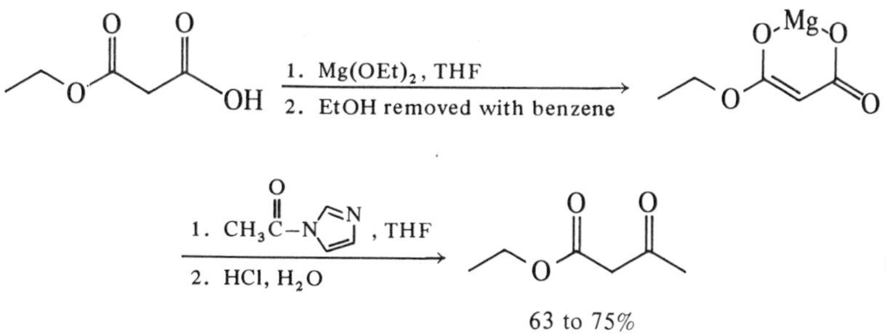

$$\text{(8)}$$

63 to 75%

acylating agent. Of course ethyl acetoacetate is readily available by the Claisen condensation, but this route was used to prepare β-^{13}C-labeled product.

Another alternative is to acylate Meldrum's acid, an acidic relative of malonic ester. The acylated products may simply be heated with an alcohol to give β-keto esters and carbon dioxide in high yield (Eq. 9) (Oikawa et al., 1978):

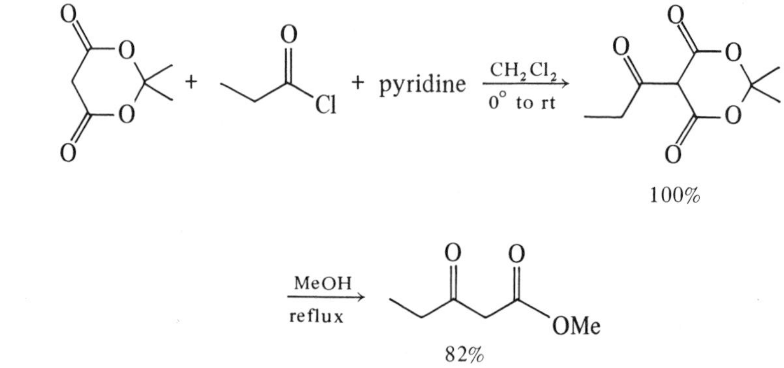

100%

$$\text{(9)}$$

82%

Finally it may be desirable to remove none of the ester groups but rather to make a triol derivative by reduction of all the carbonyl groups (Eq. 10) (Irikawa et al., 1977).

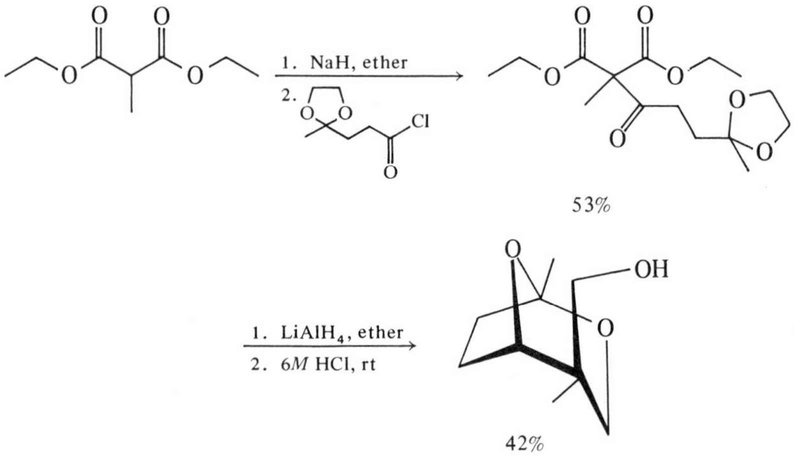

53%

42%

$$\text{(10)}$$

6.1.3 Addition-Elimination. Malonic esters or the free acid will condense with aldehydes when catalyzed by weak bases such as piperidine or pyridine. This may involve addition of the enolate anion of the malonate to the aldehyde or to an intermediate imine formed initially from the aldehyde. The reaction is generally termed the Knoevenagel or Doebner condensation (Review: G. Jones, 1967). When malonic acid is used, the isolated product is the α,β-unsaturated acid from decarboxylation (Eq. 11) (Fray et al., 1961). If the catalyst is

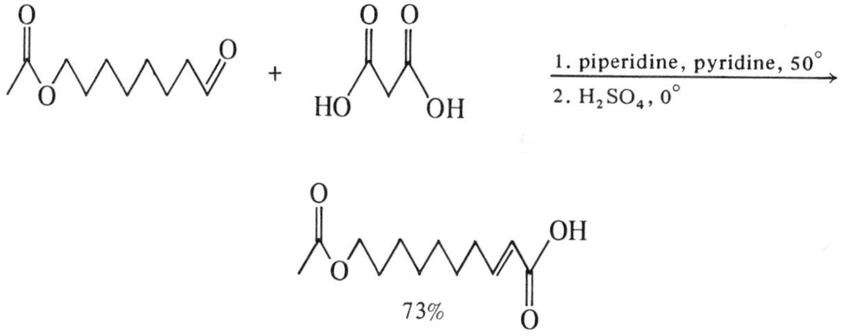

$$\tag{11}$$

73%

triethanolamine, a β,γ-unsaturated acid is formed. When an hydroxyl group is available for lactonization, decarboxylation does not occur, as in the synthesis of coumarin-3-carboxylic acids (Eq. 12) (Fukui et al., 1962).

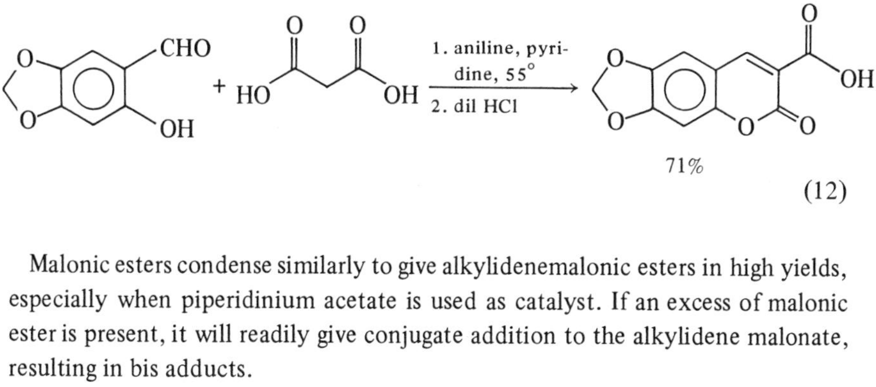

71%

$$\tag{12}$$

Malonic esters condense similarly to give alkylidenemalonic esters in high yields, especially when piperidinium acetate is used as catalyst. If an excess of malonic ester is present, it will readily give conjugate addition to the alkylidene malonate, resulting in bis adducts.

6.1.4 Conjugate Addition. The enolate anion of malonic ester gives exclusively conjugate addition to α,β-unsaturated aldehydes, ketones, nitriles, esters, and nitro compounds (Review: Bergmann et al., 1959). The high-yield addition to cyclohexenone is shown in Eq. 13 (Stetter et al., 1961). The corresponding cyclopentanone was prepared in the same manner in 85% yield. Using the preformed enolate in methanol, conjugate addition to 2-n-pentyl and 2-phenyl-

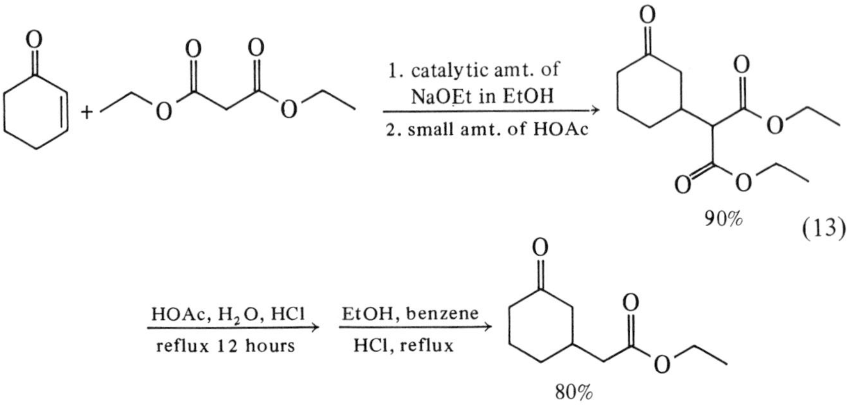

1. catalytic amt. of NaOEt in EtOH
2. small amt. of HOAc

90%   (13)

HOAc, H$_2$O, HCl / reflux 12 hours   EtOH, benzene / HCl, reflux

80%

thio-2-cyclopentenone gave the expected products (Ravid et al., 1974; Montiero, 1977).

With acyclic α,β-unsaturated ketones, the basic conditions can cause the conjugate addition product to undergo a Claisen condensation as shown in Eq. 14 (Focella et al., 1977). α-Alkylmalonic esters will also give conjugate addition.

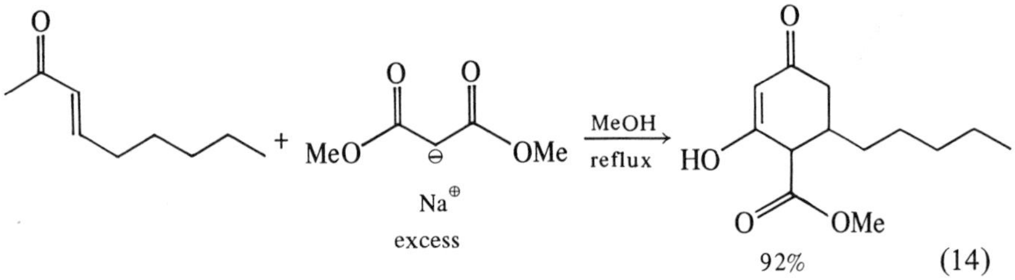

+ MeO—⊖—OMe   Na$^⊕$ excess

MeOH reflux

92%   (14)

6.2 β-KETO ESTERS

The reactions of β-keto esters are very similar to those of diethyl malonate, the difference being that after decarboxylation the products are substituted ketones. Thus the ester formation may be viewed as a temporary activating group (Section 5.1.1). The anion has been generated using NaH, NaOEt, tert-BuOK, K$_2$CO$_3$, or sodium or potassium metal.

6.2.1 Alkylation. There is an abundance of examples of alkylation at the α position of β-keto esters (Review: Henecka, 1976). An acyclic example is given Eq. 15 (E. Brown et al., 1976) where the alkylated keto ester was hydrolyzed and decarboxylated in aqueous hydroxide to give the ketone. The methyl ester of the same β-keto acid was alkylated with bromoacetone to give

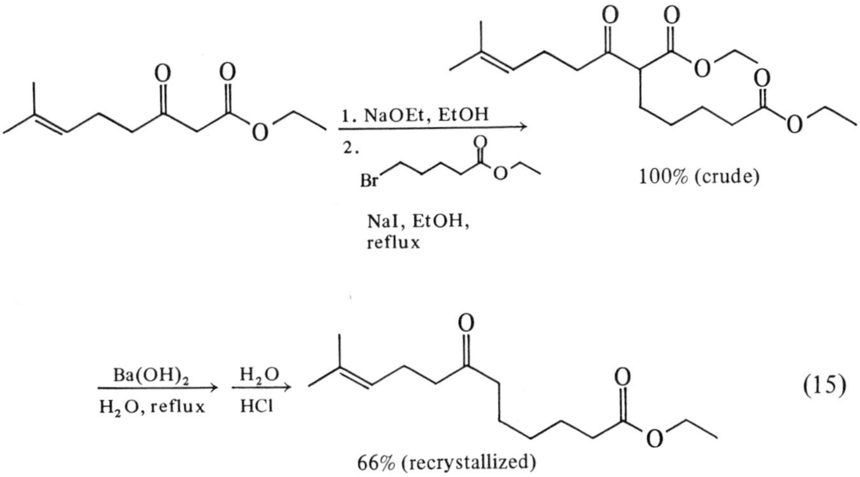

100% (crude)

NaI, EtOH, reflux

66% (recrystallized)

(15)

the diketo ester in 94% yield. This was converted to a cyclopentenone (78%) by aqueous NaOH hydrolysis, decarboxylation, and aldol cyclization (LaLonde et al., 1977).

When active halides such as allyl bromide are used in alcohol solution, there is some loss owing to attack of ethoxide on the bromide. Polar aprotic solvents favor O-alkylation, and nonpolar solvents give low yields due to the heterogeneous conditions, the enolates being insoluble. However benzene can be used as solvent if a quaternary ammonium phase transfer catalyst is used to improve the solid-liquid interaction. Thus the sodium enolate of methyl acetoacetate and allyl bromide gave the α-allylated product in 95% yield and none of the O-allylated product (Durst et al., 1974).

Concurrent enolate generation and alkylation was favorable in methylene chloride using aqueous NaOH as base with a quaternary phase transfer catalyst (Burgstahler et al., 1977a).

A second alkyl group can be attached to the α position; for example ethyl α-methylacetoacetate was alkylated with a primary bromide in benzene-DMF in 91% yield (Burgstahler et al., 1977b). The product was eventually converted to a methyl ketone by NaOH saponification and acid decarboxylation. Benzyl α-methylacetoacetate was alkylated similarly and then converted to the methyl ketone by hydrogenolysis and decarboxylation (Burgstahler et al., 1977a).

A carbomethoxycyclohexanone was alkylated by the usual procedure as shown in Eq. 16 (Grieco et al., 1977; see also Chapter 5, Eq. 7).

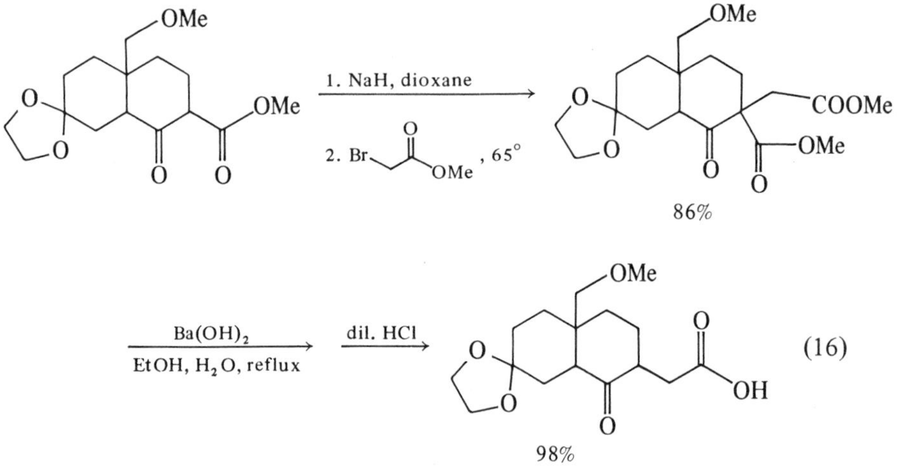

Carboalkoxycyclopentanones may be alkylated (Review: Mayer, 1963) but there is often a considerable amount of O-alkylation in competition with the desired C-alkylation, particularly with unactivated halides. Ide et al. (1976) obtained a high yield of C-alkylation in the example shown in Eq. 17 when they used a propargylic iodide, but the corresponding saturated iodide gave a

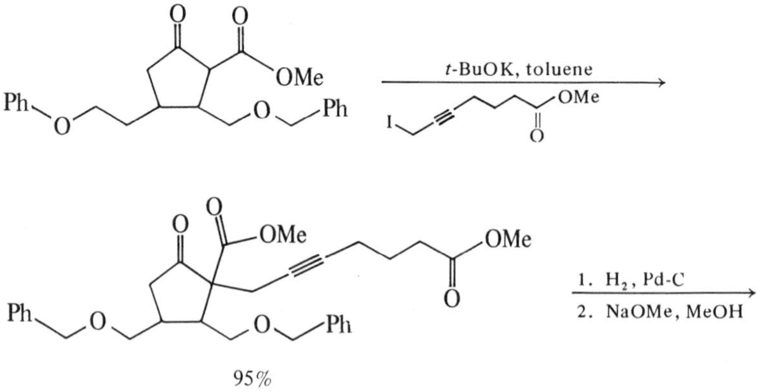

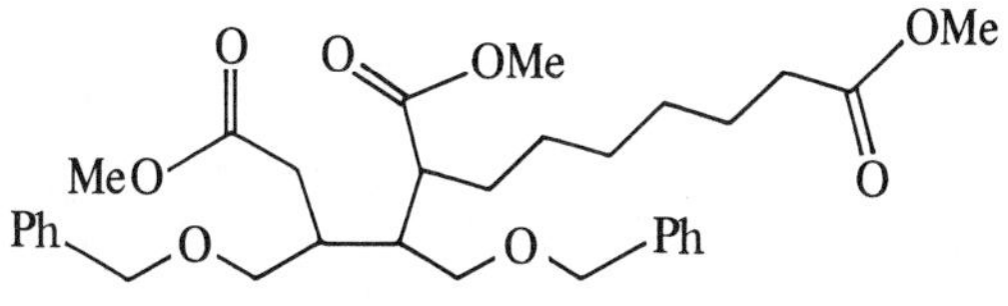

(17)

96%

large amount of *O*-alkylation. Thus the activated halide was superior even though the saturated chain was desired in the final product. Note also that in this case the alkylation was not used to prepare a ketone, but rather an ester using the retro-Claisen reaction. If the potassium enolate of a carbalkoxycyclopentanone is alkylated in DMSO at room temperature, *C*-alkylation is very much favored, even for unactivated primary and secondary halides (except cyclohexyl bromide) (Pond et al., 1967).

Carbalkoxycyclopentanones and some highly hindered acyclic α,α-disubstituted β-keto esters (Renfrow et al., 1948) show a greater tendency to give retro-Claisen condensations. Attempts to saponify these esters with even aqueous hydroxide give cleavage. In these cases hydrolysis of the ester function and decarboxylation to the ketone is preferably done in acid (Eq. 18) (Trost et al., 1978) or neutral (Eq. 19) (Torii et al., 1975) hydrolytic conditions. For other methods see also Greene et al. (1976a) and references cited therein. The

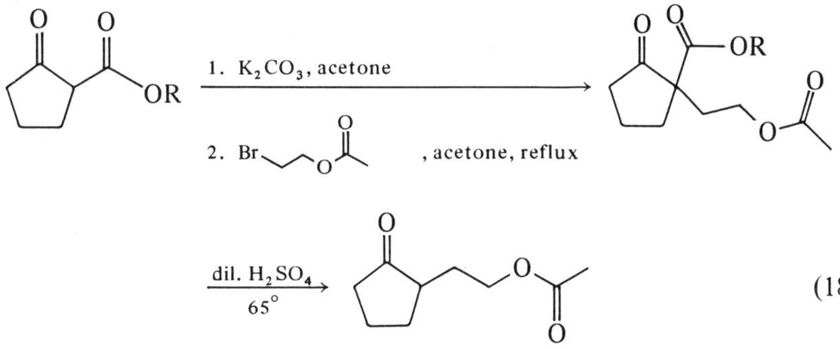

71% overall

(18)

enolate of a vinylogous keto ester (Eq. 20) gave α-, γ-, and *O*-alkylation in various solvents. Alkylation α to the ester was suppressed in DMF (Johnson et al., 1977).

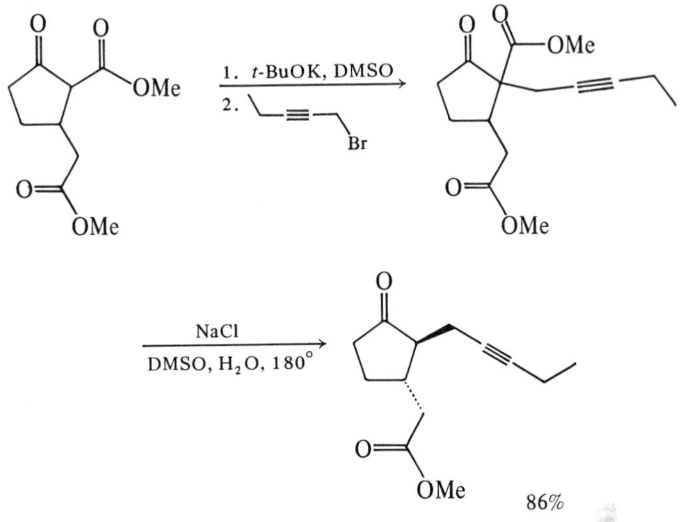

(19)

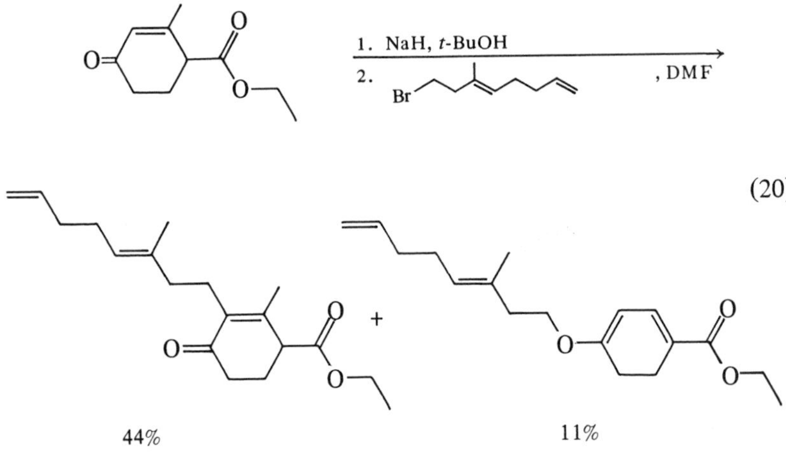

(20)

6.2.2 Acylation. β-Keto esters may be acylated with acid chlorides, affording *O*- and α-acylation. The diketo ester products cannot be hydrolyzed under acid or base conditions, the only products being the acids from the very favorable retro-Claisen condensation. The desired 1,3-diketones can be made however by hydrogenolysis of benzyl esters (Eq. 21) (Baker et al., 1952) or acid-catalyzed elimination of isobutylene from *tert*-butyl esters (Eq. 22) (Treibs et al., 1954).

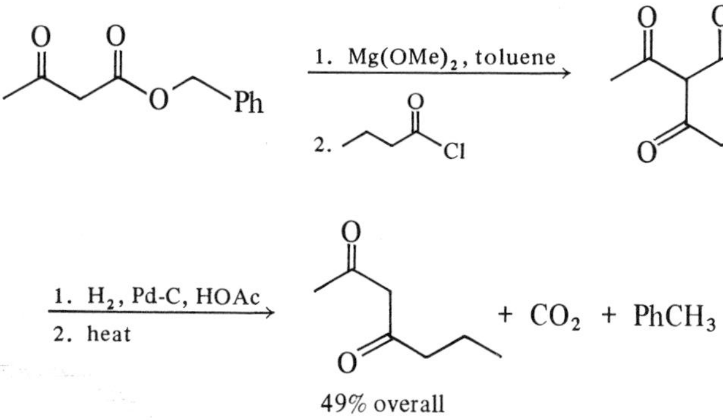

49% overall

(21)

74%   100%

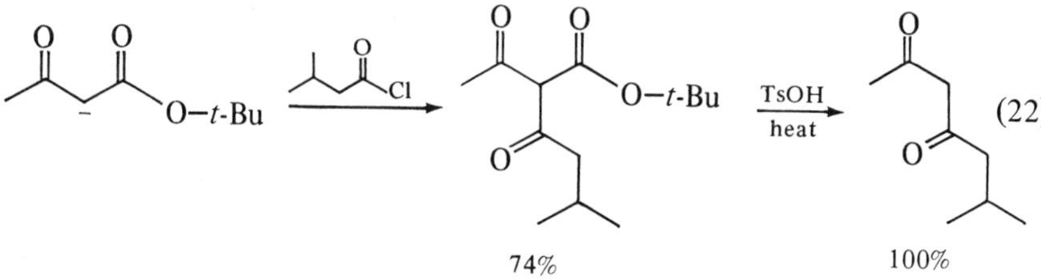

(22)

6.2.3 Addition.

The addition of a keto ester or keto acid enolate to an aldehyde followed by decarboxylation amounts to a directed aldol condensation where the nucleophilic site is determined by the extra stabilization from the carboxylate group. This has been carried out under mild conditions using zinc metal to generate the enolate (Eq. 23) (Mukaiyama et al., 1976). Other β-keto

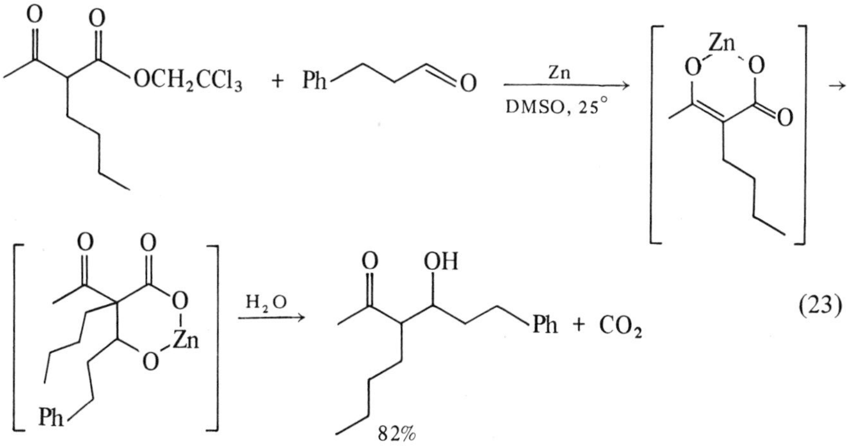

(23)

82%

acid addition reactions are carried out in buffered acidic or neutral solutions (Stiles et al., 1959; Miyano et al., 1972).

A keto ester addition-elimination combined with a conjugate addition of a β-diketone is shown in Eq. 24 (Nominé et al., 1968).

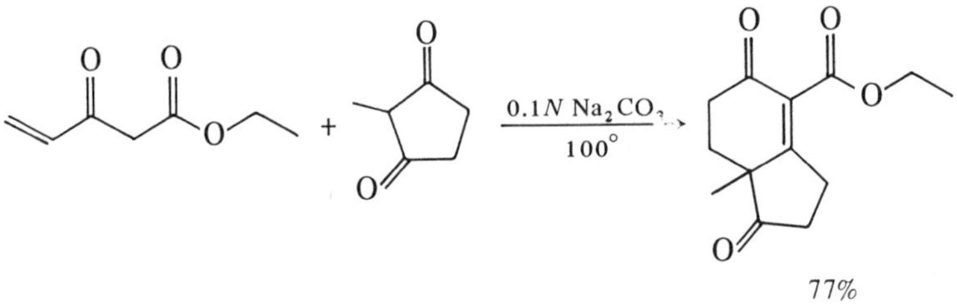

$$(24)$$

77%

6.2.4 Conjugate Addition. β-Keto esters readily undergo 1,4-addition to a variety of activated alkenes (Review: Bergmann et al., 1959); for instance the sodium enolate of ethyl acetoacetate adds to methyl vinyl ketone in 92% yield. Recently this reaction has been used to construct ring C of various diterpenoids (Eq. 25) (Meyer et al., 1977a). Here acid catalyzes removal of the *tert*-butyl ester and aldol condensation to close the ring.

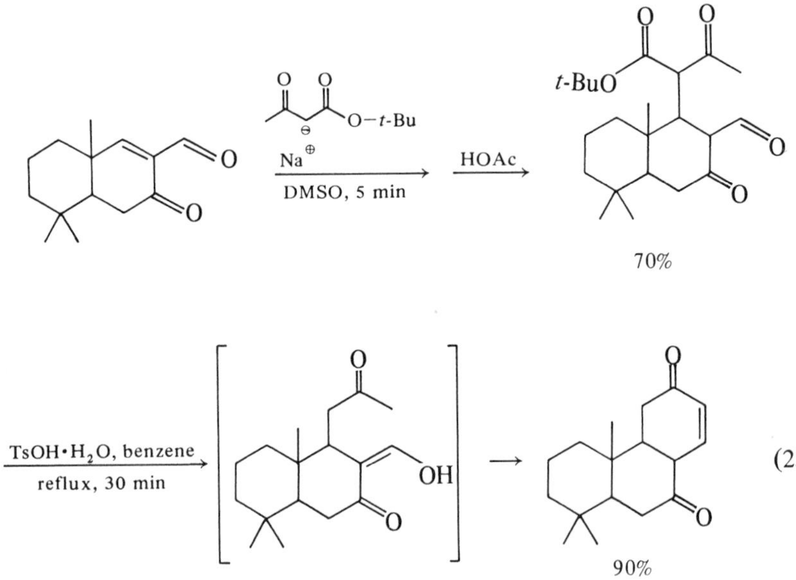

70%

$$(25)$$

90%

6.3 β-KETO ALDEHYDES AND DIKETONES

6.3.1 Alkylation (Review: Stetter, 1976). Acyclic β-diketones may be C-alkylated readily in the solvent acetone in which the alkali enolates have good solubility. For example the sodium enolate of benzoylacetone and methyl iodide in acetone gave the C-methylated product in 88% yield (Weygand et al., 1928). Pentane-2,4-dione was alkylated with methyl, ethyl, and isopropyl iodide to give the 3-alkyl diketones. The enolate was generated in situ with K_2CO_3 (Johnson et al., 1973). 2-Acetylcyclohexanone was methylated in aqueous methanolic sodium hydroxide in 79% yield (Payne, 1961). The β-diketones, which are more acidic than the foregoing classes of compounds, tend to give substantial amounts of dialkylated product and O-alkylated product. High-yield exclusive C-alkylation occurs when the thallium(I) salt of acetylacetone is treated under heterogeneous conditions with excess methyl, ethyl, or isopropyl iodide (Taylor et al., 1970).

In the enolates of cyclic β-diketones the two oxygen atoms are too far apart to both coordinate with the cation, so O-alkylation becomes predominant. Unactivated primary or secondary halides give very low yields of C-alkylation products, but the more active methyl, allylic, or propargylic halides give fair yields of C-alkylation (Eqs. 26 through 30) (Review: Stetter, 1955). Allylation

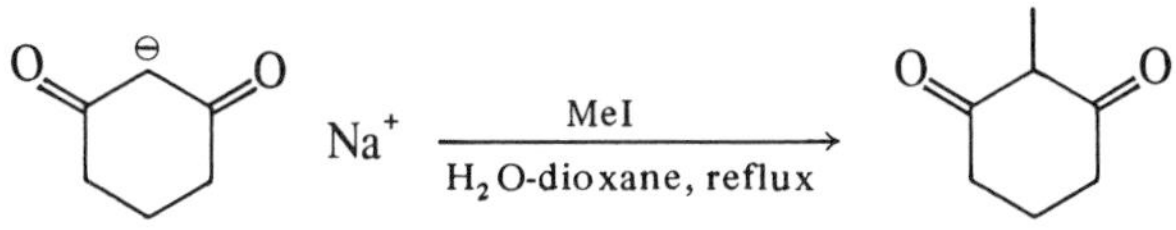

$$(26)$$

54 to 56%
(Meeker et al., 1973)

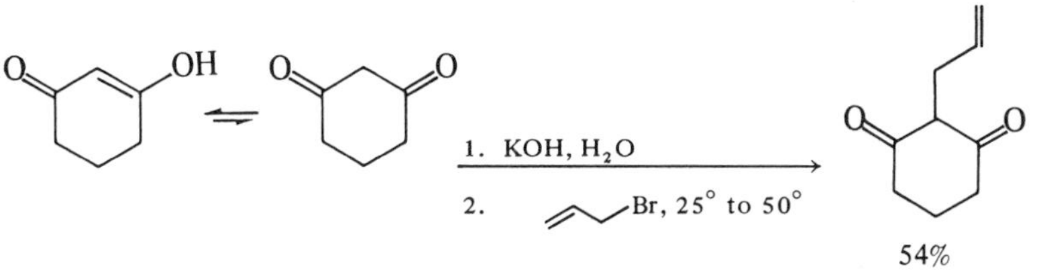

1. KOH, H_2O

2. ⌇⌇⌇Br, 25° to 50°

$$(27)$$

54%

(Bullivant et al., 1976a)

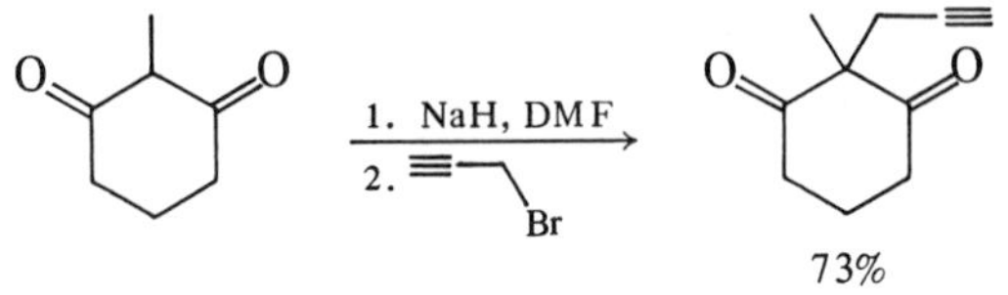

$$(28)$$

73%

(Dauben et al., 1977)

of 2-methylcyclohexane-1,3-dione with methanolic Triton B as base gave methanolysis of the allylated ring (Eq. 29) (Crispin et al., 1970). The ring

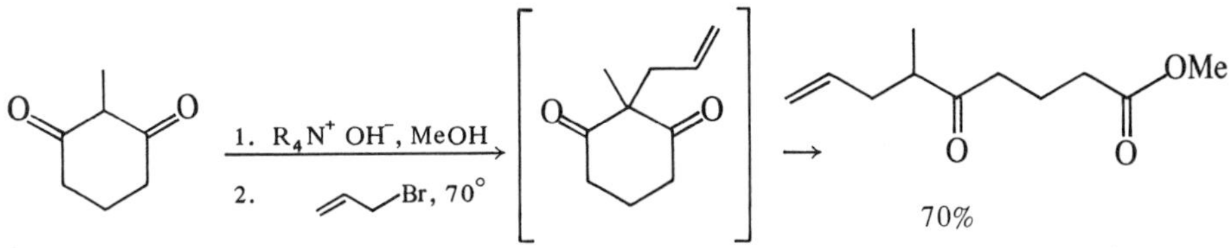

$$(29)$$

opening may be combined with a Wolff-Kishner reduction to give six-carbon homologation of halides to carboxylic acids (Eq. 30) (Mori et al., 1977; Review: Stetter, 1955).

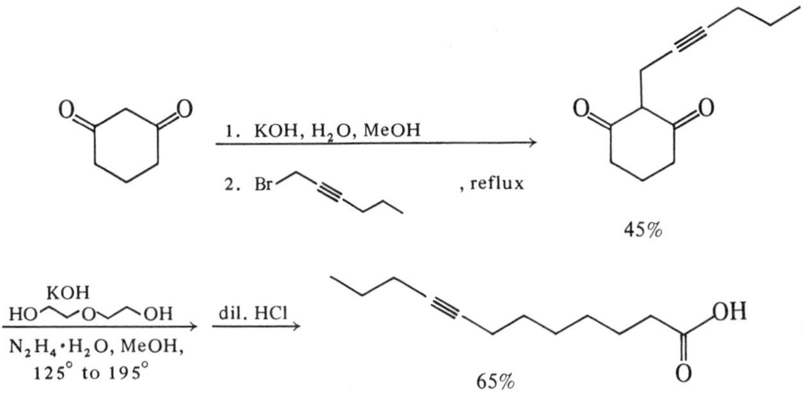

$$(30)$$

2-Methylcyclopentane-1,3-dione could be C-allylated in methanol in 73% yield (Crispin et al., 1970) or C-propargylated in DMSO in 67% yield (Hiraoka et al., 1966) or in aqueous sodium bicarbonate in 85% yield (Lansbury et al., 1978).

The alkylation of 1,3-cyclohexanediones with unactivated halides may be done indirectly using a carbanion in which the oxygens are protected as the enol ethers. The original procedure (Birch et al., 1951) involved the potassium reagent, but higher yields are obtained with lithium (Eq. 31) (Piers et al., 1977).

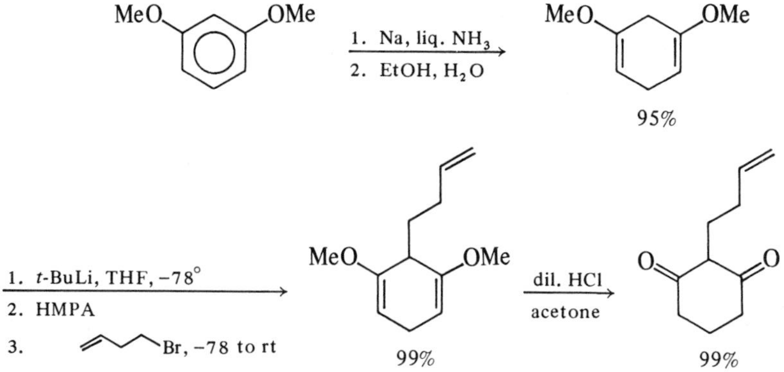

β-Keto aldehydes show the same C- versus O-alkylation characteristics as discussed for the β-diketones. Thus 2-formyl-6-methylcyclohexanone was converted to the enolate anion with potassium carbonate in acetone and alkylated with methyl iodide giving largely C-alkylation. The same process with isopropyl iodide gave an 80% yield of the enol ether (W. S. Johnson et al., 1947).

6.3.2 Addition. Cyclohexane-1,3-dione condenses readily with aldehydes, eliminating water and giving the alkylidene bis products without any catalysts. It will also self-condense as shown in Eq. 32 (Review: Stetter, 1955).

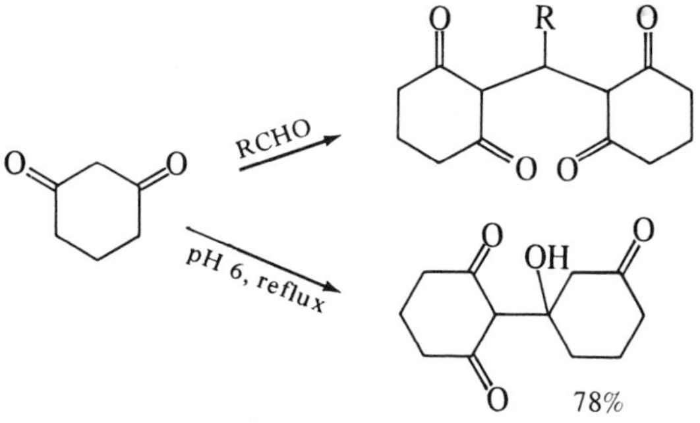

The intermediate alkoxide from the addition of oxalyl ketones to aldehydes is intercepted by the ester function to give lactones which may be thermolyzed to afford α,β-unsaturated ketones (Eq. 33). This is in effect a directed aldol condensation (Ksander et al., 1976).

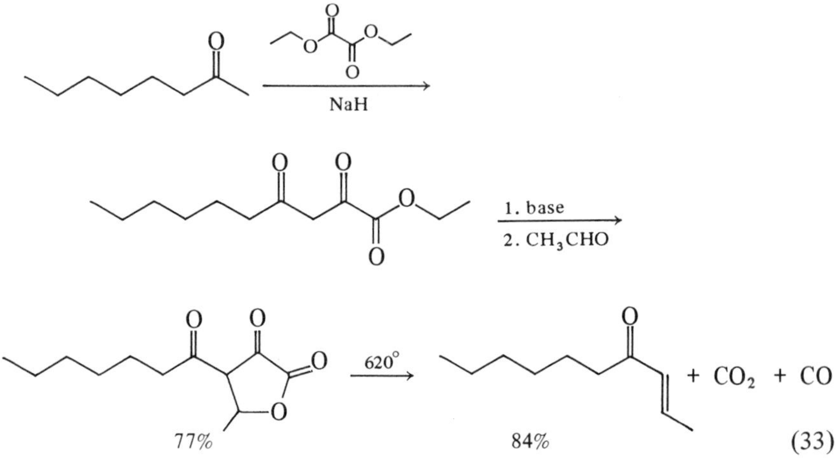

(33)

6.3.3 Conjugate Addition. Cyclic 1,3-diketones readily undergo conjugate addition to the more reactive α,β-unsaturated substrates. An addition to methyl vinyl ketone is illustrated in Eq. 34 where the triketone product is cyclized with a catalytic amount of pyrrolidine (Ramachandran et al., 1973). A similar addition of 2-methylcyclohexane- and cyclopentane-1,3-dione to a more complex

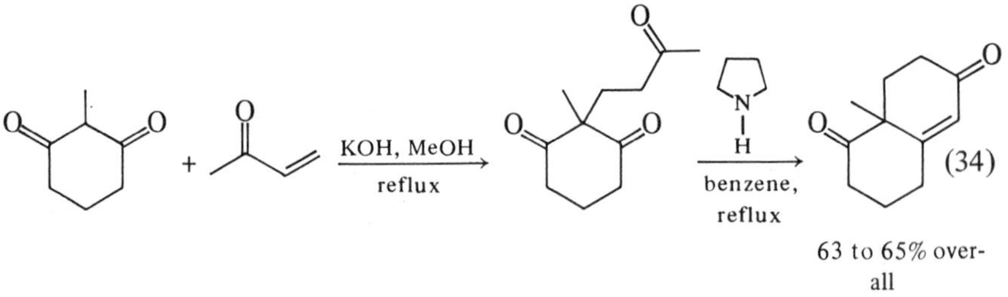

(34)

63 to 65% over-
all

α,β-unsaturated ketone was used in the synthesis of estrone derivatives and related stereoisomers (Douglas et al., 1963).

The enolate anion of 2-methylcyclopentane-1,3-dione adds readily to acrylo-nitrile (Eq. 35) (Brown et al., 1966) and to methyl α-chloroacrylate (Chapter 2, Eq. 21) (Daniewski, 1975). However the addition to ethyl acrylate requires

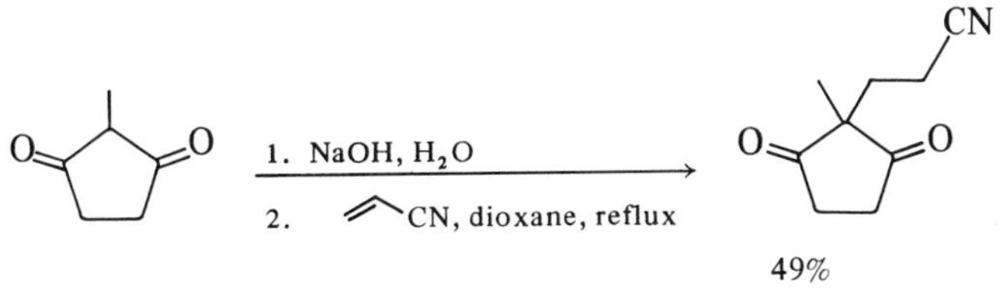

$$(35)$$

49%

refluxing DMF solvent and gives only 32% yield (Kessar et al., 1964).

β-Keto aldehydes give conjugate addition in very weakly basic solution. Ketones are often converted temporarily to β-keto aldehydes to take advantage of these mild conditions to selectively generate the one desired enolate anion in the presence of other ketones such as the acceptor methyl vinyl ketone. Thus 7-methoxy-1-tetralone was converted to a conjugate adduct via temporary formylation (Eq. 36) (Turner et al., 1956). Attempted direct condensation of the tetralone with methyl vinyl ketone using various bases failed. As a further example, methyl acrylate was the acceptor in Eq. 37 (Feutrill et al., 1976). This use of β-keto aldehydes is parallel to the use of β-keto esters, where

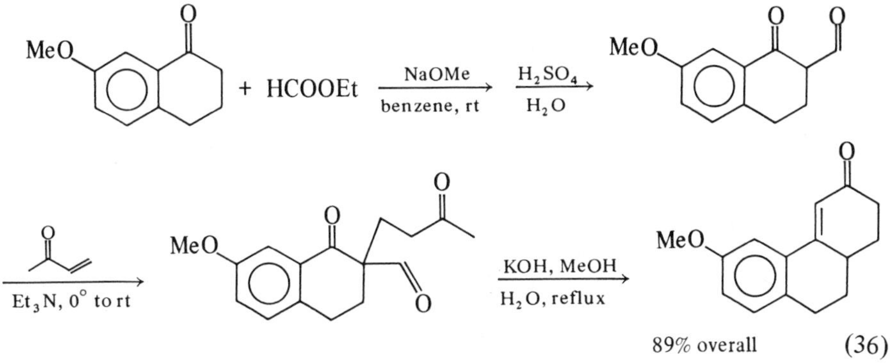

89% overall   (36)

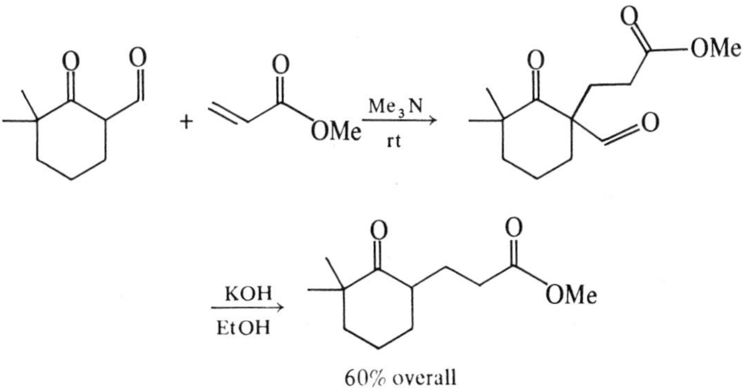

60% overall   (37)

temporary activation of a ketone for conjugate addition is likewise the object (Section 6.2.4).

Rather than remove the formyl group, de Groot et al. (1976) retained it for spirocyclization using pyrrolidine acetate as an aldol catalyst (Eq. 38).

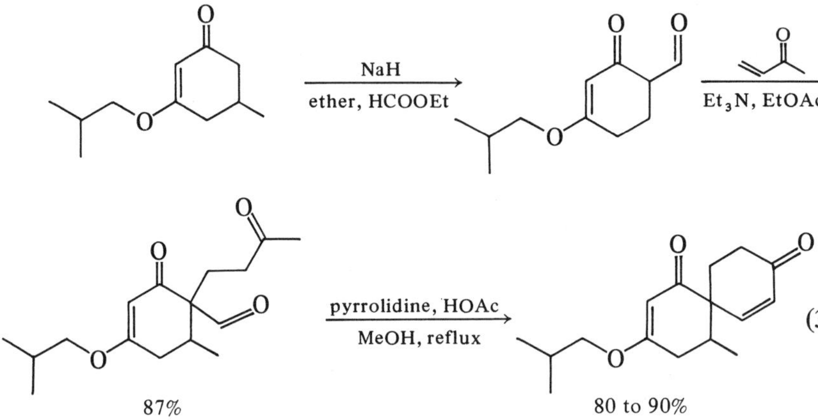

(38)

87%   80 to 90%

6.4 ACTIVATED NITRILES

Nitriles activated by another electron-delocalizing group such as an ester, ketone, amide, or a second nitrile group are readily converted to the enolate anions and then alkylated or condensed with ketones or aldehydes. The nitrile compounds are more acidic than malonic esters and tend to give more dialkylation as a by-product in monoalkylation reactions. On the other hand, if dialkylation is desired the second group enters more readily here than with malonic ester (Review: Cope et al., 1957). Dialkylation of ethyl cyanoacetate is used to prepare nitriles (Eq. 39) (Campaigne et al., 1978). The best yields of dialkylation of malononitrile are obtained in DMSO using NaH as base and

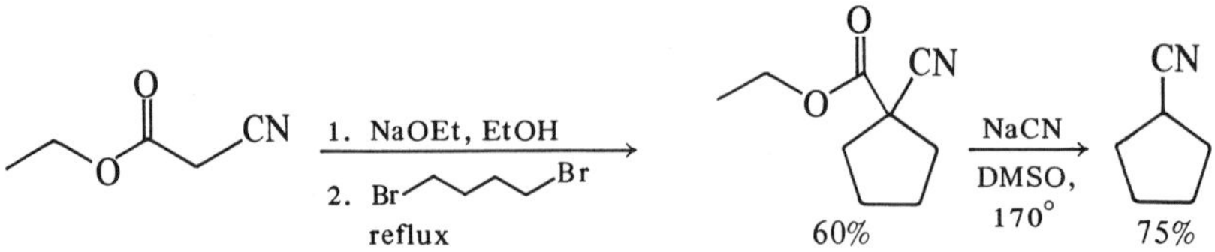

(39)

primary and secondary halides as alkylating agents. In alkoxide solutions yields are lower, in part owing to addition of alkoxide to a nitrile group (Bloomfield, 1961).

β-Keto nitriles show the usual complication of enol ether formation. Alkyl iodides are again best for C-alkylation. For example the sodium enolate of 2-cyanocyclohexanone with ethyl iodide gave a 53% yield of 2-ethyl-2-cyano-cyclohexanone plus 28% of the O-alkylation product (v. Auwers, 1928).

Cyanoacetic acid, ester, or amide will condense with carbonyl compounds (Knoevenagel reaction, Review: Jones, 1967) in good yields using catalytic amounts of weak base (Eqs. 40 through 42). The condensation

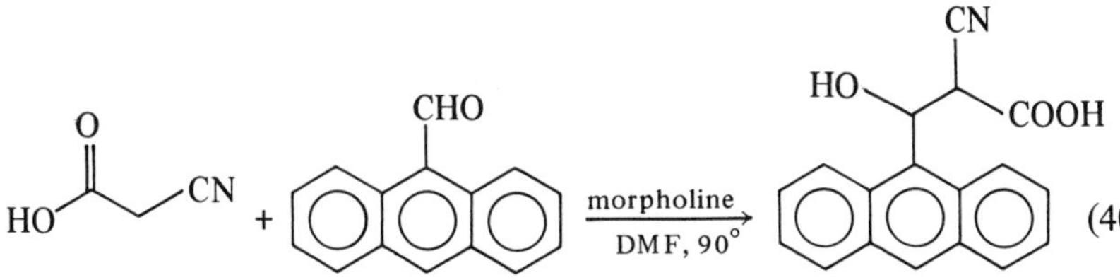

(Hann, 1974)

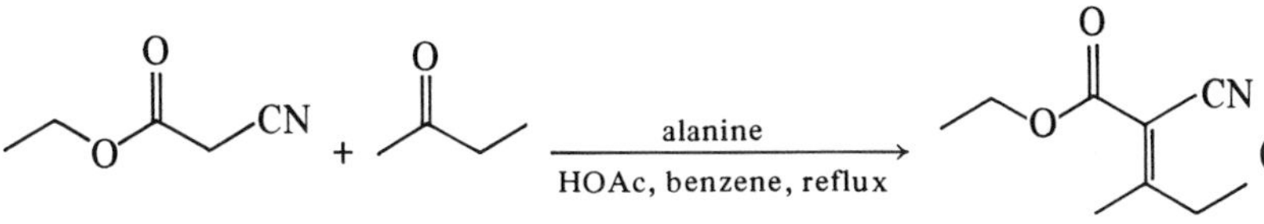

(Prout et al., 1963)

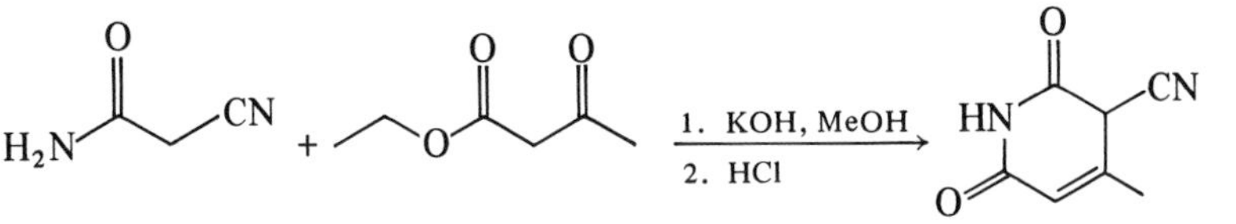

(Leete et al., 1978)

with acetaldehyde is poor so an alternative route to the alkylidene product was devised using sulfenylation, alkylation, and elimination (Eq. 43) (Bryson et al., 1976).

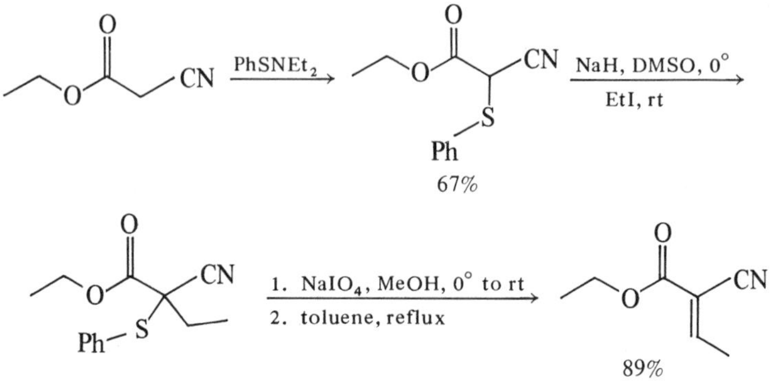

(43)

Malononitrile condenses readily with ketones and aromatic aldehydes under mild base catalysis to give the alkylidene malononitriles in good yield. Hindered or diaryl ketones often give poor yields, but the corresponding ketimates may be used instead (Eq. 44) (Campaigne et al., 1974).

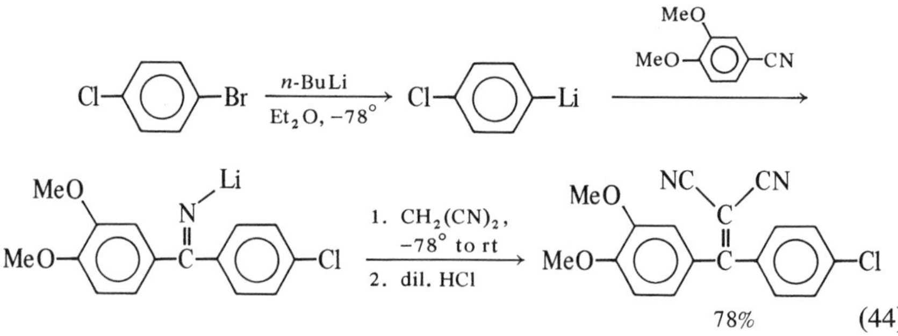

(44)

6.5 α-NITRO CARBANIONS (NITRONATE ANIONS)

Nitro compounds are sufficiently acidic so that potassium hydroxide will generate the α-anion in which the charge is largely delocalized to the two oxygen atoms. Alkylation and acylation occur mostly on oxygen, while addition reactions occur at carbon (Reviews: Erashko et al., 1966; Smith, 1966; Baer et al., 1970).

In many synthetic applications, the nitro group is ultimately converted to a keto oxygen, and thus the nitronate anion serves as the equivalent of an acyl carbanion. This conversion occurs when a nitronate ion is O-protonated and hydrolyzed with acid, and it is termed the Nef reaction. Low molecular weight cases may be done in aqueous solution, but higher ones give poor yields owing to the insolubility of the intermediates. Treatment with methanolic sodium methoxide and then methanolic sulfuric acid gives dimethyl ketals which may be hydrolyzed to the ketones in high overall yield (Jacobson, 1974). The simplest procedure is to adsorb the nitro compound on dry basic silica gel and then elute with ether (Eq. 45) (Keinan et al., 1977). Other methods include mild

$$
\underset{\text{NO}_2}{\bigcirc} \xrightarrow[\text{48 hours, rt}]{\text{basic silica gel}} \underset{99\%}{\bigcirc\!=\!\text{O}} \tag{45}
$$

oxidation or reduction conditions for preparation of the carbonyl compound (Bartlett et al., 1977; Williams et al., 1978; and references cited by Keinan, 1977).

When the nitro compound itself is desired from a nitronate anion, dilute, cold weak acids such as acetic acid should be used in order to avoid the Nef reaction. This is necessary in the alkylation of dianions and in the addition to carbonyl compounds because much of the product exists initially as the nitronate anion.

6.5.1 Alkylation. Alkylation of nitronate anions with active as well as simple alkyl halides occurs at oxygen to give an unstable nitronate ester intermediate which decomposes to an oxime and an aldehyde or ketone (Eq. 46) (Lieberman, 1955). The exceptional p-nitrobenzyl halides give mostly C-alkylation.

$$
\underset{\text{NO}_2}{\bigwedge} + \text{KOH} \underset{\text{EtOH}}{\rightleftharpoons} \left[\begin{array}{cc} O\!\!\diagdown\!\!\underset{N^{\oplus}}{\diagup}\!\!O^{\ominus} & O^{\ominus}\diagdown\!\!\underset{N^{\oplus}}{\diagup}\!\!O^{\ominus} \end{array} \right] \xrightarrow[\text{EtOH, reflux}]{n\text{-}C_{11}H_{23}Br}
$$

$$
\underset{O}{\overset{O\diagdown N\diagup O-n\text{-}C_{11}H_{23}}{\bigwedge}} \longrightarrow \underset{85\%}{\overset{N\diagup OH}{\bigwedge}} + n\text{-}C_{10}H_{21}CHO \tag{46}
$$

In the case of primary nitroalkanes, *C*-alkylation is possible if the more nucleophilic dianion is produced with a stronger base (Eq. 47) (Seebach et al., 1977). If a secondary nitro compound is treated similarly, an α,β-dianion is formed and alkylation occurs on the β-position.

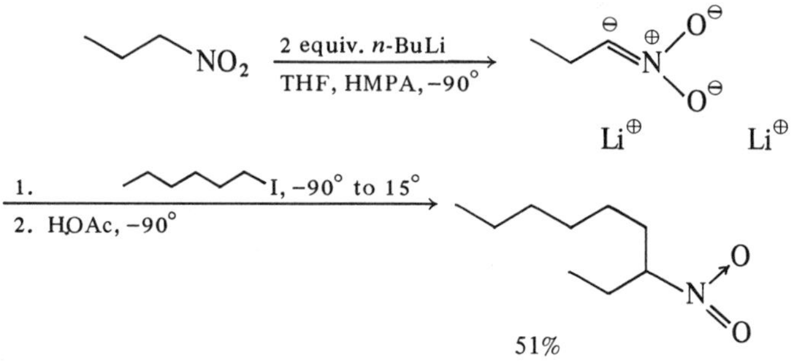

$$\tag{47}$$

The further resonance-delocalized anion from methyl nitroacetate gives *C*-alkylation in yields of 10 to 88% with primary iodides and bromides (Kaji et al., 1973; Zen et al., 1977).

6.5.2 Acylation. As mentioned earlier, acylation normally occurs at oxygen. An unusual example of *C*-acylation of nitromethane with isatoic anhydride (Eq. 48) (Gosteli, 1977) was used in a synthesis of indigo.

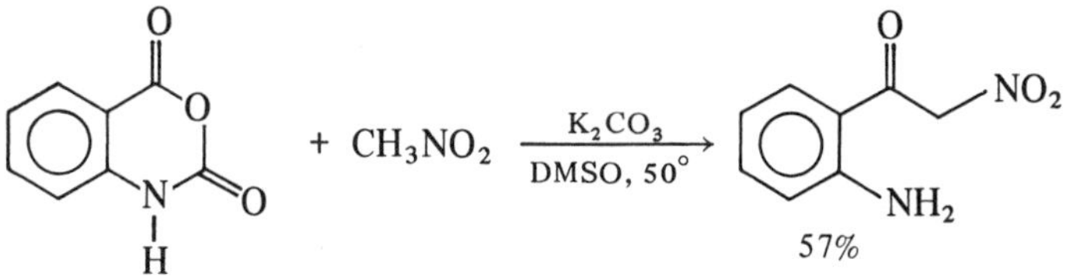

$$\tag{48}$$

The dianions may be *C*-acylated with esters in yields of 55 to 75% (Seebach et al., 1976).

6.5.3 Addition. Nitro compounds with α-hydrogens will undergo addition to ketones or aldehydes in the presence of base. This is analogous to a cross aldol condensation and is known as the Henry reaction. For example see Eq. 49 (Dauben et al., 1963). This reaction is reversible, and if distillation is attempted

$$\text{NaOEt} + \text{CH}_3\text{NO}_2 + \underset{}{\overset{O}{\bigcirc}} \xrightarrow[45°]{\text{EtOH}} \left[\overset{\text{HO} \quad \ominus \quad \text{Na}^{\oplus}}{\underset{}{\bigcirc}} {}_{\text{NO}_2} \right] \xrightarrow{\text{HOAc}} \overset{\text{HO}}{\underset{}{\bigcirc}} {}^{\text{NO}_2} \qquad (49)$$

78 to 84%

in the presence of a trace of base, starting materials will be recovered. Nitromethane and some primary nitro compounds can add to two molecules of aldehydes giving nitrodiols, depending on stoichiometry and pH.

6.5.4 Conjugate Addition. Conjugate addition of nitronate anions gives good yields with a wide variety of α,β-unsaturated carbonyl and nitrile compounds as exemplified in Eqs. 50 through 52. Conjugate addition to an α-bromo-α,β-unsaturated ketone leads to cyclopropane ring closure via a second nitronate anion (Eq. 53) (Shapiro et al., 1977).

$$\overset{}{\diagup\!\diagdown\!\diagup}\text{NO}_2 + \overset{O}{\underset{\text{OMe}}{\diagdown\!\diagup}} \xrightarrow{\text{PhCH}_2\overset{\oplus}{\text{NMe}_3} \overset{\ominus}{\text{OH}}} \underset{\underset{85\%}{\text{NO}_2}}{\overset{O}{\diagup\!\diagdown\!\diagup\!\diagdown\!\diagup\!\text{OMe}}}$$

$$\xrightarrow[\text{2. H}_2\text{SO}_4, \text{H}_2\text{O}, 5°]{\text{1. NaOH, H}_2\text{O}} \underset{\underset{65\%}{O}}{\overset{O}{\diagup\!\diagdown\!\diagup\!\diagdown\!\diagup\!\text{OMe}}} \qquad (50)$$

(Jones et al., 1968)

$$\text{CH}_3\text{NO}_2 + \overset{O}{\diagup\!\diagdown\!\diagup\!\diagdown\!\diagup\!\diagdown} \xrightarrow[\text{2. HCl, H}_2\text{O}]{\text{1. NaOMe, MeOH, 25°}}$$

$$\underset{42\%}{\overset{O}{\diagup\!\diagdown\!\diagup\!\diagdown\!\diagup\!\diagdown\!\diagup}}\text{NO}_2 \quad \xrightarrow[\text{2. Triton B,} \overset{O}{\underset{\text{OMe}}{\diagdown\!\diagup}} \text{, dioxane}]{\text{1. HO}\diagup\!\diagdown\text{OH , TsOH, benzene}}$$

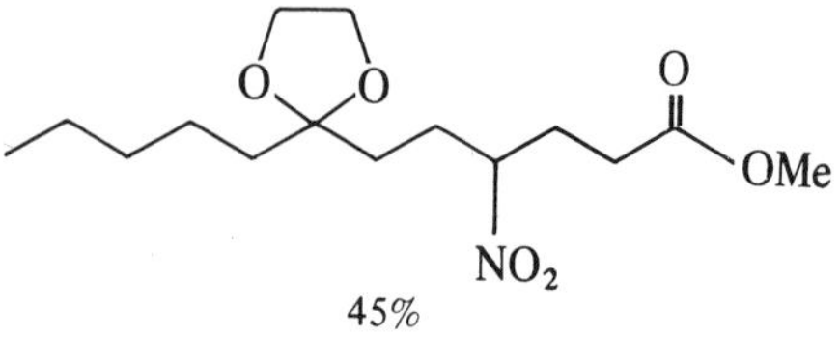

45%

(Zoretic et al., 1977b)

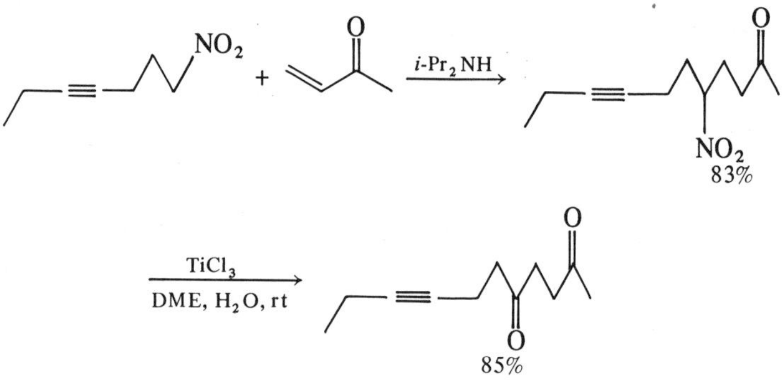

(McMurry et al., 1973)

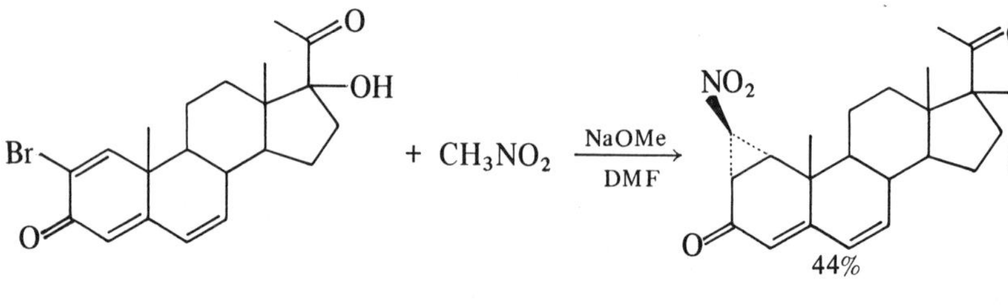

A further resonance-stabilized nitroacetate α-anion also gave conjugate addition (Eq. 54) (Richards et al., 1977).

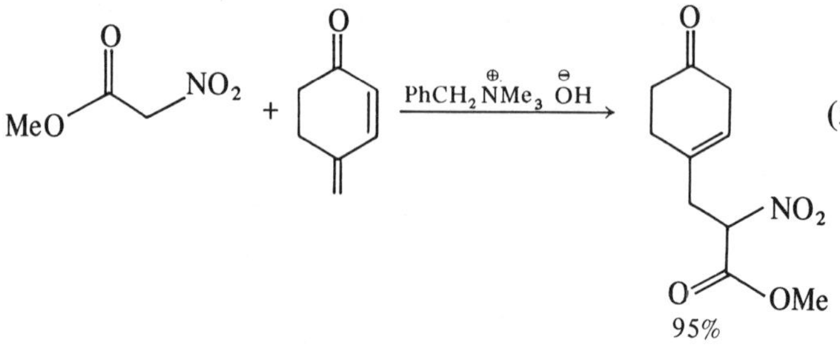

REFERENCES

Acker, R-D., *Tetrahedron Lett.*, 3407 (1977).

Akiyama, S., and J. Hooz, *Tetrahedron Lett.*, 4115 (1973).

Albarella, J. P., *J. Org. Chem.*, **42**, 2009 (1977).

Alexakis, A., G. Cahiez, J. F. Normant, and J. Villieras, *Bull. Soc. Chim. Fr.*, 693 (1977).

Anderson, R. J., and C. A. Henrick, *J. Am. Chem. Soc.*, **97**, 4327 (1975).

Anderson, W. K., E. J. LaVoie, and G. E. Lee, *J. Org. Chem.*, **42**, 1045 (1977).

Angelo, B., *Bull. Soc. Chim. Fr.*, 1848 (1970).

Angelo, B., *C.R. Acad. Sci. Paris, Ser. C*, **276**, 293 (1973).

Arco, M. J., M. H. Trammell, and J. D. White, *J. Org. Chem.*, **41**, 2075 (1976).

Armour, A. G., G. Büchi, A. Eschenmoser, and A. Storni, *Helv. Chim. Acta*, **42**, 2233 (1959).

Auerbach, J., T. Ipaktchi, and S. Weinreb, *Tetrahedron Lett.*, 4561 (1973).

Axelrod, E. H., G. M. Milne, and E. E. van Tamelen, *J. Am. Chem. Soc.*, **92**, 2139 (1970).

Babler, J. H., *Tetrahedron Lett.*, 2045 (1975).

Bachi, M. D., J. W. Epstein, Y. Herzberg-Minzly, and J. H. E. Loewenthal, *J. Org. Chem.*, **34**, 126 (1969).

Baer, H. H., and L. Urbas, *The Chemistry of the Nitro and Nitroso Groups*, H. Feuer, Ed., Part 2, Wiley-Interscience, New York, 1970, p. 75.

Baer, T. A., and R. L. Carney, *Tetrahedron Lett.*, 4697 (1976).

Bailey, E. J., D. H. R. Barton, J. Elks, and J. F. Templeton, *J. Chem. Soc.*, 1578 (1962).

Bainvel, J., B. Wojtkowiak, and R. Romanet, *Bull. Soc. Chim. Fr.*, 978 (1963).

Baker, B. R., R. E. Schaub, M. V. Querry, and J. H. Williams, *J. Org. Chem.*, **17**, 77 (1952).

Baker, R., *Chem. Rev.*, **73**, 487 (1973).

Balanson, R. D., V. M. Kobal, and R. R. Schumaker, *J. Org. Chem.*, **42**, 393 (1977).

Baldwin, J. E., G. A. Höfle, and O. W. Lever, Jr., *J. Am. Chem. Soc.*, **96**, 7125 (1974).

Baldwin, J. E., O. W. Lever, Jr., and N. R. Tzodikov, *J. Org. Chem.*, **41**, 2312 (1976).

Baldwin, J. E., S. B. Haber, C. Hoskins, and L. I. Kruse, *J. Org. Chem.*, **42**, 1239 (1977a).

Baldwin, J. E., R. C. Thomas, L. I. Kruse, and L. Silberman, *J. Org. Chem.*, **42**, 3846 (1977b).

Banerji, A., R. B. Jones, G. Mellows, L. Phillips, and K-Y. Sim, *J. Chem. Soc. Perkin I*, 2221 (1976).

Bare, T. M., and H. O. House, *Org. Syn.,* **49**, 81 (1969).

Barone, A. D., D. L. Snitman, and D. S. Watt, *J. Org. Chem.,* **43**, 2066 (1978).

Bartlett, P. A., F. R. Green, III, and T. R. Webb, *Tetrahedron Lett.,* 331 (1977).

Bartlett, P. D., S. J. Tauber, and W. P. Weber, *J. Am. Chem. Soc.,* **91**, 6362 (1969).

Bates, R. B., R. S. Cutler, and R. M. Freeman, *J. Org. Chem.,* **42**, 4162 (1977).

Bayer, O., in *Houben-Weyl, Methoden der Organischen Chemie,* Vol. 7/1, E. Müller, Ed., Georg Thieme Verlag, Stuttgart, 1954, p. 255.

Beak, P., and R. A. Brown, *J. Org. Chem.,* **42**, 1823 (1977).

Beck, A. K., M. S. Hoekstra, and D. Seebach, *Tetrahedron Lett.,* 1187 (1977).

Bellassoued, M., R. Couffignal, and M. Gaudemar, *J. Organometal. Chem.,* **61**, 9 (1973).

Bellassoued, M., and M. Gaudemar, *J. Organomet. Chem.,* **81**, 139 (1974).

Benkeser, R. A., *Synthesis,* 347 (1971).

Bergbreiter, D. E., and J. M. Killough, *J. Org. Chem.,* **41**, 2750 (1976).

Bergmann, E. D., D. Ginsburg, and R. Pappo, *Org. React.,* **10**, 179 (1959).

Bernardi, F., I. G. Csizmadia, A. Mangini, H. B. Schlegel, M-H. Whangbo, and S. Wolfe, *J. Am. Chem. Soc.,* **97**, 2209 (1975).

Beringer, F. M., and P. S. Forgione, *J. Org. Chem.,* **28**, 714 (1963).

Bertele, E., and P. Schudel, *Helv. Chim. Acta,* **50**, 2445 (1967).

Bessière-Chrétien, Y., M. M. El Gaied, and B. Meklati, *Bull. Soc. Chim. Fr.,* 1000 (1972).

Bestmann, H. J., and F. Seng, *Angew. Chem. Int. Ed.,* **1**, 116 (1962).

Bestmann, H. J., *Newer Methods of Preparative Org. Chem.,* **5**, 1 (1968).

Bestmann, H. J., W. Stransky, and O. Vostrowsky, *Chem. Ber.,* **109**, 1694, 3375 (1976).

Beumel, Jr., O. F., and R. F. Harris, *J. Org. Chem.,* **28**, 2775 (1963).

Bhanu, S., and F. Scheinmann, *J. Chem. Soc. Chem. Commun.,* 817 (1975).

Birch, A. J., and H. Smith, *J. Chem. Soc.,* 1882 (1951).

Blagoev, B., and D. Ivanov, *Synthesis,* 615 (1970).

Bloomfield, J., *J. Org. Chem.,* **26**, 4112 (1961).

Boden, R. M., *Synthesis,* 784 (1975).

Boeckman, Jr., R. K., *J. Am. Chem. Soc.,* **96**, 6179 (1974).

Boeckman, R. K., and K. J. Bruza, *Tetrahedron Lett.,* 4187 (1977).

Bohlmann, F., and J. Kocur, *Chem. Ber.,* **109**, 2969 (1976).

Bonjouklian, R., and B. Ganem, *Tetrahedron Lett.,* 2835 (1977).

Borch, R. F., A. J. Evans, and J. J. Wade, *J. Am. Chem. Soc.,* **99**, 1612 (1977).

Bordwell, F. G., W. S. Matthews, and N. R. Vanier, *J. Am. Chem. Soc.,* **97**, 442 (1975); F. G. Bordwell, J. E. Bartmess, G. E. Drucker, Z. Margolin, and W. S. Matthews, *J. Am. Chem. Soc.,* **97**, 3226 (1975); W. S. Matthews, J. E. Bares, J. E. Bartmess, F. G. Bordwell, F. J. Cornforth, G. E. Drucker, Z.

Margolin, R. J. McCallum, G. J. McCollum, and N. Vanier, *J. Am. Chem. Soc.*, **97**, 7006 (1975); F. G. Bordwell and D. Algrim, *J. Org. Chem.*, **41**, 2507 (1976); F. G. Bordwell, J. E. Bares, J. E. Bartmess, G. J. McCollum, M. Van Der Puy, N. R. Vanier, and W. S. Matthews, *J. Org. Chem.*, **42**, 321 (1977); F. G. Bordwell, D. Algrim, and N. R. Vanier, *J. Org. Chem.*, **42**, 1817 (1977).

Boutagy, J., and R. Thomas, *Chem. Rev.*, **74**, 87 (1974).

Bowman, R. E., *J. Chem. Soc.*, 322 and 325 (1950).

Brändström, A., *Ark. Kemi*, **6**, 155 (1953).

Breitholle, E. G., and A. G. Fallis, *J. Org. Chem.*, **43**, 1964 (1978).

Breslow, D. S., E. Baumgarten, and C. R. Hauser, *J. Am. Chem. Soc.*, **66**, 1286 (1944).

Brinkmeyer, R. S., E. W. Collington, and A. I. Meyers, *Org. Synth.*, **54**, 42 (1974).

Brocksom, T. J., N. Petragnani, R. Rodrigues, and H. La Scala Teixeira, *Synthesis*, 396 (1975).

Brown, C. A., *J. Am. Chem. Soc.*, **95**, 982 (1973).

Brown, C. A., *Synthesis*, 326 (1975).

Brown, C. A., and A. Yamashita, *J. Chem. Soc. Chem. Commun.* 959 (1976).

Brown, E., and R. Dhal, *J. Chem. Soc. Perkin I*, 2190 (1976).

Brown, E., J-P. Robin, and R. Dhal, *J. Chem. Soc. Chem. Commun.*, 556 (1978).

Brown, R. E., D. M. Lustgarten, R. J. Stanaback, and R. I. Meltzer, *J. Org. Chem.*, **31**, 1489 (1966).

Brownbridge, P., and S. Warren, *J. Chem. Soc. Chem Commun.*, 465 (1977).

Bryson, T. A., D. M. Donelson, R. B. Dunlap, R. R. Fisher, and P. D. Ellis, *J. Org. Chem.*, **41**, 2066 (1976).

Bruderer, H., D. Knopp, and J. J. Daly, *Helv. Chim. Acta*, **60**, 1935 (1977).

Büchi, G., and H. Wüest, *J. Org. Chem.*, **31**, 977 (1966).

Büchi, G., D. Berthet, R. Decorzant, A. Grieder, and A. Hauser, *J. Org. Chem.*, **41**, 3208 (1976).

Buhler, J. D., *J. Org. Chem.*, **38**, 904 (1973).

Bullivant, M. J., and G. Pattenden, *J. Chem. Soc. Perkin I*, 249 (1976a).

Bullivant, M. J., and G. Pattenden, *J. Chem. Soc. Perkin I*, 256 (1976b).

Burford, C., F. Cooke, E. Ehlinger, and P. Magnus, *J. Am. Chem. Soc.*, **99**, 4536 (1977).

Burgstahler, A. W., M. E. Sanders, C. G. Shaefer, and L. O. Weigel, *Synthesis*, 405 (1977a).

Burgstahler, A. W., L. O. Weigel, M. E. Sanders, C. G. Schaefer, W. J. Bell, and S. B. Vuturo, *J. Org. Chem.*, **42**, 566 (1977b).

Burness, D. M., *Org. Synth.*, **Coll. Vol. 4**, 649 (1963).

Byon, C., G. Büyüktür, P. Choay, and M. Gut, *J. Org. Chem.*, **42**, 3619 (1977).

Cahiez, G., D. B. Bernard, and J. F. Normant, *Synthesis*, 245 (1976a).

Cahiez, G., A. Masuda, D. Bernard, and J. F. Normant, *Tetrahedron Lett.*, 3155 (1976b).

Cahiez, G., D. Bernard, and J. F. Normant, *Synthesis*, 130 (1977a).

Cahiez, G., and J. F. Normant, *Tetrahedron Lett.*, 3383 (1977b).

Caine, D., and F. N. Tuller, *J. Org. Chem.*, **34**, 222 (1969).

Caine, D., and S. T. Chao, *Org. Synth.*, **56**, 52 (1977).

Cainelli, G., N. Tangari, and A. U. Ronchi, *Tetrahedron*, **28**, 3009 (1972).

Campaigne, E., D. Mais, and E. M. Yokley, *Synth. Commun.*, **4**, 379 (1974).

Campaigne, E., and R. A. Forsch, *J. Org. Chem.*, **43**, 1044 (1978).

Cardillo, G.; M. Contento, and S. Sandri, *Tetrahedron Lett.*, 2215 (1974).

Cardillo, G., D. Savoia, and A. Umani-Ronchi, *Synthesis*, 453 (1975).

Casey, C. P., and D. F. Marten, *Tetrahedron Lett.*, 925 (1974).

Cason, J., K. L. Rinehart, Jr., and S. D. Thornton, Jr., *J. Org. Chem.*, **18**, 1594 (1953).

Cason, J., and F. S. Prout, *Org. Synth.* Coll. Vol. 3, 601 (1955).

Cavill, G. W. K., B. S. Goodrich, and D. G. Laing, *Aust. J. Chem.*, **23**, 83 (1970).

Chamberlin, A. R., J. E. Stemke, and F. T. Bond, *J. Org. Chem.*, **43**, 147 (1978).

Chan, K., N. Cohen, J. P. De Noble, A. C. Specian, Jr., and G. Saucy, *J. Org. Chem.*, **41**, 3497 (1976).

Chastrette, M., R. Amouroux, and M. Subit, *J. Organomet. Chem.*, **99**, C41 (1975).

Chavdarian, C. G., and C. H. Heathcock, *J. Am. Chem. Soc.*, **97**, 3822 (1975).

Chen, T. S., J. Wolinska-Mocydlarz, and L. C. Leitch, *J. Labelled Comp.* **6**, 285 (1971).

Ciochetto, L. J., D. E. Bergbreiter, and M. Newcomb, *J. Org. Chem.*, **42**, 2948 (1977).

Clark, R. D., and C. H. Heathcock, *J. Org. Chem.*, **41**, 1396 (1976).

Clive, D. L. J., *Tetrahedron*, **34**, 1049 (1978).

Coates, R. M., H. D. Pigott, and J. Ollinger, *Tetrahedron Lett.*, 3955 (1974).

Cohen, T., D. A. Bennett, and A. J. Mura, Jr., *J. Org. Chem.*, **41**, 2506 (1976).

Colon, I., G. W. Griffin, and E. J. O'Connell, Jr., *Org. Synth.*, **52**, 33 (1972).

Cook, D., *J. Org. Chem.*, **41**, 2173 (1976).

Cooke, R. G., and I. J. Rainbow, *Aust. J. Chem.*, **30**, 2241 (1977).

Cookson, R. C., and P. J. Parsons, *J. Chem. Soc. Chem. Commun.*, 990 (1976).

Cooper, G. K., and L. J. Dolby, *Tetrahedron Lett.*, 4675 (1976).

Cope A. C., H. L. Holmes, and H. O. House, *Org. React.*, **9**, 107 (1957).

Corey, E. J., and M. Chaykovsky, *J. Am. Chem. Soc.*, **84**, 866 (1962).

Corey, E. J., and M. Chaykovsky, *J. Am. Chem. Soc.*, **86**, 1639 (1964).

Corey, E. J., and M. Chaykovsky, *J. Am. Chem. Soc.*, **87**, 1353 (1965).

Corey, E. J., and D. Seebach, *J. Org. Chem.*, **31**, 4097 (1966).

Corey, E. J., and G. H. Posner, *J. Am. Chem. Soc.*, **89**, 3911 (1967).

Corey, E. J., and G. H. Posner, *J. Am. Chem. Soc.*, **90**, 5615 (1968a).

Corey, E. J., and M. Jautelat, *Tetrahedron Lett.*, 5787 (1968b).

Corey, E. J., N. M. Weinshenker, T. K. Schaff, and W. Huber, *J. Am. Chem. Soc.*, **91**, 5675 (1969).

Corey, E. J., and H. Yamamoto, *J. Am. Chem. Soc.*, **92**, 226 (1970a).

Corey, E. J., and D. Seebach, *Org. Synth.*, **50**, 72 (1970b).

Corey, E. J., B. W. Erickson, and R. Noyori, *J. Am. Chem. Soc.*, **93**, 1724 (1971a).

Corey, E. J., J. A. Katzenellenbogen, S. A. Roman, and N. W. Gilman, *Tetrahedron Lett.*, 1821 (1971b).

Corey, E. J., and B. W. Erickson, *J. Org. Chem.*, **36**, 3553 (1971c).

Corey, E. J., and I. Kuwajima, *Tetrahedron Lett.*, 487 (1972a).

Corey, E. J., and D. J. Beames, *J. Am. Chem. Soc.*, **94**, 7210 (1972b).

Corey, E. J., and M. Chaykovsky, *Org. Synth.*, Coll. **5**, 755 (1973).

Corey, E. J., and P. Helquist, *Tetrahedron Lett.*, 4091 (1975a).

Corey, E. J., M. Petrzilka, and Y. Ueda, *Tetrahedron Lett.*, 4343 (1975b).

Corey, E. J., and G. N. Widiger, *J. Org. Chem.*, **40**, 2975 (1975c).

Corey, E. J., and D. Enders, *Tetrahedron Lett.*, 3, 11 (1976a).

Corey, E. J., D. Enders, and M. G. Bock, *Tetrahedron Lett.*, 7 (1976b).

Corey, E. J., M. Shibasaki, K. C. Nicolaou, C. L. Malmsten, and B. Samuelsson, *Tetrahedron Lett.*, 737 (1976c).

Corey, E. J., and S. Knapp, *Tetrahedron Lett.*, 3667, 4687 (1976d).

Corey, E. J., M. Shibasaki, J. Knolle, and T. Sugahara, *Tetrahedron Lett.*, 785 (1977a).

Corey, E. J., and D. R. Williams, *Tetrahedron Lett.*, 3847 (1977b).

Courtois, G., and L. Miginiac, *J. Organomet. Chem.*, **69**, 1 (1974).

Creary, X., *J. Org. Chem.*, **41**, 3740 (1976).

Creger, P. L., *J. Org. Chem.*, **37**, 1907 (1972).

Cregge, R. J., J. L. Herrmann, C. S. Lee, J. E. Richman, and R. H. Schlessinger, *Tetrahedron Lett.*, 2425 (1973).

Crispin, D. J., A. E. Vanstone, and J. S. Whitehurst, *J. Chem. Soc.* (C), 10 (1970).

Crombie, L., P. Hemesley, and G. Pattenden, *J. Chem. Soc. C*, 1016 (1969).

Curé, J., and M. Gaudemar, *Bull. Soc. Chim. Fr.*, 2471 (1969).

Cuvigny, T., M. Larchevêque, and H. Normant, *Liebigs Ann.*, 719 (1975).

Cuvigny, T., J. F. Le Borgne, M. Larchevêque, and H. Normant, *Synthesis*, 237 (1976).

Damon, R. E., and R. H. Schlessinger, *Tetrahedron Lett.*, 4551 (1975).

Damon, R. E., R. H. Schlessinger, and J. F. Blount, *J. Org. Chem.*, **41**, 3772 (1976).

d'Angelo, J., *Tetrahedron*, **32**, 2979 (1976).

Daniewski, A. R., *J. Org. Chem.*, **40**, 3135 (1975).

Danishefsky, S., K. Nagaswara, and N. Wang, *J. Org. Chem.*, **40**, 1989 (1975).

Danishefsky, S., and A. Zimmer, *J. Org. Chem.*, **41**, 4059 (1976).

Dauben, Jr., H. J., H. J. Ringold, R. H. Wade, D. L. Pearson, and A. G. Anderson, Jr., *Org. Synth.*, **Coll. Vol. 4**, 221 (1963).

Dauben, W. G., and A. P. Kozikowski, *Tetrahedron Lett.*, 3711 (1973a).

Dauben, W. G., and J. Ipaktschi, *J. Am. Chem. Soc.*, **95**, 5088 (1973b).

Dauben, W. G., G. H. Beasley, M. D. Broadhurst, B. Muller, D. J. Peppard, P. Pesnelle, and C. Suter, *J. Am. Chem. Soc.*, **97**, 4973 (1975).

Dauben, W. G., and D. J. Hart, *J. Org. Chem.*, **42**, 3787 (1977).

Daviaud, G., and P. Miginiac, *Tetrahedron Lett.*, 3345 (1973).

Davis, F. A., and P. A. Mancinelli, *J. Org. Chem.*, **42**, 398 (1977).

Dean, R. T., and H. Rapoport, *J. Org. Chem.*, **43**, 2115 (1978).

Debal, A., T. Cuvigny, and M. Larchevêque, *Synthesis*, 391 (1976).

DeBoer, C. D., *J. Org. Chem.*, **39**, 2426 (1974).

de Groot, Ae. and B. J. M. Jansen, *Tetrahedron Lett.*, 2709 (1976).

Dehmlow, E. V., *Angew. Chem. Int. Ed.*, **16**, 493 (1977).

Deprès, J-P., A. E. Greene, and P. Crabbe, *Tetrahedron Lett.*, 2191 (1978).

Derguini-Boumechal, F., and G. Linstrumelle, *Tetrahedron Lett.*, 3225 (1976).

Derguini-Boumechal, F., R. Lorne, and G. Linstrumelle, *Tetrahedron Lett.*, 1181 (1977).

DiBiase, S. A., and G. W. Gokel, *Synthesis*, 629 (1977).

Dietl, H. K., and K. C. Brannock, *Tetrahedron Lett.*, 1273 (1973).

D'Incan, E., and J. Seyden-Penne, *Synthesis*, 516 (1975).

Dockx, J., *Synthesis*, 441 (1973).

Dolak, T. M., and T. A. Bryson, *Tetrahedron Lett.*, 1961 (1977).

Douglas, G. H., J. M. H. Graves, D. Hartley, G. A. Hughes, B. J. McLoughlin, J. Siddall, and H. Smith, *J. Chem. Soc.*, 5072 (1963).

Drake, N. L., and P. Allen, Jr., *Org. Synth.*, **Coll. Vol. 1**, 77 (1932).

Dreger, E. E., Org. Synth., **Coll. Vol. 1**, 306 (1932).

Dubois, J. E., S. Molnarfi, F. Hennequin, M. H. Durand, and R. Fellous, *Bull. Soc. Chim. Fr.*, 1491 (1963).

Durst, H. D., and L. Liebeskind, *J. Org. Chem.*, **39**, 3271 (1974).

Durst, T., *J. Am. Chem. Soc.*, **91**, 1034 (1969).

Durst, T., and K-C. Tin, *Tetrahedron Lett.*, 2369 (1970).

Durst T., and M. J. LeBelle, *Can. J. Chem.*, **50**, 3196 (1972).

Dyke, S. F., *The Chemistry of Enamines,* Cambridge University Press, Cambridge, 1973.

Eftax, D. S. P., and A. P. Dunlop, *J. Org. Chem.,* **30,** 1317 (1965).

Eicher, T., *The Chemistry of the Carbonyl Group,* S. Patai, Ed., Interscience, New York, 1966, Chap. 13.

Eisch, J. J., and A. M. Jacobs, *J. Org. Chem.,* **28,** 2145 (1963).

Eiter, K., E. Truscheit, and M. Boness, *Ann. Chem.,* **709,** 29 (1967).

Elliott, W. J., and J. Fried, *J. Org. Chem.,* **41,** 2475 (1976).

Ellison, R. A., W. D. Woessner, and C. C. Williams, *J. Org. Chem.,* **37,** 2757 (1972).

Enders, D., and H. Eichenauer, *Angew. Chem. Int. Ed.,* **15,** 549 (1976); *Tetrahedron Lett.,* 191 (1977).

Erashko, V. I., S. A. Shevelev, and A. A. Fainzil'berg, *Russ. Chem. Rev.,* **35,** 719 (1966).

Erickson, A. S., and N. Kornblum, *J. Org. Chem.,* **42,** 3764 (1977).

Evans, D. A., G. C. Andrews, T. T. Fujimoto, and D. Wells, *Tetrahedron Lett.,* 1385, 1389 (1973).

Evans, D. A., and G. C. Andrews, *Acc. Chem. Res.,* **7,** 147 (1974a).

Evans, D. A., G. C. Andrews, and B. Buckwalter, *J. Am. Chem. Soc.,* **96,** 5560 (1974b).

Ewing, G. D., and L. A. Paquette, *J. Org. Chem.,* **40,** 2965 (1975).

Fayos, J., J. Clardy, L. J. Dolby, and T. Farnham, *J. Org. Chem.,* **42,** 1349 (1977).

Feutrill, G. I., and R. N. Mirrington, *J. Chem. Soc. Chem. Commun.,* 589 (1976).

Fitt, J. J., and H. W. Gschwend, *J. Org. Chem.,* **41,** 4029 (1976).

Fitt, J. J., and H. W. Gschwend, *J. Org. Chem.,* **42,** 2639 (1977).

Floyd, J. C., *Tetrahedron Lett.,* 2877 (1974).

Focella, A., S. Teitel, and A. Brossi, *J. Org. Chem.,* **42,** 3456 (1977).

Fornasier, R., F. Montanari, G. Podda, and P. Tundo, *Tetrahedron Lett.,* 1381 (1976).

Fortunato, J. M., and B. Ganem, *J. Org. Chem.,* **41,** 2194 (1976).

Fouquet, G., and M. Schlosser, *Angew. Chem. Int. Ed.,* **13,** 82 (1974).

Fray, G. I., R. H. Jaeger, E. D. Morgan, R. Robinson, and A. D. B. Sloan, *Tetrahedron,* **15,** 18 (1961).

Freeman, P. K., and L. L. Hutchinson, *Tetrahedron Lett.,* 1849 (1976).

Freiesleben, W., *Angew. Chem. Int. Ed.,* **2,** 396 (1963).

Fukui, K., and M. Nakayama, *Bull. Chem. Soc. Japan,* **35,** 1321 (1962).

Furukawa, J., and N. Kawabata, *Adv. Organomet. Chem.,* **12,** 83 (1974).

Gall, M., and H. O. House, *Org. Synth.,* **52,** 39 (1972).

Ganem, B., and J. M. Fortunato, *J. Org. Chem.,* **40,** 2846 (1975).

Garbers, C. F., J. A. Steenkamp, and H. E. Visagie, *Tetrahedron Lett.*, 3753 (1975).

Gassman, P. G., and B. L. Fox, *J. Org. Chem.*, **31**, 982 (1966a).

Gassman, P. G., and G. D. Richmond, *J. Org. Chem.*, **31**, 2355 (1966b).

Gassman, P. G., J. T. Lumb, and F. V. Zalar, *J. Am. Chem. Soc.*, **89**, 946 (1967).

Gassman, P. G., D. P. Gilbert, and S. M. Cole, *J. Org. Chem.*, **42**, 3233 (1977a).

Gassman, P. G., and R. J. Balchunis, *J. Org. Chem.*, **42**, 3236 (1977b).

Gaudemar, M., and J. Cure, *C.R. Acad. Sci., Ser. C*, **262**, 213 (1966).

Gawley, R. E., *Synthesis*, 777 (1976).

Gersmann, H. R., and A. F. Bickel, *J. Chem. Soc. B*, 2230 (1971).

Gilbert, J. C., and K. R. Smith, *J. Org. Chem.*, **41**, 3883 (1976).

Gilman, H., and R. D. Gorisch, *J. Org. Chem.*, **23**, 550 (1958).

Glaze, W. H., and P. C. Jones, *J. Chem. Soc. Chem. Commun.*, 1434 (1969).

Glaze, W. H., J. E. Hanicak, D. J. Berry, and D. P. Duncan, *J. Organomet. Chem.*, **44**, 49 (1972).

Glaze, W. H., D. P. Duncan, and D. J. Berry, *J. Org. Chem.*, **42**, 694 (1977).

Gokel, G. W., and H. D. Durst, *Synthesis*, 168 (1976a).

Gokel, G. W., and H. D. Durst, *Synthesis*, 182 (1976b).

Gokel, G. W., S. A. DiBiase, and B. A. Lipisko, *Tetrahedron Lett.*, 3495 (1976c).

Gosteli, J., *Helv. Chim. Acta*, **60**, 1980 (1977).

Greene, A. E., A. Cruz, and P. Crabbé, *Tetrahedron Lett.*, 2707 (1976a).

Greene, A. E., and P. Crabbé, *Tetrahedron Lett.*, 4867 (1976b).

Greenwald, R., M. Chaykovsky, and E. J. Corey, *J. Org. Chem.*, **28**, 1128 (1963).

Grieco, P. A., and R. S. Finkelhor, *Tetrahedron Lett.*, 3781 (1972).

Grieco, P. A., and R. S. Finkelhor, *J. Org. Chem.*, **38**, 2909 (1973).

Grieco, P. A., D. Boxler, and C. S. Pogonowski, *Chem. Commun.* 497 (1974a).

Grieco, P. A., and Y. Masaki, *J. Org. Chem.*, **39**, 2135 (1974b).

Grieco, P. A., and Y. Masaki, *J. Org. Chem.*, **40**, 150 (1975).

Grieco, P. A., J. A. Noguez, and Y. Masaki, *J. Org. Chem.*, **42**, 495 (1977).

Gröbel, B-T., and D. Seebach, *Synthesis*, 357 (1977).

Gschwend, H. W., and A. Hamdan, *J. Org. Chem.*, **40**, 2008 (1975).

Hamon, A., B. Lacoume, A. Olivier, and W. R. Pilgrim, *Tetrahedron Lett.*, 4481 (1975); *ibid.*, 644 (1976).

Hampton, K. G., T. M. Harris, and C. R. Hauser, *J. Org. Chem.*, **29**, 3511 (1964).

Hampton, K. G., T. M. Harris, and C. R. Hauser, *Org. Synth.*, **Coll. Vol. 5**, 848 (1973).

Hann, R. A., *J. Chem. Soc. Perkin I*, 1379 (1974).

Hansen, J. F., K. Kamata, and A. I. Meyers, *J. Heterocycl. Chem.*, **10**, 711 (1973).

Hansen, J. F., and S. Wang, *J. Org. Chem.*, **41**, 3635 (1976).

Harris, T. M., and C. M. Harris, *Org. React.*, **17**, 155 (1969).

Hart, H., B. Chen, and C. Peng, *Tetrahedron Lett.*, 3121 (1977).

Hauser, C. R., and W. B. Renfrow, Jr., *J. Am. Chem. Soc.*, **59**, 1823 (1937).

Hauser, C. R., and B. E. Hudson, Jr., *Org. React.*, **1**, 266 (1942).

Hauser, C. R., F. W. Swamer, and J. T. Adams, *Org. React.*, **8**, 59 (1954).

Hauser, F. M., and R. P. Rhee, *J. Org. Chem.*, **43**, 178 (1978).

Hegedus, L. S., P. M. Kendall, S. M. Lo, and J. R. Sheats, *J. Am. Chem. Soc.*, **97**, 5448 (1975).

Heissler, D., F. Jung, J. P. Vevert, and J. J. Riehl, *Tetrahedron Lett.*, 4879 (1976).

Henecka, H., in *Houben-Weyl, Methoden der Organischen Chemie*, Vol. 8, E. Müller, Ed., Georg Thieme Verlag, Stuttgart, 1952, pp. 600-628, a.

Henecka, H., ibid., p. 567, b.

Henecka, H., ibid., p. 573, c.

Henecka, H., in *Houben-Weyl, Methoden der Organischen Chemie* Vol. 7/2b, E. Müller, Ed., Georg Thieme Verlag, Stuttgart, 1976, p. 1435.

Heng, K. K., and R. A. J. Smith, *Tetrahedron Lett.*, 589 (1975).

Henrick, C. A., *Tetrahedron*, **33**, 1845 (1977).

Hergrueter, C. A., P. D. Brewer, J. Tagat, and P. Helquist, *Tetrahedron Lett.*, 4145 (1977).

Herrmann, J. L., and R. H. Schlessinger, *J. Chem. Soc. Chem. Commun.*, 711 (1973a).

Herrmann, J. L., and R. H. Schlessinger, *Tetrahedron Lett.*, 2429 (1973b).

Herrmann, J. L., G. R. Kieczykowski, and R. H. Schlessinger, *Tetrahedron Lett.*, 2433 (1973c).

Herrman, J. L., J. E. Richman, and R. H. Schlessinger, *Tetrahedron Lett.*, 3271, 3275 (1973d).

Herrmann, J. L., J. E. Richman, P. J. Wepplo, and R. H. Schlessinger, *Tetrahedron Lett.*, 4707 (1973e).

Hill, E. A., H. G. Richey, Jr., and T. C. Rees, *J. Org. Chem.*, **28**, 2161 (1963).

Hirai, K., Y. Iwano, and Y. Kishida, *Tetrahedron Lett.*, 2677 (1977).

Hiraoka, T., and I. Iwai, *Chem. Pharm. Bull. (Tokyo)*, **14**, 262 (1966).

Ho, T-L., H. C. Ho, and C. M. Wong, *J. Chem. Soc. Chem. Commun.*, 791 (1972); *Can. J. Chem.*, **50**, 3740 (1972).

Hori, I., T. Hayashi, and H. Midorikawa, *Synthesis*, 705 (1974).

House, H. O., L. J. Czuba, M. Gall, and H. D. Olmstead, *J. Org. Chem.*, **34**, 2324 (1969).

House, H. O., M. Gall, and H. D. Olmstead, *J. Org. Chem.,* **36,** 2361 (1971).

House, H. O., *Modern Synthetic Reactions,* 2nd ed., W. A. Benjamin, Menlo Park, California, 1972, Chap. 9.

House, H. O., R. A. Auerbach, M. Gall, and N. P. Peet, *J. Org. Chem.,* **38,** 514 (1973a).

House, H. O., and M. J. Umen, *J. Org. Chem.,* **38,** 3893 (1973b).

House, H. O., C. Chu, J. M. Wilkins, and M. J. Umen, *J. Org. Chem.,* **40,** 1460 (1975).

House, H. O., *Acc. Chem. Res.,* **9,** 59 (1976).

House, H. O., and J. M. Wilkins, *J. Org. Chem.,* **41,** 4031 (1976).

House, H. O., C. Chu, W. V. Phillips, T. S. B. Sayer, and C. Yau, *J. Org. Chem.,* **42,** 1709 (1977a).

House, H. O., and E. J. Zaiko, *J. Org. Chem.,* **42,** 3780 (1977b).

House, H. O., W. V. Phillips, T. S. B. Sayer, and C-C. Yau, *J. Org. Chem.,* **43,** 700 (1978).

Huckin, S. N., and L. Weiler, *J. Am. Chem. Soc.,* **96,** 1082 (1974).

Hudson, Jr., B. E., and C. R. Hauser, *J. Am. Chem. Soc.,* **63,** 3156 (1941).

Huet, F., G. Emptoz, and A. Jubier, *Tetrahedron,* **29,** 479 (1973).

Huffman, J. W., and P. G. Arapakos, *J. Org. Chem.,* **30,** 1604 (1965).

Huffman, J. W., and P. G. Harris, *Syn. Commun.,* **7,** 137 (1977a).

Huffman, J. W., and P. G. Harris, *J. Org. Chem.,* **42,** 2357 (1977b).

Hünig, S., and G. Wehner, *Synthesis,* 180 (1975a).

Hünig, S., and G. Wehner, *Synthesis,* 391 (1975b).

Hunsdiecker, H., *Chem. Ber.,* **75,** 455 (1942).

Huston, R. G., and A. H. Agett, *J. Org. Chem.,* **6,** 123 (1941).

Ibuka, T., Y. Mori, and Y. Inubushi, *Tetrahedron Lett.,* 3169 (1976).

Ide, J., and K. Sakai, *Tetrahedron Lett.,* 1367 (1976).

Ireland, R. E., and J. A. Marshall, *J. Org. Chem.,* **27,** 1615, 1620 (1962).

Ireland, R. E., and P. W. Schiess, *J. Org. Chem.,* **28,** 6 (1963).

Ireland, R. E., and G. Pfister, *Tetrahedron Lett.,* 2145 (1969).

Irikawa, H., T. Ishikura, and Y. Okumura, *Bull. Chem. Soc. Japan,* **50,** 2811 (1977).

Isobe, M., H. Iio, T. Kawai, and T. Goto, *Tetrahedron Lett.,* 703 (1977).

Ito, Y., T. Konoike, T. Harada, and T. Saegusa, *J. Am. Chem. Soc.,* **99,** 1487 (1977).

Ivanov, D., G. Vassilev, and I. Panayotov, *Synthesis,* 83, (1975).

Izzo, P. T., and S. R. Safir, *J. Org. Chem.,* **24,** 701 (1959).

Jacobson, R. M., *Tetrahedron Lett.,* 3215 (1974).

Johns, G., C. J. Ransom, and C. B. Reese, *Synthesis,* 515 (1976).

Johnson, A. W., *Ylid Chemistry,* Academic Press, New York, 1966.

Johnson, A. W., E. Markham, and R. Price, *Org. Synth.*, **Coll. Vol. 5**, 785 (1973).

Johnson, C. R., and E. R. Janiga, *J. Am. Chem. Soc.*, **95**, 7692 (1973a).

Johnson, C. R., and G. A. Dutra, *J. Am. Chem. Soc.*, **95**, 7777 (1973b).

Johnson, P. Y., and D. J. Kerkman, *J. Org. Chem.*, **41**, 1768 (1976).

Johnson, R. R. and J. A. Nicholson, *J. Org. Chem.*, **30**, 2918 (1965).

Johnson, W. S., and H. Posvic, *J. Am. Chem. Soc.*, **69**, 1361 (1947).

Johnson, W. S., M. I. Dawson, and B. E. Ratcliffe, *J. Org. Chem.*, **42**, 153 (1977).

Jones, G., *Org. React.*, **15**, 204 (1967).

Jones, J. B., and J. M. Young, *J. Med. Chem.*, **11**, 1176 (1968).

Jones, J. B., and J. D. Leman, *Can. J. Chem.*, **49**, 2420 (1971).

Jones, J. B., and R. Grayshan, *Can. J. Chem.*, **50**, 1414 (1972).

Jones, J. R., *The Ionization of Carbon Acids*, Academic Press, New York, 1973.

Jones, R. G., and H. Gilman, *Org. React.*, **6**, 339 (1951).

Julia, M., and P. Ward, *Bull. Soc. Chim. Fr.*, 3065 (1973a).

Julia, M., and J-M. Paris, *Tetrahedron Lett.*, 4833 (1973b).

Julia, M., and B. Badet, *Bull. Soc. Chim. Fr.*, 1363 (1975).

Julia, M., and B. Badet, *Bull. Soc. Chim. Fr.*, 525 (1976).

Jung, M. E., *Tetrahedron*, **32**, 3 (1976).

Jung, M. E., and R. B. Blum, *Tetrahedron Lett.*, 3791 (1977).

Kaiser, E. M., and C. R. Hauser, *J. Org. Chem.*, **33**, 3402 (1968).

Kaiser, E. M., J. D. Petty, and P. L. A. Knutson, *Synthesis*, 509 (1977).

Kaji, E., and S. Zen, *Bull. Chem. Soc. Japan*, **46**, 337 (1973).

Kajiwara, T., Y. Odake, and A. Hatanaka, *Agr. Biol. Chem.*, **39**, 1617 (1975).

Karrer, P., B. Shibata, A. Wettstein, and L. Jacubowicz, *Helv. Chim. Acta*, **13**, 1292 (1930).

Keinan, E., and Y. Mazur, *J. Am. Chem. Soc.*, **99**, 3861 (1977).

Kergomard, A., and H. Veschambre, *Tetrahedron Lett.*, 4069 (1976).

Kessar, S. V., A. L. Rampal, K. Kumar, and R. R. Jogi, *Ind. J. Chem.*, **2**, 240 (1964).

Kharasch, M. S., and O. Reinmuth, *Grignard Reactions of Nonmetallic Substances*, Prentice-Hall, New York, 1954.

Kieczykowski, G. R., R. H. Schlessinger, and R. B. Sulsky, *Tetrahedron Lett.*, 4647 (1975).

Kieczykowski, G. R., R. H. Schlessinger, and R. B. Sulsky, *Tetrahedron Lett.*, 597 (1976).

Kitatani, K., T. Hiyama, and H. Nozaki, *Bull. Chem. Soc. Japan*, **50**, 3288 (1977).

Klein, J., and E. Gurfinkel, *Tetrahedron,* **26**, 2127 (1970).

Knox, I., S-C. Chang, and A. H. Andrist, *J. Org. Chem.,* **42**, 3981 (1977).

Kobayashi, Y., T. Taguchi, and E. Tokuno, *Tetrahedron Lett.,* 3741 (1977).

Köbrich, G., *Angew. Chem. Int. Ed.,* **6**, 41 (1967); *ibid.,* **11**, 473 (1972).

Kocienski, P. J., and G. J. Cernigliaro, *J. Org. Chem.,* **41**, 2927 (1976).

Kocienski, P. J., G. Cernigliaro, and G. Feldstein, *J. Org. Chem.,* **42**, 353 (1977).

Kodama, M., Y. Matsuki, and S. Itô, *Tetrahedron Lett.,* 3065 (1975).

Kondo, K., and D. Tunemoto, *Tetrahedron Lett.,* 1007 (1975a).

Kondo, K., E. Saito, and D. Tunemoto, *Tetrahedron Lett.,* 2275 (1975b).

Kondo, K., and D. Tunemoto, *Tetrahedron Lett.,* 1397 (1975c).

Kondo, K., and M. Matsumoto, *Tetrahedron Lett.,* 391 (1976).

Kornblum, N., and L. Cheng, *J. Org. Chem.,* **42**, 2944 (1977).

Korte, D. E., L. S. Hegedus, and R. K. Wirth, *J. Org. Chem.,* **42**, 1329 (1977).

Korte, F., and K. H. Büchel, *Chem. Ber.,* **93**, 1025 (1960).

Kraft, W. M., *J. Am. Chem. Soc.,* **70**, 3569 (1948).

Krapcho, A. P., D. S. Kashdan, and E. G. E. Jahngen, Jr., *J. Org. Chem.,* **42**, 1189 (1977).

Krapcho, A. P., J. F. Weimaster, J. M. Eldridge, E. G. E. Jahngen, Jr., A. J. Lovey, and W. P. Stephens, *J. Org. Chem.,* **43**, 138 (1978).

Kraus, G. A., and B. Roth, *Tetrahedron Lett.,* 3129 (1977).

Ksander, G. M., and J. E. McMurry, *Tetrahedron Lett.,* 4691 (1976).

Kühlein, K., A. Linkies, and D. Reuschling, *Tetrahedron Lett.,* 4463 (1976).

Kuwajima, I., and M. Uchida, *Tetrahedron Lett.,* 649 (1972).

Kuwajima, I., and E. Nakamura, *J. Am. Chem. Soc.,* **97**, 3257 (1975).

Kuwajima, I., T. Sato, M. Arai, and N. Minami, *Tetrahedron Lett.,* 1817 (1976a).

Kuwajima, I., N. Minami, and T. Sato, *Tetrahedron Lett.,* 2253 (1976b).

Kyrides, L. P., *J. Am. Chem. Soc.,* **55**, 3431 (1933).

LaLonde, R. T., N. Muhammad, and C. F. Wong, *J. Org. Chem.,* **42**, 2113 (1977).

Lansbury, P. T., and A. K. Serelis, *Tetrahedron Lett.,* 1909 (1978).

Larchevêque, M., P. Mulot, and T. Cuvigny, *J. Organomet. Chem.,* **57**, C33 (1973).

Larchevêque, M., G. Valette, and T. Cuvigny, *Synthesis,* 424 (1977).

Lawson, J. A., W. T. Colwell, J. I. DeGraw, R. H. Peters, R. L. Dehn, and M. Tanabe, *Synthesis,* 729 (1975).

Leake, W. W., and R. Levine, *J. Am. Chem. Soc.,* **81**, 1169, 1627 (1959).

Le Borgne, J. F., T. Cuvigny, M. Larchevêque, and H. Normant, *Tetrahedron Lett.,* 1379 (1976).

Leete, E., *J. Org. Chem.,* **41**, 3438 (1976).

Leete, E., and S. A. S. Leete, *J. Org. Chem.*, **43**, 2122 (1978).

LeGoff, E., S. E. Ulrich, and D. B. Denney, *J. Am. Chem. Soc.*, **80**, 622 (1958).

Lever, Jr., O. W., *Tetrahedron*, **32**, 1943 (1976).

Levine, R., E. Baumgarten, and C. R. Hauser, *J. Am. Chem. Soc.*, **66**, 1230 (1944).

Levy, A. B., P. Talley, and J. A. Dunford, *Tetrahedron Lett.*, 3545 (1977).

Lewis, R. G., D. H. Gustafson, and W. F. Erman, *Tetrahedron Lett.*, 401 (1967).

Liebermann, S. V., *J. Am. Chem. Soc.*, **77**, 1114 (1955).

Liedtke, R. J., and C. Djerassi, *J. Org. Chem.*, **37**, 2111 (1972).

Linstrumelle, G., J. K. Krieger, and G. M. Whitesides, *Org. Synth.*, **55**, 103 (1976).

Lipnick, R. L., *J. Am. Chem. Soc.*, **96**, 2941 (1974).

Liu, W-S., and G. I. Glover, *J. Org. Chem.*, **43**, 754 (1978).

Loewenthal, H. J. E., and S. Schatzmiller, *J. Chem. Soc. Perkin I*, 944 (1976).

Logue, M. W., *J. Org. Chem.*, **39**, 3455 (1974).

Lompa-Krzymien, L., and L. C. Leitch, *Synthesis*, 124 (1976).

Longone, D. T., and W. D. Wright, *Tetrahedron Lett.*, 2859 (1969).

Maercker, A., *Org. React.*, **14**, 270 (1965).

Magnus, P. D., *Tetrahedron*, **33**, 2019 (1977).

Magnus, P., and G. Roy, *J. Chem. Soc. Chem. Commun.*, 297 (1978).

Magnusson, G., *Tetrahedron Lett.*, 2713 (1977).

Majetich, G., P. A. Grieco, and M. Nishizawa, *J. Org. Chem.*, **42**, 2327 (1977).

Makosza, M., and A. Jończyk, *Org. Synth.*, **55**, 91 (1976a).

Makosza, M., J. Czyzewski, and M. Jawdosiuk, *Org. Synth.*, **55**, 99 (1976b).

Makosza, M., and E. Bialecka, *Tetrahedron Lett.*, 183 (1977).

Malone, G. R., and A. I. Meyers, *J. Org. Chem.*, **39**, 623 (1974).

Marino, J. P., and D. M. Floyd, *J. Am. Chem. Soc.*, **96**, 7138 (1974).

Marino, J. P., and D. M. Floyd, *Tetrahedron Lett.*, 3897 (1975a).

Marino, J. P., and J. S. Farina, *Tetrahedron Lett.*, 3901 (1975b).

Marino, J. P., and L. J. Browne, *J. Org. Chem.*, **41**, 3629 (1976).

Marshall, J. A., N. H. Andersen, and P. C. Johnson, *J. Org. Chem.*, **35**, 186 (1970).

Marshall, J. A., and P. G. Wuts, *J. Org. Chem.*, **42**, 1794 (1977a).

Marshall, J. A., and R. Bierenbaum, *J. Org. Chem.*, **42**, 3309 (1977b).

Martin, S. F., and D. R. Moore, *Tetrahedron Lett.*, 4459 (1976).

Masaki, M., and M. Ohta, *Bull. Chem. Soc. Japan*, **34**, 1257 (1961).

Masure, D., R. Sauvetre, J. F. Normant, and J. Villieras, *Synthesis*, 761 (1976).

Mayer, R., *Newer Methods of Preparative Org. Chem.*, **2**, W. Foerst, Ed., Academic Press, New York, 1963, p. 101.

McConnell, O. J., and W. Fenical, *Tetrahedron Lett.*, 1851 (1977).

McMorris, T. C., and S. R. Schow, *J. Org. Chem.*, **41**, 3759 (1976).

McMurry, J. E., and S. J. Isser, *J. Am. Chem. Soc.*, **94**, 7132 (1972).

McMurry, J. E., and J. Melton, *J. Org. Chem.*, **38**, 4367 (1973).

McMurry, J. E., and J. H. Musser, *J. Org. Chem.*, **40**, 2556 (1975).

Meekler, A. B., S. Ramachandran, S. Swaminathan, and M. S. Newman, *Org. Synth.*, **Coll. Vol. 5**, 743 (1973).

Merz, A., and G. Märkl, *Angew. Chem. Int. Ed.*, **12**, 845 (1973).

Meyer, W. L., D. D. Cameron, and W. S. Johnson, *J. Org. Chem.*, **27**, 1130 (1962).

Meyer, W. L., R. A. Manning, P. G. Schroeder, and D. C. Shew, *J. Org. Chem.*, **42**, 2754 (1977a).

Meyer, W. L., T. E. Goodwin, R. J. Hoff, and C. W. Sigel, *J. Org. Chem.*, **42**, 2761 (1977b).

Meyers, A. I., and D. L. Temple, Jr., *J. Am. Chem. Soc.*, **92**, 6644 (1970).

Meyers, A. I., and R. C. Strickland, *J. Org. Chem.*, **37**, 2579 (1972a).

Meyers, A. I., and E. M. Smith, *J. Org. Chem.*, **37**, 4289 (1972b).

Meyers, A. I., and N. Nazarenko, *J. Am. Chem. Soc.*, **94**, 3243 (1972c).

Meyers, A. I., A. Nabeya, H. W. Adickes, I. R. Politzer, G. R. Malone, A. C. Kovelesky, R. L. Nolen, and R. C. Portnoy, *J. Org. Chem.*, **38**, 36 (1973a).

Meyers, A. I., and N. Nazarenko, *J. Org. Chem.*, **38**, 175 (1973b).

Meyers, A. I., *Heterocycles in Organic Synthesis*, Wiley, New York, 1974.

Meyers, A. I., E. D. Mihelich, and R. L. Nolen, *J. Org. Chem.*, **39**, 2783 (1974a).

Meyers, A. I., and G. Knaus, *J. Am. Chem. Soc.*, **96**, 6508 (1974b).

Meyers, A. I., and E. D. Mihelich, *J. Org. Chem.*, **40**, 1186 (1975a).

Meyers, A. I., and J. L. Durandetta, *J. Org. Chem.*, **40**, 2021 (1975b).

Meyers, A. I., J. L. Durandetta, and R. Munavu, *J. Org. Chem.*, **40**, 2025 (1975c).

Meyers, A. I., and C. E. Whitten, *J. Am. Chem. Soc.*, **97**, 6266 (1975d).

Meyers, A. I., and E. D. Mihelich, *Angew. Chem. Int. Ed.*, **15**, 270 (1976a).

Meyers, A. I., and K. Kamata, *J. Am. Chem. Soc.*, **98**, 2290 (1976b).

Midland, M. M., *J. Org. Chem.*, **40**, 2250 (1975).

Miginiac, L., and B. Mauzé, *Bull. Soc. Chim. Fr.*, 3832 (1968).

Millard, A. A., and M. W. Rathke, *J. Org. Chem.*, **43**, 1834 (1978).

Millon, J., R. Lorne, and G. Linstrumelle, *Synthesis*, 434 (1975).

Minami, N., and I. Kuwajima, *Tetrahedron Lett.*, 1423 (1977).

Miyano, M., C. R. Dorn, and R. A. Mueller, *J. Org. Chem.*, **37**, 1810 (1972).

Miyano, M., and M. A. Stealey, *J. Org. Chem.*, **40**, 2840 (1975).

Moersch, G. W., and A. R. Burkett, *J. Org. Chem.*, **36**, 1149 (1971).

Monteiro, H. J., *J. Org. Chem.*, **42**, 2324 (1977).

Mori, K., H. Hashimoto, Y. Takenaka, and T. Takigawa, *Synthesis*, 720 (1975).

Mori, K., M. Uchida, and M. Matsui, *Tetrahedron*, **33**, 385 (1977).

Morizur, J. P., G. Bidan, and J. Kossanyi, *Tetrahedron Lett.*, 4167 (1975).

Mousseron, M., R. Jacquier, and H. Christol, *Bull. Soc. Chim. Fr.*, 346 (1957).

Muchmore, D. C., *Org. Synth.*, **52**, 109 (1972).

Mukaiyama, T., M. Araki, and H. Takei, *J. Am. Chem. Soc.*, **95**, 4763 (1973).

Mukaiyama, T., T. Sato, S. Suzuki, T. Inoue, and H. Nakamura, *Chem. Lett.*, 95 (1976).

Murphy, W. S., P. J. Hamrick, and C. R. Hauser, *Org. Synth.*, **Coll. Vol. 5**, 523 (1973).

Näf, F., R. Decorzant, and W. Thommen, *Helv. Chim. Acta*, **60**, 1196 (1977).

Nagata, W., and Y. Hayase, *J. Chem. Soc. C*, 460 (1969).

Nagata, W., T. Wakabayashi, and Y. Hayase, *Org. Synth.*, **53**, 44, 104 (1973).

Nakai, T., and M. Okawara, *Chem. Lett.*, 731 (1974).

Narasaka, K., T. Sakashita, and T. Mukaiyama, *Bull. Chem. Soc. Japan*, **45**, 3724 (1972).

Nedelec, L., J. C. Gasc, and R. Bucourt, *Tetrahedron*, **30**, 3263 (1974).

Needles, H. L., and R. H. Whitfield, *J. Org. Chem.*, **31**, 989 (1966).

Neef, G., and U. Elder, *Tetrahedron Lett.*, 2825 (1977).

Nelson, D. J., and E. A. Uschak, *J. Org. Chem.*, **42**, 3308 (1977).

Neumann, S. M., and J. K. Kochi, *J. Org. Chem.*, **40**, 599 (1975).

Newkome, G. R., *Synthesis*, 517, 808 (1975).

Newman, M. S., and A. S. Smith, *J. Org. Chem.*, **13**, 592 (1948).

Newman, M. S., and B. J. Magerlein, *Org. React.*, **5**, 414 (1949).

Nielsen, A. T., and W. J. Houlihan, *Org. React.*, **16**, 1 (1968).

Nishihata, K., and M. Nishio, *Tetrahedron Lett.*, 1695 (1976).

Niznik, G. E., W. H. Morrison, III, and H. M. Walborsky, *J. Org. Chem.*, **39**, 600 (1974).

Nominé, G., G. Amiard, and V. Torelli, *Bull. Soc. Chim. Fr.*, 3664 (1968).

Normant, H., *Adv. Org. Chem.*, **2**, 1 (1960a).

Normant, H., and B. Angelo, *Bull. Soc. Chim. Fr.*, 354 (1960b).

Normant, J. F., G. Cahiez, M. Bourgain, C. Chuit, and J. Villieras, *Bull. Soc. Chim. Fr.*, 1656 (1974); J. F. Normant, G. Cahiez, C. Chuit, and J. Villieras, *J. Organomet. Chem.*, **77**, 269, 281 (1974).

Normant, J. F., A. Commercon, M. Bourgain, and J. Villieras, *Tetrahedron Lett.*, 3833 (1975).

Nützel, K., in *Houben-Weyl, Methoden der Organischen Chemie*, Vol. 13/2a, E. Müller, Ed., Georg Thieme Verlag, Stuttgart, 1973, p. 54 (a).

Nützel, K., *ibid.*, p. 197 (b).

Nützel, K., *ibid.*, p. 570 (c).

Ogura, K., and G. Tsuchihashi, *Tetrahedron Lett.*, 3151 (1971).

Ogura, K., and G. Tsuchihashi, *Tetrahedron Lett.*, 2681 (1972).

Ogura, K., M. Yamashita, and G. Tsuchihashi, *Tetrahedron Lett.*, 759 (1976).

Ohloff, G., and W. Giersch, *Helv. Chim. Acta*, **60**, 1496 (1977).

Ohta, H., T. Kobori, and T. Fujisawa, *J. Org. Chem.*, **42**, 1231 (1977).

Oikawa, Y., K. Sugano, and O. Yonemitsu, *J. Org. Chem.*, **43**, 2087 (1978).

Olofson, R. A., and C. M. Dougherty, *J. Am. Chem. Soc.*, **95**, 581, 582 (1973).

Olson, G. L., H. Cheung, K. D. Morgan, C. Neukom, and G. Saucy, *J. Org. Chem.*, **41**, 3287 (1976).

Oppolzer, W., and R. L. Snowden, *Tetrahedron Lett.*, 4187 (1976).

Ortiz de Montellano, P. R., and C. K. Hsu, *Tetrahedron Lett.*, 4215 (1976).

Oshima, K., K. Shimoji, H. Takahashi, H. Yamamoto, and H. Nozaki, *J. Am. Chem. Soc.*, **95**, 2694 (1973a).

Oshima, K., H. Yamamoto, and H. Nozaki, *J. Am. Chem. Soc.*, **95**, 4446 (1973b).

Oshima, K., H. Takahashi, H. Yamamoto, and H. Nozaki, *J. Am. Chem. Soc.*, **95**, 2693 (1973c).

Ostrowski, P. C., and V. V. Kane, *Tetrahedron Lett.*, 3549 (1977).

Owens, F. H., R. P. Fellman, and F. E. Zimmerman, *J. Org. Chem.*, **25**, 1808 (1960).

Paine, III, J. B., W. B. Kirshner, and D. W. Moskowitz, *J. Org. Chem.*, **41**, 3857 (1976).

Palmer, D. C., and M. J. Strauss, *Chem. Rev.*, **77**, 13 (1977).

Paquette, L. A., M. J. Kukla, and J. C. Stowell, *J. Am. Chem. Soc.*, **94**, 4920 (1972).

Parham, W. E., L. D. Jones, and Y. A. Sayed, *J. Org. Chem.*, **41**, 1184 (1976a).

Parham, W. E., and L. D. Jones, *J. Org. Chem.*, **41**, 1187, 2704 (1976b).

Parham, W. E., and D. W. Boykin, *J. Org. Chem.*, **42**, 260 (1977).

Patel, K. M., H. J. Pownall, J. D. Morrisett, and J. T. Sparrow, *Tetrahedron Lett.*, 4015 (1976).

Pattison, F. L. M., and R. E. A. Dear, *Can. J. Chem.*, **41**, 2600 (1963).

Payne, G. B., *J. Org. Chem.*, **26**, 4793 (1961).

Pearce, G. T., W. E. Gore, and R. M. Silverstein, *J. Org. Chem.*, **41**, 2797 (1976).

Pearson, D. E., and C. A. Buehler, *Chem. Rev.*, **74**, 45 (1974).

Peterson, D. J., *J. Org. Chem.*, **32**, 1717 (1967).

Pfeffer, P. E., L. S. Silbert, and J. M. Chirinko, Jr., *J. Org. Chem.*, **37**, 451 (1972).

Piers, E., and I. Nagakura, *J. Org. Chem.*, **40**, 2694 (1975).

Piers, E., and J. R. Grierson, *J. Org. Chem.*, **42**, 3755 (1977).

Pirkle, W. H., and C. W. Boeder, *J. Org. Chem.*, **43**, 2091 (1978).

Pittman, Jr., C. U., and R. M. Hanes, *J. Org. Chem.*, **42**, 1194 (1977).

Pitzele, B. S., J. S. Baran, and D. H. Steinman, *J. Org. Chem.*, **40**, 269 (1975).

Plantema, O. G., H. de Koning, and H. O. Huisman, *Tetrahedron Lett.*, 2945 (1975).

Poisel, H., and U. Schmidt, *Chem. Ber.*, **105**, 625 (1972).

Pond, D. M., and R. L. Cargill, *J. Org. Chem.*, **32**, 4064 (1967).

Poonia, N. S., K. Chhabra, C. Kumar, and V. W. Bhagwat, *J. Org. Chem.*, **42**, 3311 (1977).

Poos, G. I., W. F. Johns, and L. H. Sarett, *J. Am. Chem. Soc.*, **77**, 1026 (1955).

Posner, G. H., *Org. React.*, **19**, 1 (1972).

Posner, G. H., *Org. React.*, **22**, 253 (1975).

Posner, G. H., C. E. Whitten, and P. E. McFarland, *J. Am. Chem. Soc.*, **94**, 5106 (1972).

Posner, G. H., C. E. Whitten, and J. J. Sterling, *J. Am. Chem. Soc.*, **95**, 7788 (1973).

Posner, G. H., D. J. Brunelle, and L. Sinoway, *Synthesis*, 662 (1974).

Posner, G. H., J. J. Sterling, C. E. Whitten, C. M. Lentz, and D. J. Brunelle, *J. Am. Chem. Soc.*, **97**, 107 (1975).

Posner, G. H., and C. E. Whitten, *Org. Synth.*, **55**, 122 (1976).

Postis, J. de, *C. R. Acad. Sci., Paris*, **222**, 398 (1946).

Prout, F. S., R. J. Hartman, E. P-Y. Huang, C. J. Korpics, and G. R. Tichelaar, *Org. Synth.*, **Coll. Vol. 4**, 93 (1963).

Quesada, M. L., and R. H. Schlessinger, *J. Org. Chem.*, **43**, 346 (1978).

Raggio, M. L., and D. S. Watt, *J. Org. Chem.*, **41**, 1873 (1976).

Ramachandran, S., and M. S. Newman, *Org. Synth.*, **Coll. Vol. 5**, 486 (1973).

Ramamurthy, V., and R. S. H. Liu, *J. Org. Chem.*, **40**, 3460 (1975).

Ramanathan, V., and R. Levine, *J. Org. Chem.*, **27**, 1216 (1962).

Ranade, A. C., R. S. Mali, S. R. Bhide, and S. R. Mehta, *Synthesis*, 123 (1976).

Rathke, M. W., *J. Am. Chem. Soc.*, **92**, 3222 (1970).

Rathke, M. W., *Org. React.*, **22**, 423 (1975).

Rathke, M. W., and A. Lindert, *J. Org. Chem.*, **35**, 3966 (1970).

Rathke, M. W., and J. Deitch, *Tetrahedron Lett.*, 2953 (1971a).

Rathke, M. W., and A. Lindert, *Tetrahedron Lett.*, 3995 (1971b).

Rathke, M. W., and A. Lindert, *J. Am. Chem. Soc.*, **93**, 2318 (1971c).

Rathke, M. W., and D. F. Sullivan, *J. Am. Chem. Soc.*, **95**, 3050 (1973).

Rausch, M. D., and A. J. Sarnelli, *Adv. Chem. Ser.* No. **130**, 248 (1974).

Rautenstrauch, V., *Helv. Chim. Acta*, **57**, 496 (1974).

Ravid, U., and R. Ikan, *J. Org. Chem.*, **39**, 2637 (1974).

Ravid, U., and R. M. Silverstein, *Tetrahedron Lett.*, 423 (1977).

Reich, H. J., and S. K. Shah, *J. Am. Chem. Soc.*, **97**, 3250 (1975a).

Reich, H. J., J. M. Renga, and I. L. Reich, *J. Am. Chem. Soc.*, **97**, 5434 (1975b).

Reiff, H., *Newer Methods of Preparative Organic Chemistry*, Vol. 6, W. Foerst, Ed., Academic Press, New York, 1971, p. 48.

Renfrow, W. B., and G. B. Walker, *J. Am. Chem. Soc.*, **70**, 3957 (1948).

Reutrakul, V., and W. Kanghae, *Tetrahedron Lett.*, 1225 (1977a).

Reutrakul, V., and W. Kanghae, *Tetrahedron Lett.*, 1377 (1977b).

Reynolds, G. A., W. J. Humphlett, F. W. Swamer, and C. R. Hauser, *J. Org. Chem.*, **16**, 165 (1951).

Richards, K. D., A. J. Kolar, A. Srinivasan, R. W. Stephenson, and R. K. Olsen, *J. Org. Chem.*, **41**, 3674 (1976).

Richards, R. W., J. L. Rodwell, and K. J. Schmalzl, *J. Chem. Soc. Chem. Commun.*, 849 (1977).

Richman, J. E., J. L. Herrmann, and R. H. Schlessinger, *Tetrahedron Lett.*, 3267 (1973).

Rieke, R. D., S. J. Uhm, and P. M. Hudnall, *J. Chem. Soc. Chem. Commun.*, 269 (1973).

Rieke, R. D., and S. J. Uhm, *Synthesis*, 452 (1975).

Rieke, R. D., *Acc. Chem. Res.*, **10**, 301 (1977).

Riemschneider, R., and W. Grunow, *Monatsh. Chem.*, **92**, 1191 (1961).

Ringold, H. J., and S. K. Malhorta, *Tetrahedron Lett.*, 669 (1962).

Rose, A. F., J. A. Pettus, Jr., and J. J. Sims, *Tetrahedron Lett.*, 1847 (1977).

Rosenblum, L. D., R. J. Anderson, and C. A. Henrick, *Tetrahedron Lett.*, 419 (1976).

Rossi, R., *Synthesis*, 817 (1977).

Ruppert, J. F., and J. D. White, *J. Org. Chem.*, **39**, 269 (1974).

Ruppert, J. F., and J. D. White, *J. Org. Chem.*, **41**, 550 (1976).

Santaniello, E., and A. Manzocchi, *Synthesis*, 698 (1977).

Sasson, I., and J. Labovitz, *J. Org. Chem.*, **40**, 3670 (1975).

Savignac, P., and F. Mathey, *Tetrahedron Lett.*, 2829 (1976).

Savoia, D., C. Trombini, and A. Umani-Ronchi, *Tetrahedron Lett.*, 653 (1977).

Schaap, A. and J. F. Arens, *Rec. Trav. Chim. Pays-Bas*, **87**, 1249 (1968).

Schaefer, J. P., and J. J. Bloomfield, *Org. React.*, **15**, 1 (1967).

Schatz, P. F., *J. Chem. Ed.*, **55**, 468 (1978).

Schill, G., and C. Merkel, *Synthesis*, 387 (1975).

Schlosser, M., and K. F. Christmann, *Liebigs Ann. Chem.*, **708**, 1 (1967).

Schlosser, M., and G. Heinz, *Chem. Ber.*, **102**, 1944 (1969).

Schlosser, M., *Top. Stereochem.*, **5**, 1 (1970).

Schlosser, M., K. F. Christmann, and A. Piskala, *Chem. Ber.*, **103**, 2814 (1970).

Schmidt, F., in *Houben-Weyl, Methoden der Organischen Chemie,* Vol. 6/2, E. Müller, Ed., Georg Thieme Verlag, Stuttgart, 1963, p. 5.

Schöllkopf, U., in *Houben-Weyl, Methoden der Organischen Chemie,* Vol. 13/1, E. Müller, Ed., Georg Thieme Verlag, Stuttgart, 1970, p. 353. (a).

Schöllkopf, U., *ibid.,* p. 148, (b).

Schultz, A. G., and Y. K. Yee, *J. Org. Chem.,* **41**, 4044 (1976).

Schwarz, M., and R. M. Waters, *Synthesis,* 567 (1972).

Scilly, N. F., *Synthesis,* 160 (1973).

Scopes, D. I. C., A. F. Kluge, and J. A. Edwards, *J. Org. Chem.,* **42**, 376 (1977).

Seebach, D., and A. K. Beck, *Org. Synth.,* **51**, 76 (1971).

Seebach, D., and M. Teschner, *Tetrahedron Lett.,* 5113 (1973).

Seebach, D., and H. Neumann, *Chem. Ber.,* **107**, 847 (1974a).

Seebach, D., and M. Kolb, *Chem. Ind.,* 687 (1974b).

Seebach, D., and E. J. Corey, *J. Org. Chem.,* **40**, 231 (1975a).

Seebach, D., H. F. Leitz, and V. Ehrig, *Chem. Ber.,* **108**, 1924 (1975b).

Seebach, D., and F. Lehr, *Angew. Chem. Int. Ed.,* **15**, 505 (1976).

Seebach, D., R. Henning. F. Lehr, and J. Gonnermann, *Tetrahedron Lett.,* 1161 (1977).

Selikson, S. J., and D. S. Watt, *J. Org. Chem.,* **40**, 267 (1975).

Semmelhack, M. F., *Org. React.,* **19**, 115 (1972).

Semmelhack, M. F., and P. M. Helquist, *Org. Synth.,* **52**, 115 (1972).

Seyferth, D., and M. A. Weiner, *J. Am. Chem. Soc.,* **83**, 3583 (1961).

Shapiro, E. L., M. J. Gentles, L. Weber, and G. Page, *Tetrahedron Lett.,* 3557 (1977).

Sharpless, K. B., R. F. Lauer, and A. Y. Teranishi, *J. Am. Chem. Soc.,* **95**, 6137 (1973).

Sher, F. T., and G. A. Berchtold, *J. Org. Chem.,* **42**, 2569 (1977).

Shirley, D. A., *Org. React.,* **8**, 28 (1954).

Sih, C. J., J. B. Heather, G. P. Peruzzotti, P. Price, R. Sood, and L. H. Lee, *J. Am. Chem. Soc.,* **95**, 1676 (1973).

Sih, C. J., J. B. Heather, R. Sood, P. Price, G. Peruzzotti, L. F. H. Hee, and S. S. Lee, *J. Am. Chem. Soc.,* **97**, 865 (1975).

Slocum, D. W., and D. I. Sugarman, *Adv. Chem. Ser.* No. **130**, 222 (1974).

Smith, III, A. B., S. J. Branca, and B. H. Toder, *Tetrahedron Lett.,* 4225 (1975).

Smith, H. A., B. J. L. Huff, W. J. Powers, III, and D. Caine, *J. Org. Chem.,* **32**, 2851 (1967).

Smith, P. A. S., *The Chemistry of Open-Chain Organic Nitrogen Compounds,* Vol. 2, W. A. Benjamin, New York, 1966, p. 391.

Smith, W. N., *Adv. Chem. Ser.* No. **130**, 23 (1974).

Sondheimer, F., and R. Mechoulam, *J. Am. Chem. Soc.,* **79**, 5029 (1957).

Sosnovsky, G., and J. H. Brown, *Chem. Rev.*, **66**, 529 (1966).

Sowerby, R. L., and R. M. Coates, *J. Am. Chem. Soc.*, **94**, 4758 (1972).

Sprague, J. M., L. J. Beckham, and H. Adkins, *J. Am. Chem. Soc.*, **56**, 2665 (1934).

Staab, H. A., and W. Rohr, *Newer Methods of Prep. Org. Chem.*, **5**, 61 (1968).

Starks, C. M., *J. Am. Chem. Soc.*, **93**, 195 (1971).

Stephenson, L. M., and D. L. Mattern, *J. Org. Chem.*, **41**, 3614 (1976).

Stetter, H., *Angew. Chem.*, **67**, 769 (1955).

Stetter, H., I. Krüger-Hansen, and M. Rizk, *Chem. Ber.*, **94**, 2702 (1961).

Stetter, H., and E. Reske, *Chem. Ber.*, **103**, 643 (1970).

Stetter, H. in *Houben-Weyl, Methoden der Organischen Chemie*, Vol. 7/2b, E. Müller, Ed., Georg Thieme Verlag, Stuttgart, 1976, p. 1423.

Stiles, M., D. Wolf, and G. V. Hudson, *J. Am. Chem. Soc.*, **81**, 628 (1959).

Still, W. C., and T. L. Macdonald, *J. Am. Chem. Soc.*, **96**, 5561 (1974).

Still, W. C., *Tetrahedron Lett.*, 2115 (1976).

Still, W. C., and T. L. Macdonald, *J. Org. Chem.*, **41**, 3620 (1976).

Stork, G., and S. R. Dowd, *J. Am. Chem. Soc.*, **85**, 2178 (1963).

Stork, G., and L. Maldonado, *J. Am. Chem. Soc.*, **93**, 5286 (1971).

Stork, G., G. A. Kraus, and G. A. Garcia, *J. Org. Chem.*, **39**, 3459 (1974a).

Stork, G., L. D. Cama, and D. R. Coulson, *J. Am. Chem. Soc.*, **96**, 5268 (1974b).

Stork, G., and J. F. Cohen, *J. Am. Chem. Soc.*, **96**, 5270 (1974c).

Stork, G., and L. Maldonado, *J. Am. Chem. Soc.*, **96**, 5272 (1974d).

Stork, G., and J. Singh, *J. Am. Chem. Soc.*, **96**, 6181 (1974e).

Stork, G., and S. R. Dowd, *Org. Synth.*, **54**, 46 (1974f).

Stork, G., A. Y. W. Leong, and A. M. Touzin, *J. Org. Chem.*, **41**, 3491 (1976).

Stork, G., and T. Takahashi, *J. Am. Chem. Soc.*, **99**, 1275 (1977a).

Stork, G., and J. Benaim, *Org. Synth.*, **57**, 69 (1977b).

Stotter, P. L., and K. A. Hill, *J. Org. Chem.*, **38**, 2576 (1973).

Streitwieser, Jr., A., and R. G. Lawler, *J. Am. Chem. Soc.*, **87**, 5388 (1965).

Streitwieser, Jr., A., W. B. Hollyhead, A. H. Pudjaatmaka, P. H. Owens, T. L. Kruger, P. A. Rubenstein, R. A. MacQuarrie, M. L. Brokaw, W. K. C. Chu, and H. M. Niemeyer, *J. Am. Chem. Soc.*, **93**, 5088 (1971).

Sullivan, D. F., R. P. Woodbury, and M. W. Rathke, *J. Org. Chem.*, **42**, 2038 (1977).

Taber, D. F., *J. Org. Chem.*, **41**, 2649 (1976).

Tagaki, W., I. Inoue, Y. Yano, and T. Okonogi, *Tetrahedron Lett.*, 2587 (1974).

Taguchi, H., K. Shimoji, H. Yamamoto, and H. Nozaki, *Bull. Chem. Soc. Japan*, **47**, 2529 (1974a).

Taguchi, H., H. Yamamoto, and H. Nozaki, *J. Am. Chem. Soc.*, **96**, 3010 (1974b).

Taguchi, H., H. Yamamoto, and H. Nozake, *Tetrahedron Lett.*, 2617 (1976).

Tamura, M., and J. Kochi, *Synthesis*, 303 (1971).

Tanaka, K., R. Tanikaga, and A. Kaji, *Chem. Lett.*, 917 (1976).

Tanaka, T., S. Kurozumi, T. Toru, M. Kobayashi, S. Miura, and S. Ishimoto, *Tetrahedron Lett.*, 1535 (1975).

Taylor, E. C., and A. McKillop, *Acc. Chem. Res.*, 3, 338 (1970).

Taylor, E. C., and J. L. LaMattina, *J. Org. Chem.*, 43, 1200 (1978).

Thomas, M. T., and A. G. Fallis, *Tetrahedron Lett.*, 4687 (1973).

Tokuda, M., T. Taguchi, O. Nishio, and M. Itoh, *J. Chem. Soc. Chem. Commun.*, 606 (1976).

Torii, S., H. Tanaka, and T. Mandai, *J. Org. Chem.*, 40, 2221 (1975).

Torii, S., K. Uneyama, and K. Hamada, *Bull. Chem. Soc. Japan*, 50, 2503 (1977).

Touzin, A. M., *Tetrahedron Lett.*, 1477 (1975).

Treibs, A., and K. Hintermeier, *Chem. Ber.*, 87, 1163 (1954).

Trost, B. M., and R. A. Kunz, *J. Org. Chem.*, 39, 2475 (1974).

Trost, B. M., and Y. Tamaru, *Tetrahedron Lett.*, 3797 (1975a).

Trost, B. M., and C. H. Miller, *J. Am. Chem. Soc.*, 97, 7182 (1975b).

Trost, B. M., and L. S. Melvin, Jr., *Sulfur Ylides*, Academic Press, New York, 1975, (c).

Trost, B. M., H. C. Arndt, P. E. Strege, and T. R. Verhoeven, *Tetrahedron Lett.*, 3477 (1976a).

Trost, B. M., T. N. Salzmann, and K. Hiroi, *J. Am. Chem. Soc.*, 98, 4887 (1976b).

Trost, B. M., and L. H. Latimer, *J. Org. Chem.*, 43, 1031 (1978).

Turner, R. B., D. E. Nettleton, Jr., and R. Ferebee, *J. Am. Chem. Soc.*, 78, 5923 (1956).

Tyman, J. H. P., *Synth., Commun.*, 5, 21 (1975).

Van Ende, D., W. Dumont, and A. Krief, *Angew. Chem. Int. Ed.*, 14, 700 (1975).

van Tamelen, E. E., G. M. Milne, M. I. Suffness, M. C. R. Chauvin, R. J. Anderson, and R. S. Achini, *J. Am. Chem. Soc.*, 92, 7202 (1970).

v. Auwers, K., *Chem. Ber.*, 61, 408 (1928).

Vedejs, E., and P. L. Fuchs, *J. Org. Chem.*, 36, 366 (1971).

Vedejs, E., *J. Am. Chem. Soc.*, 96, 5944 (1974).

Vedejs, E., and J. E. Telschow, *J. Org. Chem.*, 41, 740 (1976).

Vedejs, E., and W. T. Stolle, *Tetrahedron Lett.*, 135 (1977).

Vedejs, E., D. A. Engler, and J. E. Telschow, *J. Org. Chem.*, 43, 188 (1978).

Villieras, J., P. Perriot, and J. F. Normant, *Bull. Soc. Chim. Fr.*, 765 (1977).

Vlattas, I., L. D. Vecchia, and A. O. Lee, *J. Am. Chem. Soc.*, 98, 2008 (1976).

Von, I., and E. C. Wagner, *J. Org. Chem.*, **9**, 155 (1944).

Wadsworth, Jr., W. S., and W. D. Emmons, *J. Am. Chem. Soc.*, **83**, 1733 (1961).

Wadsworth, Jr., W. S., *Org. React.*, **25**, 73 (1977).

Wagner, R. B., and H. D. Zook, *Synthetic Organic Chemistry*, Wiley, New York, 1953, p. 424.

Wakefield, B. J., *The Chemistry of Organolithium Compounds*, Pergamon Press, Elmsford, New York, 1974.

Walton, D. R. M., in *Protecting Groups in Organic Chemistry*, J. F. W. McOmie, Ed., Plenum Press, London, 1973, p. 28.

Wannagat, U., and H. Niederprüm, *Chem. Ber.*, **94**, 1540 (1961).

Wasserman, H. H., and B. H. Lipshutz, *Tetrahedron Lett.*, 1731 (1975).

Wasserman, H. H., and B. H. Lipshutz, *Tetrahedron Lett.*, 4613 (1976).

Watanabe, Y., M. Shiono, and T. Mukaiyama, *Chem. Lett.*, 871 (1975).

Watanabe, T., Y. Nakashita, S. Katayama, and M. Yamauchi, *J. Chem. Soc. Chem. Commun.*, 493 (1977).

Watt, D. S., *Tetrahedron Lett.*, 707 (1974).

Weizmann, C., E. Bergmann, and M. Sulzbacher, *J. Org. Chem.*, **15**, 918 (1950).

Welch, S. C., and T. A. Valdes, *J. Org. Chem.*, **42**, 2108 (1977).

Westmijze, H., J. Meijer, H. J. T. Bos, and P. Vermeer, *Rec. Trav. Chim. Pays-Bas*, **95**, 299, 304 (1976).

Weygand, C., and C. Bischoff, *Chem. Ber.*, **61**, 687 (1928).

Whitesides, G. M., W. F. Fischer, Jr., J. San Filippo, Jr., R. W. Bashe, and H. O. House, *J. Am. Chem. Soc.*, **91**, 4871 (1969).

Williams, J. R., L. R. Unger, and R. H. Moore, *J. Org. Chem.*, **43**, 1271 (1978).

Wilson, S. R., and L. R. Phillips, *Tetrahedron Lett.*, 3047 (1975).

Wingler, F., in *Houben-Weyl, Methoden der Organischen Chemie*, Vol. 7/2a, E Müller, Ed., Georg Thieme Verlag, Stuttgart, 1973, p. 603.

Wittig, G., and A. Hesse, *Org. Synth.*, **50**, 66 (1970).

Wittig, G., and U. Schoellkopf, *Org. Synth.*, Coll. Vol. 5, 751 (1973).

Wollenberg, R. H., K. F. Albizati, and R. Peries, *J. Am. Chem. Soc.*, **99**, 7365 (1977).

Woodbury, R. P., and M. W. Rathke, *J. Org. Chem.*, **42**, 1688 (1977).

Wroble, R. R., and D. S. Watt, *J. Org. Chem.*, **41**, 2939 (1976).

Yamada, S., N. Oh-hashi, and K. Achiwa, *Tetrahedron Lett.*, 2557 (1976).

Yates, P., N. Yoda, W. Brown, and B. Mann, *J. Am. Chem. Soc.*, **80**, 202 (1958).

Zen, S., and E. Kaji, *Org. Synth.*, **57**, 60 (1977).

Ziegler, F. E., and J. A. Schwartz, *Tetrahedron Lett.*, 4643 (1975).

Ziegler, F. E., and J. A. Schwartz, *J. Org. Chem.*, **43**, 985 (1978).

Zoretic, P. A., and F. Barcelos, *Tetrahedron Lett.*, 529 (1977a).

Zoretic, P. A., and J. Chiang, *J. Org. Chem.*, **42**, 2103 (1977b).

Carbanion
Equivalent Index

Many carbanions that contain protecting or delocalizing groups give products (after removal of the groups) which formally would arise from other simpler but often hypothetical carbanions. Those simpler carbanions are grouped below in order of increasing numbers of carbon atoms bearing or separating functionally, and the page numbers indicate the coverage of the protected and/or delocalized equivalents thereof. In making selections from this index, consider that more substitution of R groups is often possible or, inversely, that H may replace R.

$$\overset{O}{\overset{\|}{H C}} CH_2 CH_2 \overset{O}{\overset{\|}{C}} CH_2^- \qquad 94$$

$$\overset{O}{\overset{\|}{H C}} CH_2 CH_2 CH_2 \overset{O}{\overset{\|}{C}}^- \qquad 87$$

$$C_6 \quad HO \overset{O}{\overset{\|}{C}} CH_2 CH_2 CH_2 CH_2 CH_2^- \qquad 206$$

$$RO \overset{O}{\overset{\|}{C}} CH_2 CH_2 CH_2 \overset{O}{\overset{\|}{C}} \overset{-}{C} HCH_3 \qquad 206$$

General Index